CANADIAN BULLETIN OF FISHERIES AND AQUATIC SCIENCES 218

Pacific Oyster Culture in British Columbia

D. B. Quayle

Department of Fisheries and Oceans
Fisheries Research Branch
Pacific Biological Station
Nanaimo, British Columbia V9R 5K6

Scientific Excellence
Resource Protection & Conservation
Benefits for Canadians

DEPARTMENT OF FISHERIES AND OCEANS
Ottawa 1988

The *Canadian Bulletins of Fisheries and Aquatic Sciences* are designed to interpret current knowledge in scientific fields pertinent to Canadian fisheries and aquatic environments.

The *Canadian Journal of Fisheries and Aquatic Sciences* is published in annual volumes of monthly issues. *Canadian Special Publications of Fisheries and Aquatic Sciences* are issued periodically. These series are available from authorized bookstore agents and other bookstores, or you may send your prepaid order to the Canadian Government Publishing Centre, Supply and Services Canada, Ottawa, Ont. K1A 0S9. Make cheques or money orders payable in Canadian funds to the Receiver General for Canada.

Communications Directorate

Nicole Deschênes	Director General
John Camp	Director
Gerald J. Neville	Editorial and Publishing Services

Typesetter: Typo Litho composition inc.
Printer: Kromar Printing Ltd., Winnipeg, Manitoba
Cover Design: Diane Dufour

Available from authorized bookstore agents, other bookstores or you may send your prepaid order to the Canadian Government Publishing Centre Supply and Services Canada, Ottawa, Ont. K1A 0S9.

Make cheques or money orders payable in Canadian funds to the Receiver General for Canada.

A deposit of this publication is also available for reference in public libraries across Canada.

Canada $15.00, Other countries $18.00
+ shipping & handling on orders

under $25	=	$1.55
$25.01 to $75	=	$2.25
$75.01 to $200	=	$4.50
$200.01 and over	=	4% dollar value of order

Cat. No. Fs 94-218E
ISBN 0-660-12869-1
ISSN 0706-6503
DFO/3798

Price subject to change without notice

Correct citation for this publication:

QUAYLE, D.B. 1988. Pacific oyster culture in British Columbia. Can. Bull. Fish. Aquat. Sci. 218: 241 p.

Contents

Abstract

QUAYLE, D. B. 1988. Pacific oyster culture in British Columbia. Can. Bull. Fish. Aquat. Sci. 218: 241 p.

Of the three species of oysters (*Crassostrea gigas, Crassostrea virginica*, and *Ostrea lurida*) that occur in British Columbia, only the exotic Japanese species (*C. gigas*) the Pacific oyster, is now cultured. The anatomy of *C. gigas* is described in some detail as well as the reproductive process. Successful breeding of the species in most British Columbia waters is erratic except for certain areas such as Pendrell Sound, Hotham Sound, Ladysmith Harbour in Georgia Strait, Pipestem Inlet and Nootka Sound on the west coast of Vancouver Island. Collection of seed, particularly in Pendrell Sound is described in some detail. Methods of culture include intertidal and subtidal bottom culture, suspended string, stake, tray, and both suspended and intertidal stick culture. Harvesting, processing, and storage methods are described. The problems of Pacific oyster culture include industrial and sewage pollution, paralytic shellfish poisoning along with predators and disease. How these may be dealt with is discussed. Administrative aspects of this industry are reviewed.

Résumé

QUAYLE, D. B. 1988. Pacific oyster culture in British Columbia. Can. Bull. Fish. Aquat. Sci. 218: 241 p.

Des trois espèces d'huître (*Crassostrea gigas, Crassostrea virginica* et *Ostrea lurida*) peuplant les eaux de la Colombie-Britannique, seule l'huître creuse du Pacifique (*C. gigas*) fait l'objet d'une culture. On présente certains détails sur l'anatomie et la reproduction de cette espèce. Dans la plus grande partie des eaux de la Colombie-Britannique, la culture de l'huître creuse du Pacifique obtient un succès erratique sauf à certains endroits comme les baies Pendrell et Hotham, le havre de Ladysmith dans le détroit de Géorgie, ainsi que l'inlet Pipestem et la baie Nootka sur la côte ouest de l'île Vancouver. On présente certains détails du captage du naissain, surtout dans la baie Pendrell. Les méthodes de culture comprennent la culture à plat en eau intertidale et infratidale, la culture sur claies, la culture surélevée et la culture suspendue sur filière flottante ou sur pilotis fixes en milieu intertidal. On décrit aussi les méthodes de prélèvement, de traitement et d'entreposage. Les problèmes auxquels est exposée l'ostriculture en Colombie-Britannique comprennent la pollution par les eaux industrielles et usées, l'intoxication paralysante par les mollusques, les prédateurs et les maladies. On examine certaines solutions à ces problèmes et on traite des aspects administratifs de cette industrie.

Introduction

The first edition of Bulletin 169 was a revision and expansion of a 1933 Bulletin of the Biological Board of Canada (later Fisheries Research Board of Canada) entitled *Oysters in British Columbia* by Dr. C. R. Elsey. Although long out of print, Bulletin 169 is still in demand by both the industry and teaching institutions. Although our understanding of the basic biology of oysters has not altered greatly, significant developments in culture practices have warranted a revision in factual and source material. Bulletin 178 entitled *Pacific Oyster Raft Culture in British Columbia* is incorporated in this version.

History

Native Oyster

The native oyster (*Ostrea lurida*), known in the State of Washington as the Olympia oyster, is the species on which the original industry in British Columbia was based. The main centers of production were Boundary Bay and Ladysmith Harbour where now only occasional specimens may be found. It occurs throughout the coast in lagoons or sloughs where the oysters are not, or only briefly, exposed to drying. The species is susceptible to extremes of temperature, particularly freezing. This characteristic led to the development of the shallow dike system of culture. Further details are given in Appendix F. Production of native oysters, essentially from a fishery rather than a culture, ceased about 1940.

There is still a small production in Puget Sound only possible because of the present high value and appeal to oyster gourmets. The flavour is comparable to that of the highly esteemed European flat oyster, also an *Ostrea*. Slow growth of 4–5 years to market, high mortality, specialized culture requirements, and small size make this species a dubious candidate for culture under present circumstances. There is no known attempt to culture native oysters in suspension.

Eastern Oyster

Production records show that the native oyster (*Ostrea lurida*) was marketed in British Columbia as early as 1884, although undoubtedly it was used before then. This was the only oyster available until 1903, at which time the eastern or Atlantic species (*Crassostrea virginica*) (Fig. 1) was introduced into Boundary Bay, Esquimalt and Ladysmith harbours, also in Hammond and Nanoose bays (Fig. 2). Only limited success was obtained with this oyster at Vancouver Island sites, but in Boundary Bay results were better and varying numbers were planted until about 1940. At first they were grown to a marketable size from seed procured from the Atlantic coast but high mortality rates soon discouraged this practice, so, 3- or 4-years-olds were relaid for a year or so for a half-shell (raw oyster) market chiefly on transpacific Empress liners. Even with these older oysters, mortality still amounted to about 25% in one year of relaying.

Breeding of the Atlantic oyster was limited to the estuaries of the Serpentine and Nicomekl rivers which flow into Boundary Bay. By 1930 there were probably 75 tonnes (t) on the bars in these rivers. However, populations have declined markedly, and now only a few are to be found there, and these are being threatened by real estate developments.

Pacific Oyster

The Japanese oyster (*Crassostrea gigas*) was first introduced into British Columbia waters about 1912 or 1913 when a few were planted in Ladysmith Harbour and, it is believed, in Fanny Bay. By 1925 there was evidence that some breeding had occurred

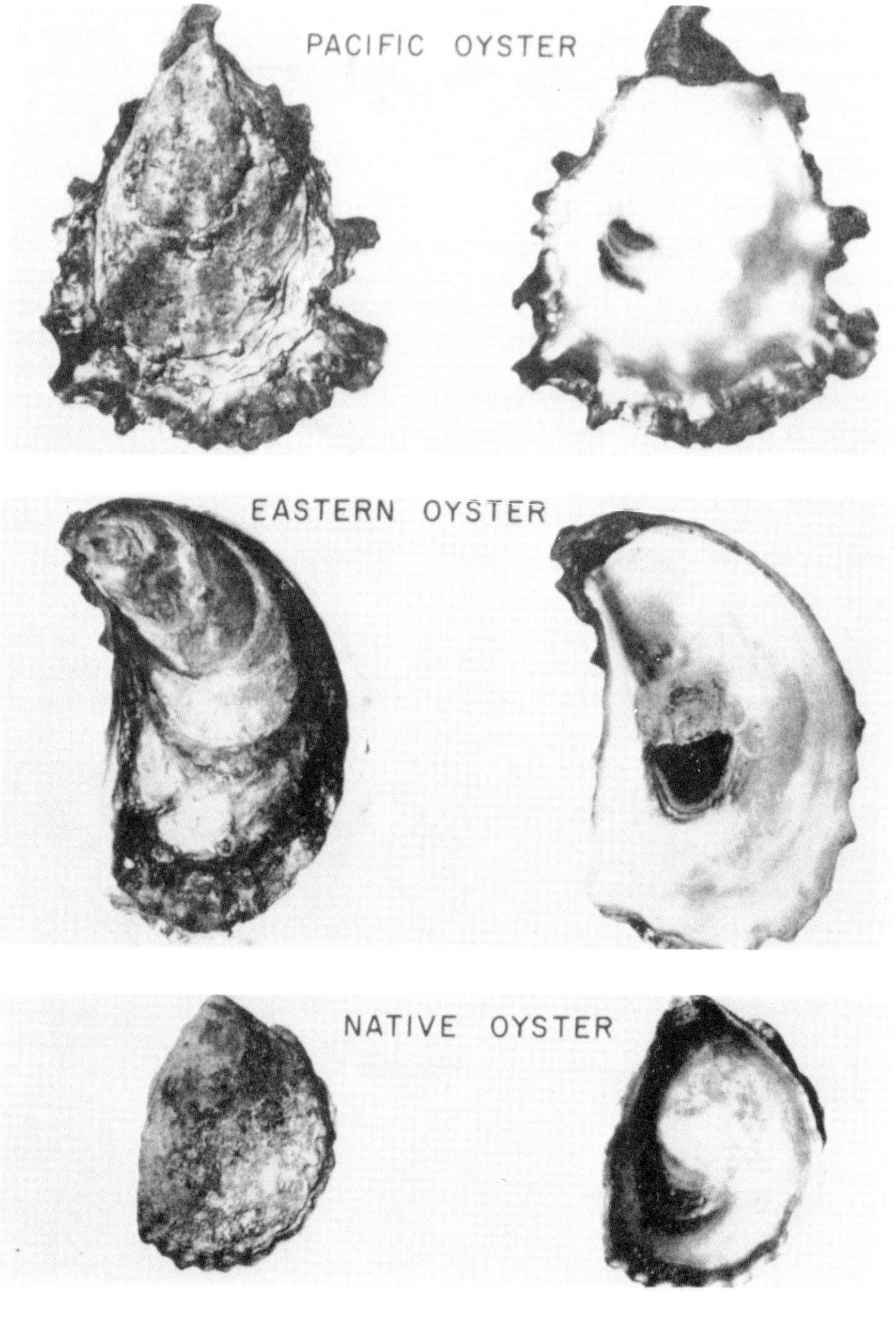

FIG. 1. The three species of oyster that occur in British Columbia: Pacific oyster (*Crassostrea gigas*), ×0.5; Eastern oyster (*Crassostrea virginica*), ×1; native or Olympia oyster (*Ostrea lurida*), ×0.75.

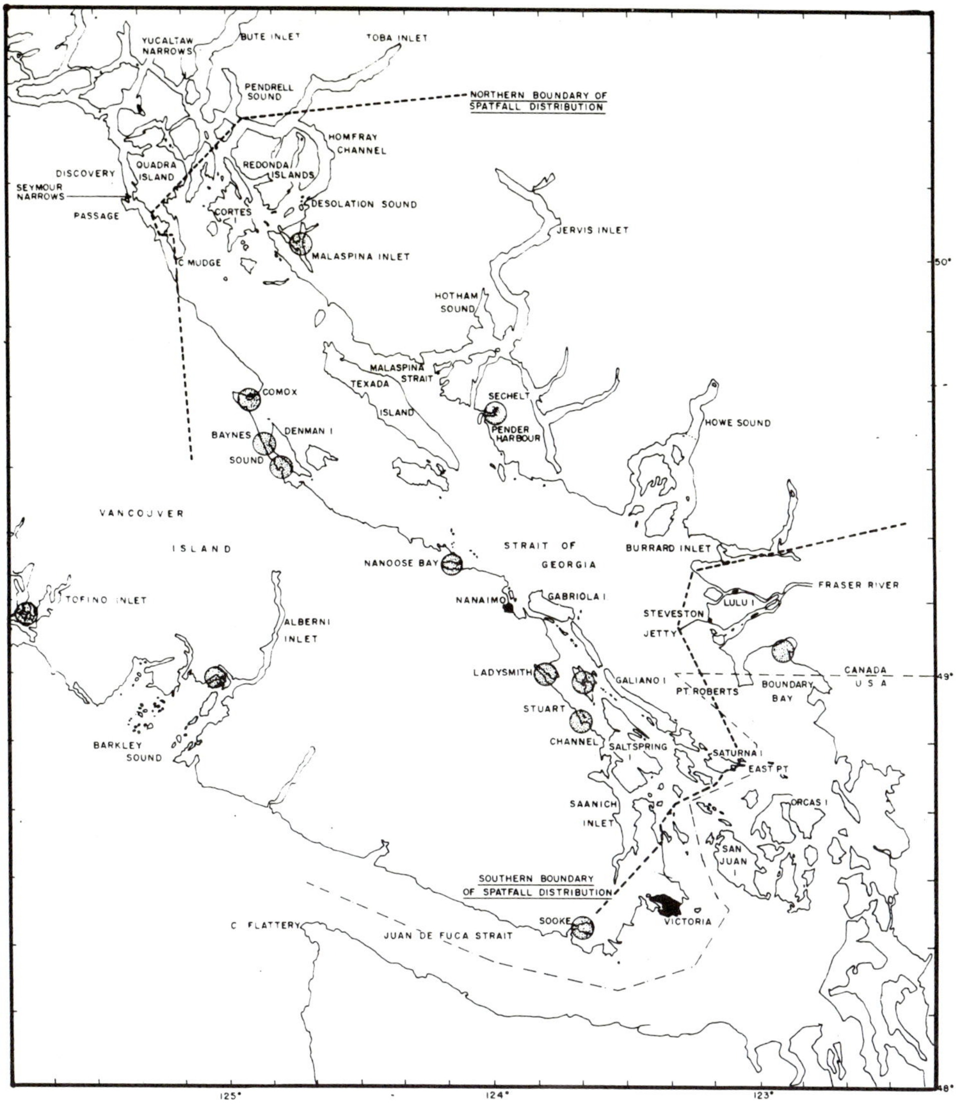

FIG. 2. British Columbia coast with oyster-growing areas marked in circles. The heavy broken line shows the limits of the 1958 spatfall distribution.

in Ladysmith Harbour, for a few oysters were found attached to large rocks. The first significant importation of Japanese oysters took place in 1926 when about 2000, 2–3-year-old oysters were obained from Samish Bay in Washington, and 20 cases of seed came directly from Japan. These averaged one year in age and were spatted on brush.

Between 1929 and 1932, approximately 4 million seed oysters were imported. Also small numbers of naturally set oysters were found as a result of spawnings in Ladysmith Harbour in 1926, 1930, and 1931. The first major spawning there occured in 1932 and this was the beginning of a series of breedings of some magnitude in this harbour.

Subsequently, Japanese oysters were introduced by planting imported seed into a number of areas in the Strait of Georgia such as Pender Harbour, Boundary Bay, and Baynes Sound. In 1942, there was a general breeding in all of these areas, and many beaches, particularly in the northern part of the Strait of Georgia and in the Gulf Islands, were heavily seeded. These "wild" oysters, as they were called, provided a supply that replaced, in part, normal requirements for Japanese seed, the importation of which was suspended during World War II.

In 1958 there was prolific and widespread breeding of Pacific oysters (as the Japanese species came to be called) and this had a significant effect on the industry. It used these wild stocks to a considerable extent and as a consequence replaced the need for Japanese seed, the importation of which virtually ceased about 1961. Also, the increased supply of oysters demonstrated a market potential greater than previously realized.

Another major event in the history of oyster culture in British Columbia was development of Pendrell Sound as a seed-producing area and a decision to prohibit leasing there. The Order-in-Council designating the reserve there is shown in Appendix A.

The introduction in 1949 of the first comprehensive set of regulations for sanitary control of the shellfish industry was a significant milestone and changed oystering from a relatively disorganized business into a stable, responsible industry.

Beginning, in 1963, with closures and restrictions on certain oyster beds, and followed by others in 1964 and 1965, sewage pollution became a serious problem for the oyster industry as did several outbreaks of paralytic shellfish poisoning since 1957. Attendant publicity, which had an effect on sales, brought an awareness of this problem to the industry which was not previously realized. As a result, the dormant British Columbia Oyster Growers Association was reorganized and is now an active force in the affairs of the oyster industry. A further important event was formation of a Shellfish Marketing Board in 1964.

Since publication of Bulletin 169 in 1969, the development of hatchery techniques has altered the pattern of seed procurement, but it is unlikely that it will completely supplant collection of natural seed. Production of hatchery seed and its proper utilization is still in a state of evolution.

A market for half shell oysters produced by tray culture is being developed and additional forms of cultch are undergoing trials. Suspended culture is only now becoming an accepted form of oyster production although it was demonstrated as a method for British Columbia conditions many years ago. Participation of Federal and Provincial governments in development, control and research in the oyster industry continues to evolve. An attempt to concentrate sales efforts and processing for the industry came to naught. Somewhat easier financing is now possible and some of the benefits of land farmers are now available.

While problems, particularly sewage pollution, beset the industry, there is no reason why these cannot be solved with ingenuity and determination. A growing population will inevitably increase demand for oysters and as there is only a limited amount of growing ground, the industry must utilize available ground more effectively than it has in the past and be prepared to accept changing concepts, and other methods of culture.

Taxonomy

The species with which this bulletin is concerned is the Pacific oyster. Earlier it was referred to as the Japanese oyster because of its origin. The scientific name, *Crassostrea gigas* (Thunberg), is used to distinguish it from the many species of oysters that occur in Japan and other parts of the world.

The two main groups of commercial oysters are readily separable on the basis of anatomical and breeding differences. In the *Ostrea* type, to which the British Columbia native oyster belongs, eggs are discharged into the inhalant chamber where fertilization, and the initial stages of development occur. In the genus *Crassostrea*, which includes Pacific and eastern oysters, eggs are discharged into the open water where fertilization and all subsequent development takes place. This form also has an additional exhalant water outlet, the promyal chamber, which does not occur in *Ostrea*. The latter is considered to be adapted to clear water environments while *Crassostrea* is better adapted to muddy estuarine waters.

The name of the Pacific oyster (*Crassostrea gigas*), when analysed, means *Crass*, thick + *ostrea*, oyster + *gigas*, huge. The name of the native oyster (*Ostrea lurida*) means the lurid or yellow oyster.

About 100 named species of oysters occur throughout the world but there may, in fact, be fewer, for so many of them have been named mainly on the basis of shell characteristics, which are now known to be extremely variable.

Originally, all North American oysters were called *Ostrea* but, in 1950, it was suggested there should be separation of the two basic types and now both *Crassostrea* and *Ostrea* are generally accepted. The name *Crassostrea* was validated in 1955 in accordance with the rules of the International Commission on Zoological Nomenclature.

Gryphaea, which was the generic name applied to the Portuguese oyster, is now used only for a certain group of fossil oysters and the living oyster has been renamed *Crassostrea angulata*.

TABLE 1. Comparative characteristics of oyster species found in British Columbia.

Scientific name:	*Ostrea lurida*	*Crassostrea virginica*	*Crassostrea gigas*
Common name:	native or Olympia	eastern or Atlantic	Pacific or Japanese
Anatomy:	without promyal chamber	with promyal chamber	with promyal chamber
Adductor	without pigment	dark purple or brown	mauve or white
Muscle scar:	clearly outlined	sharply outlined	not clearly outlined
Size:	5 cm	7–10 cm	10–15 cm
Concentric growth:	indistinct	flat but clear	projecting with flutings
Colour inside shell:	yellow-green iridescent	white	white
Colour outside shell:	gray with dark purple	yellow brown	gray with purple
Hinge area denticles (chomata):	present	not present	not present
Radial grooves:	not apparent	barely apparent	generally deep

Descriptions of the three species of oysters found in British Columbia are given below, and Table 1 and Fig. 1 show comparative characteristics.

Native Oyster

Ostrea lurida (Carpenter) is the native oyster and was named by the conchologist P. P. Carpenter in 1864 from specimens taken in Willapa Harbor, Washington. This oyster is generally circular in outline and small in size, seldom exceeding a diameter of 5 cm in British Columbia. Externally, the shell is gray and seldom fluted. Internally, the shell is quite nacreous and often iridescent olive-green or yellow. The shell edges near the hinge have a series of tiny teeth called chomata not present in *Crassostrea* types. The species is distributed from Alaska to lower California, chiefly in estuaries and saltwater lagoons. It occurs in numerous places in British Columbia, but nowhere in abundance. Some areas where it may now be found are the Gorge near Victoria, Ladysmith, Van Donop Creek on Cortes Island, Pendrell Sound, Toquart Bay on the west coast of Vancouver Island as well as in most other inlets there, Blunden Harbour in Queen Charlotte Sound, and Campbell Island near Namu.

Eastern Oyster

Crassostrea virginica (Gmelin) is the Virginia, eastern, or Atlantic oyster and was named by Gmelin in 1792. This is a fairly large oyster, up to 10 cm in length and normally somewhat pear-shaped in outline. Externally, the shell is fairly smooth but with definite concentric sculpture and is light brown or yellow. Internally, the shell is smooth with the black, brown, or dark blue scar of the adductor muscle quite deeply impressed. On the Atlantic coast, it ranges from the Gulf of St. Lawrence to the Gulf of Mexico and the West Indies. On the Pacific coast, it was introduced to San Francisco Bay, Willapa Harbor, and Puget Sound in Washington, and to Boundary Bay, Esquimalt Harbour, and Ladysmith Harbour in British Columbia. In general, it may be said that it failed to become established, although a small breeding population still exists in the river estuaries of Boundary Bay.

Pacific Oyster

Crassostrea gigas (Thunberg) is the Japanese, Pacific, or Miyagi oyster named by Thunberg in 1795. As the name indicates, this is a large oyster and exceptionally may attain a length of 30 cm. The shape is irregular and depends on type of substrate on which it is grown, as well as on degree of crowding. The external surface may be either smooth or highly fluted (Fig. 3). The colour, particularly the flutings, is brown or purple but mainly gray. The upper flat right valve is smaller than the lower cupped left valve. The interior of the shell is normally pure white with a smooth polished surface, often with irregularly shaped chalky areas. The adductor muscle scar is not as well defined as that of *C. virginica* and is usually light mauve.

In Japan, *C. gigas* exists in several forms or races in different parts of the country. On the northernmost island of Japan, there occurs the Hokkaido type which is characterized by fast growth, large shallow shells, and by its grey-white colour. Further south is the Miyagi type with colour and shell depth intermediate between the Hokkaido and more southerly Hiroshima type which has still slower growth, a smaller quite deep shell, and is blackish purple and brown. Finally, in the extreme south of Japan a stunted form of *C. gigas* occurs called the Kumamoto oyster, the name of its main production area.

The only Japanese seed imported into British Columbia was the Miyagi type which originates from Miyagi Prefecture near Sendai on northern Honshu Island (Fig. 4). Small quantities of Kumamoto-type seed are imported into the State of Washington where the small size makes it a useful substitute for native oysters. For many years importation of Kumamoto oysters into British Columbia was prohibited because it was not known whether it would cross-breed with the Miyagi type to produce a less desirable oyster. However, experience in the State of Washington indicates this possibility is unlikely and there should be no objection to its use in this province.

FIG. 3. Three basic shell types of the Pacific oyster: A — round fluted type from hard gravel. ×0.5; B — smooth type grown on fairly soft ground. ×0.5; C — long smooth type grown crowded on muddy ground. ×0.3.

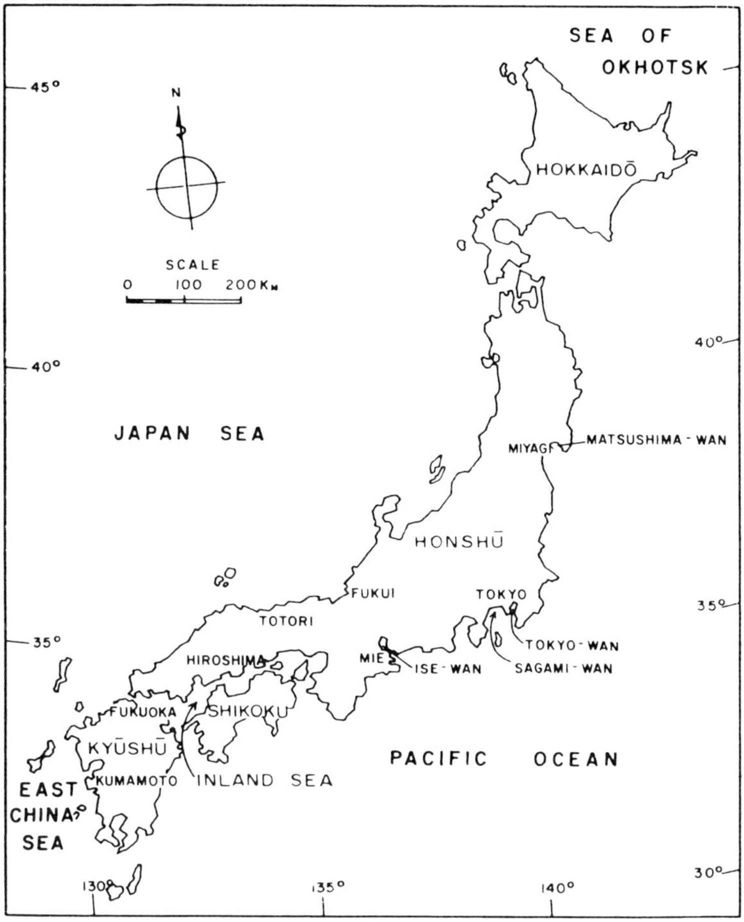

FIG. 4. Japan showing place names of the main oyster growing areas.

Hybrids

A number of attempts to cross-breed various species of oysters have been made. The most detailed attempt was made in Japan where various races [Kumamoto (south), Hiroshima (central), Miyagi and Hokkaido (north)] of oysters from different sections of the country were cross-bred to the third generation. It was shown that the characters of cross-bred oysters maintained a position intermediate betwcen the two parents in shell dimensions, shell flatness, and colouration. The hybrid oysters showed greater adaptability to a variety of environmental conditions than did the parent strains.

Although cross-fertilization between *C. gigas* and several species such as *C. virginica, C. echinata, C. rivularis*, and *C. angulata* could be achieved, only that with *C. angulata*, the Portuguese oyster, resulted in viable hybrids. It would therefore seem possible to breed oysters selectively for certain specific characteristics, an area where oyster hatcheries may be of use.

In addition to the Miyagi type *Crassostrea gigas* there is also interest in the United States in another species of Japanese oyster known as the Sumino (*Crassostrea rivularis*) which rates high in culturability, flavour and market acceptability.

The Hiroshima variety is midway in size between the Miyagi and Kumamoto and has been crossed with the latter variety to form what is called a Gigamoto to give an oyster with characteristics of both parents. It is not difficult to produce in hatcheries and has potential as a commercial oyster.

Anatomy

The anatomy of an oyster and functions of the various organs are described in some detail, for this will enable oyster growers to understand the influence on the oyster of various factors in the external environment, such as pollution, silting, and temperature changes.

Shell

Since an oyster is an invertebrate animal, it has no backbone or true internal skeleton. Its shell, however, forms a protective outer covering for the soft body. The two valves which comprise one shell are different in form, and the shell is said to be inequivalve; what is known as the lower or left valve is somewhat larger and more deeply cupped than the upper, flatter, right valve. One end of the valves is pointed (where they are hinged) and is designated as the head, anterior end, or umbonal area, for this is the region where the mouth is located. If an oyster is placed on edge with the anterior end directed to the right, and the cupped valve away from the observer, then the lower part of the shell and body is the ventral or belly section and the upper shell edge and body is the dorsal or back section (Fig. 5).

Shells of oysters are extremely variable in shape and surface sculpture. Shape is generally a result of the type of ground on which they are grown or on degree of crowding, particularly on the mother shell to which oyster were originally attached. If grown on hard, pebbly ground without crowding, they tend to become round, deeply cupped with a high degree of fluting on the outer surface, particularly on the right valve. If grown in water of low salinity, the shells are soft, and in high salinity, hard and brittle. If crowded, they grow long and narrow (Fig. 3). There are generally markings on the shell showing growth increments, but whereas it is generally possible to determine age of a clam from these marks or checks which are the result of cessation of growth, this is seldom possible with oysters because of the frequency of what are known as disturbance checks. Thus age of most oysters can be determined only from a knowledge of the time when it was spatted.

The shell itself is composed of three main layers common to most molluscan shells (Fig. 6). Externally, there is a thin horny layer known as the periostracum. Initially, this completely covers the whole outside of the shell, but in many species it is so thin and delicate that it is soon worn away. However, in such molluscs as razor and horse clams,

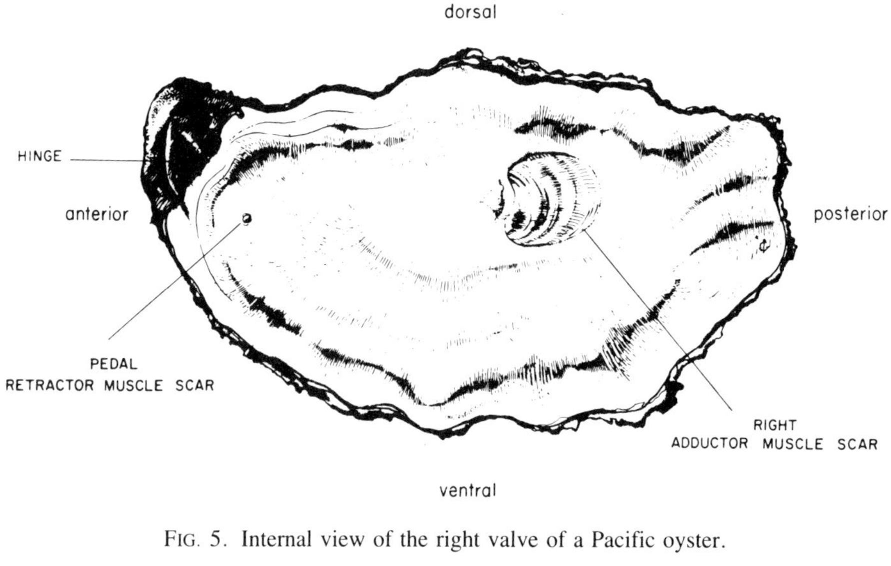

FIG. 5. Internal view of the right valve of a Pacific oyster.

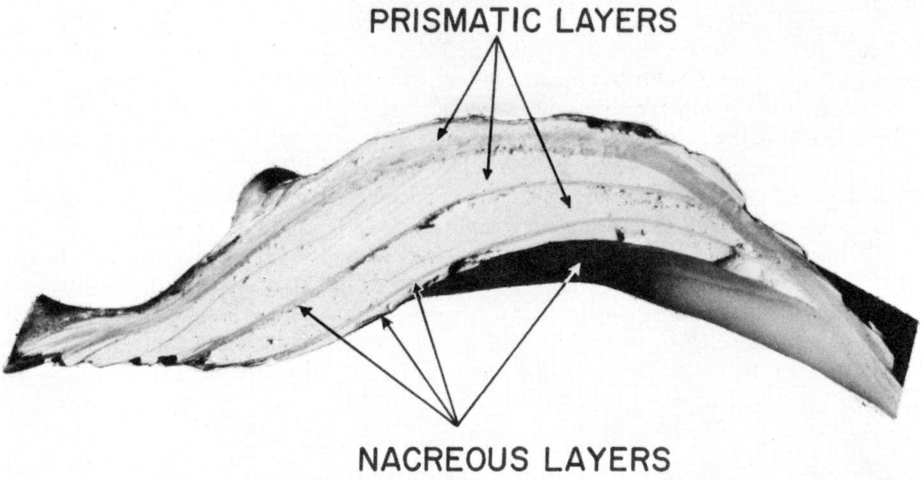

FIG. 6. Section of the shell of a Pacific oyster cut through the hinge to show the shell layers.

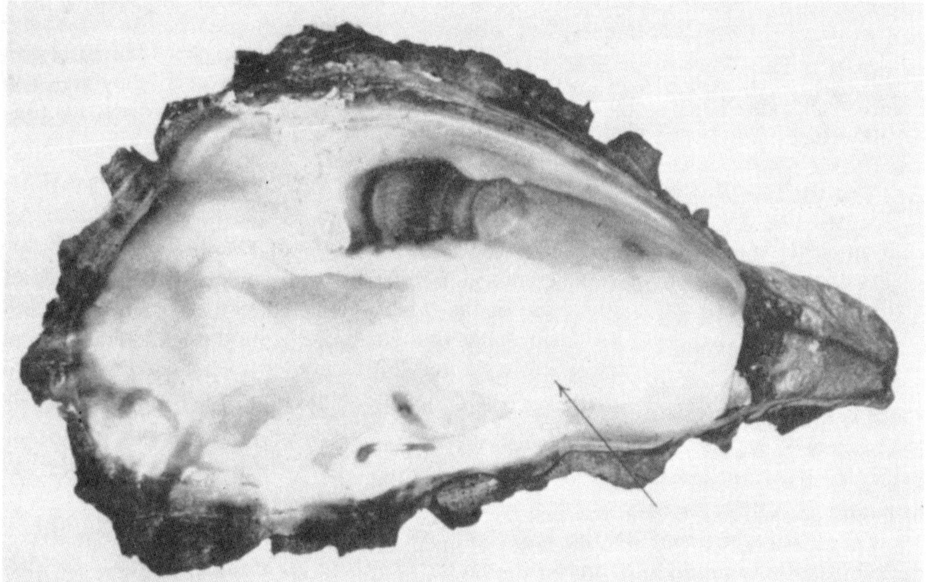

FIG. 7. Interior of Pacific oyster shell to show the white chalky shell deposits (indicated by arrow).

the periostracum is quite thick and is retained to a considerable extent on the adult shell. Immediately under the periostracum lies a layer the framework of which is made up of conchyolin, which is largely composed of vertical prisms of a chalky material called calcite. The third and innermost layer of shell, which lies next to the body is composed largely of hard aragonite, but there the plates lie flat. This is the nacreous layer and forms the mother-of-pearl from which pearls are formed.

Often superimposed on the nacreous layer are irregularly shaped deposits of chalky shell soft enough to be marked with a fingernail (Fig. 7). These are thought to be attempts by the oyster to contour the inner surface. Sometimes very definite chambers are formed by secreting well away from the nacreous surface, an additional thin layer of nacreous material that contains very little calcium. The chamber may be filled with sea water or sometimes a jellylike material. This condition, called chambering, is believed to be caused by shrinkage of the mantle tissue, which secretes the shell. A probable function of the process is to reduce the internal shell volume. Shrinkage of mantle tissue may be caused by drastic changes in salinity, by injury, or by decimation of the whole body.

The hinge ligament, located at the pointed umbonal region (Fig. 5), is quite different in structure and function from the shell. The purpose of the hinge is to spring the valves apart, acting in opposition to the adductor muscle which holds the valves together and adjusts the distance they are allowed to separate. The hinge is formed of horny conchyolin similar to that of the main shell but the amount of calcification is much reduced, resulting in an elastic, compressible material. The action of the hinge of a bivalve mollusc is more readily seen in clams such as the horse clam (*Tresus*) with an internal ligament or a butter clam (*Saxidomus*) with an external ligament.

When an oyster dies and the body is removed without damaging the hinge, the two valves tend to spread apart and such an oyster is called a ''gaper'' although this term often refers to an open oyster with at least some meat remaining. Completely empty shells are known as ''boxes'' or ''cluckers''. As an oyster grows, the hinge ligament also grows by addition to the edges and to under surfaces. At the same time exposed areas of the ligament are being eroded away.

Pearls

Pearls are essentially molluscan shell formed by mantle tissue to protect the oyster from an irritant such as a sand grain or encysted parasitic worm.

Although *Crassostrea* is only distantly related to the true pearl oyster (which, in fact, is not an oyster) it can and occasionally does produce pearls. However, they are of no commercial value because the nacre of this species of oyster is soft and not lustrous. Quality pearls are formed from hard lustrous nacre.

Body

If an oyster is opened with the hinge toward the right and the cupped valve down, the exposed right side of the body takes the same orientation designations as given for the shell (Fig. 8). Nearly the whole edge of the body is covered by two skirts or mantles that are separated except for a short distance in the anterior or head region and at a point just above the adductor muscle where right and left mantle edges are fused. This latter junction serves to separate the area of the exhalant or outgoing water current from the inhalant as shown by the heavy arrows.

Figure 9 shows an oyster in the same position with sections of the mantle cut away along the dotted lines. This reveals a pair of right labial palps just ventral and posterior to the hinge, and just behind these is the right side pair of gills which extend from the palps in a posterior direction to the point of mantle fusion. There is a corresponding pair of palps and gills on the left side of the body, the function of which will be described later. Just above the gills and forward of the point of mantle fusion lies an adductor muscle which holds the two valves together and immediately above it is the opening of the anus at the end of the digestive tract. The space below the gills is known as the inhalant chamber and the space above as the exhalant.

If an oyster is now turned over to show the left side, with the anterior or head region to the left and as much mantle as possible cut away, the asymmetry, or difference in structure between the right and left halves, is demonstrated by the fact that part of the exhalant chamber exists only on the right side, while the inhalant chamber is fully open on both sides. Also, the inhalant chamber is unobstructed throughout its length, while the exhalant chamber is divided by the adductor muscle, with water being discharged through two passages, the cloacal chamber posterior to, and a promyal chamber anterior to, the adductor muscle. This chamber is absent in oysters of the *Ostrea* genus.

Contact or fusion of the body and shell occurs mainly at the adductor muscle, but there is another point of contact, less obvious, just back of the hinge, the position of which may be determined from the tiny scar less than 2 mm in diameter. If an oyster is opened carefully, the mantle may be observed to adhere slightly to the shell at this point. This is the pedal muscle-scar, thought to represent the remnant of the gill retractor muscles

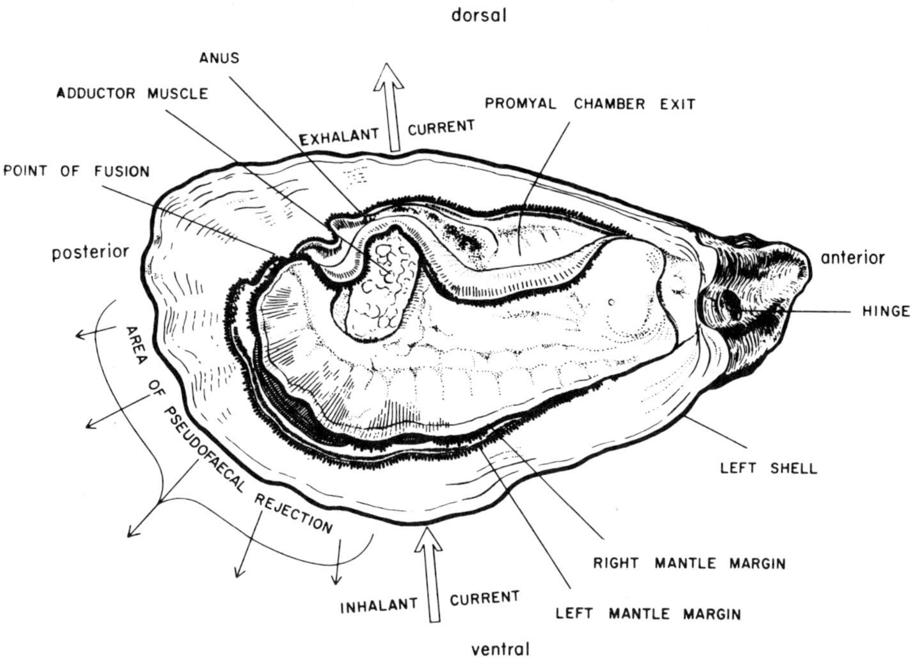

8. Pacific oyster with right valve removed.

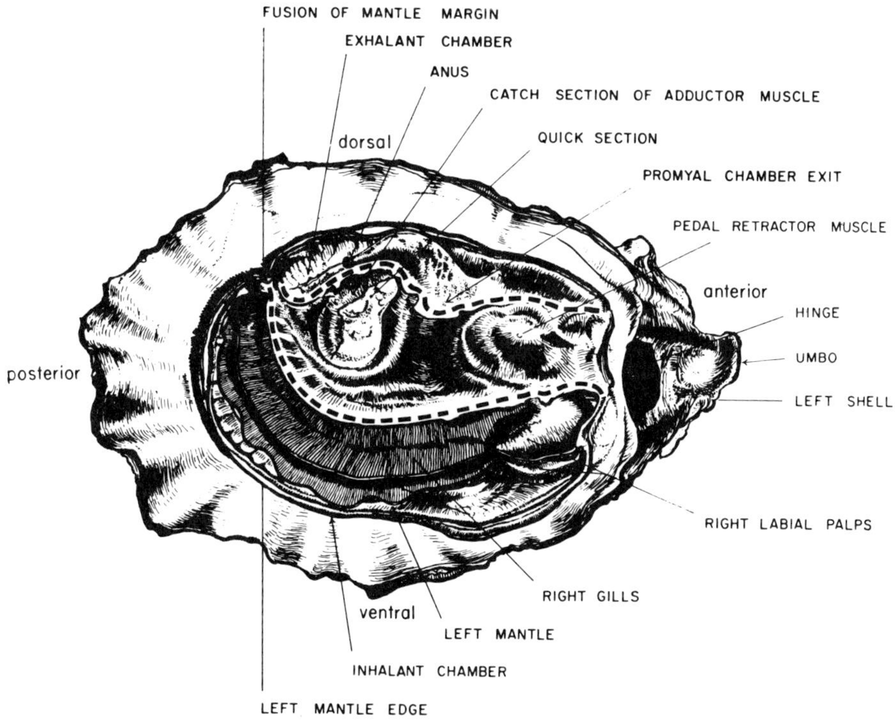

FIG. 9. Pacific oyster with right valve and part of mantle removed.

of the clam-like ancestor of the oyster and sometimes called Quenstedt's muscle; the two muscle scars are shown in Fig. 5.

Mantle

As indicated previously, the mantle of an oyster is the general surface covering of the body and is formed of two lobes, one covering the right half and the other the left half of the body. The two lobes are fused in the anterior region to form a hood over the mouth area; at the posterior end they are joined together at a single point (fusion of mantle margins; Fig. 9) to form a separation point between inhalant and exhalant water chambers. The mantle, while appearing to be formed of a thin skin over most of its area, is threaded with muscles, particularly in the posterior regions.

The mantle edge is thickened, and throughout its full length this border is divided into three parallel folds or ridges (Fig. 10) which are of great importance. The outer fold, next to the shell edge, is the area of shell formation, the calcareous shell originating from the outer surface of this fold, while the periostracum originates from its inner surface. The middle fold is heavily tentacled; it has a sensory function and is able to detect changes in chemical composition of water which flows over it. Each fourth or fifth tentacle is larger than the intervening ones. The inner fold is largest and is strongly muscled with powerful, equally sized tentacles. This fold is sometimes referred to as the pallial curtain and tentacles of the right and left lobes can interlock to form a barrier to the passage of water, either over the whole length of the mantle edge or in certain areas. The mantle edge, with three folds, is often pigmented black in the Pacific oyster.

The whole mantle surface is involved in formation of the greater part of the shell, specifically the calcite-ostracum. The region of adductor muscle attachment is composed of a special material known as hypostracum. Since the mantle is in constant contact with

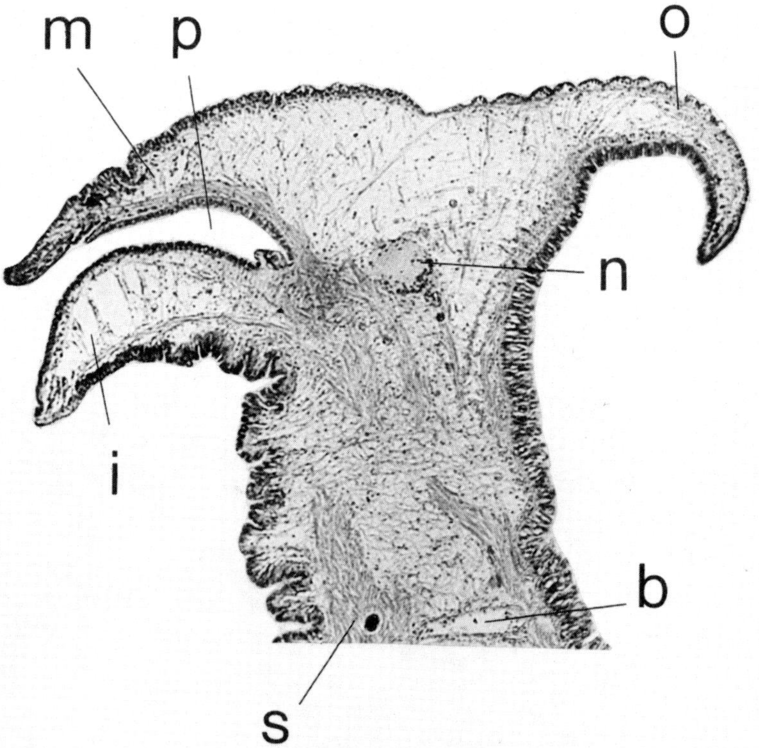

FIG. 10. Cross section of the mantle edge of the Pacific oyster. ×35. *b* = blood vessel; *i* = inner lobe; *m* = middle lobe; *n* = nerve; *o* = outer lobe; *p* = periostracal groove; *s* = muscle.

the inner shell surface deposition of new shell is possible in this area throughout life, whereas the shell-secreting mantle lobes are only able to lay down shell layers for which they are responsible, at or near the shell margin. The mantle, particularly in young oysters, also takes part in respiration.

Gills

While the mantle is involved to some extent in respiration, the main organs responsible for this function are gills (properly called ctenidia). These are lined, gray-green or brown, leaflike organs lying in the mantle cavity just below the visceral mass (Fig. 11). The gills are formed of eight leaves or lamellae, four on each side of the body. The four leaves of each group are arranged to form a "W" with each of the two lamellae joined at the ventral edge but separated dorsally to form a demibranch (Fig. 12). The dorsal edge of the outer left and outer right lamellae are joined to the mantle. In the mouth region, the dorsal edges of all gills are joined and there is a connection with the visceral mass between the bases of the palps. Towards the posterior, these lines of attachment separate, although they still remain connected with the visceral mass. Thus, four distinct chambers are formed, and since these are above the gills, they are termed suprabranchial (or epibranchial) chambers. These four chambers are reduced to two at a point just anterior to the adductor muscle because the connection between the two lateral axes of the gills and visceral mass is lost. Still further back, the two remaining chambers merge into a single suprabranchial cavity termed the cloacal chamber. On the right side, anterior to the level of the pericardial cavity, the two suprabranchial chambers fuse to form an additional discharge area called the promyal chamber. This chamber does not occur in the genus *Ostrea* and forms one of the major anatomical differences between *Ostrea* and *Crassostrea*. At this time, asymmetry of *C. gigas* may be best seen by observing, particularly in cross sections, how the visceral mass is displaced to the left. This asymmetry may also be seen in the position of the mouth, oral hood, palps, and heart, as well as in the gill system. Furthermore, the promyal chamber on the right side gives a general impression of asymmetry.

The account of the gills given so far describes only gross structures. Each demibranch encloses a V-shaped water space and to strengthen the system there are V-shaped cross

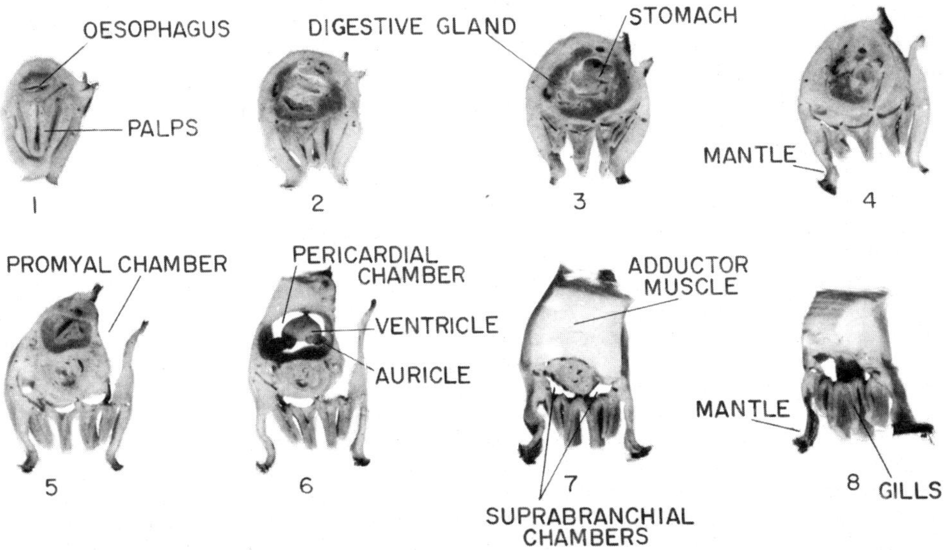

FIG. 11. Several cross sections of a Pacific oyster. Section 1 at the anterior end, slightly posterior to the mouth; section 8 near posterior end of adductor muscle.

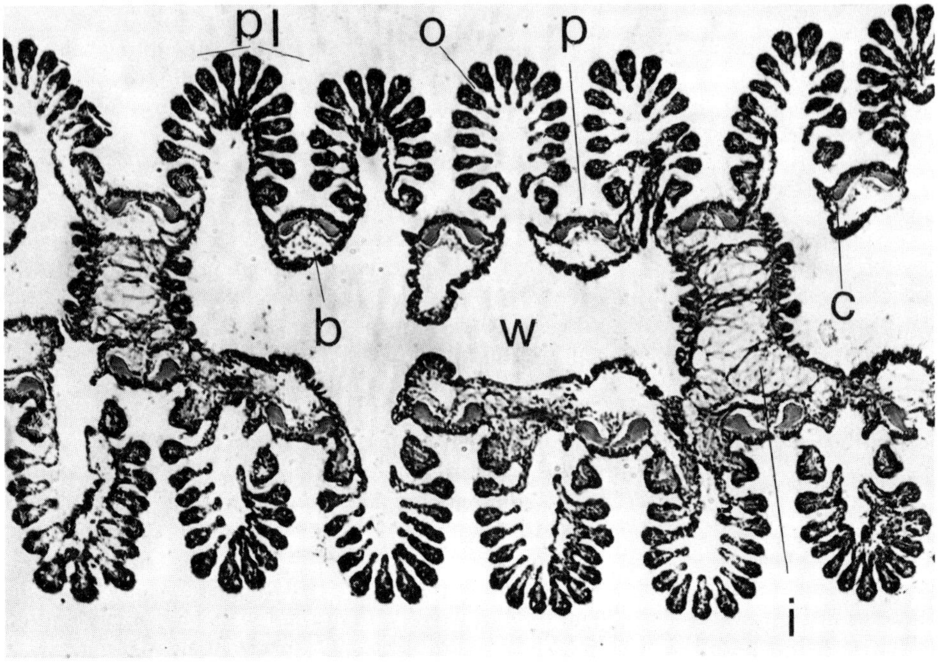

A

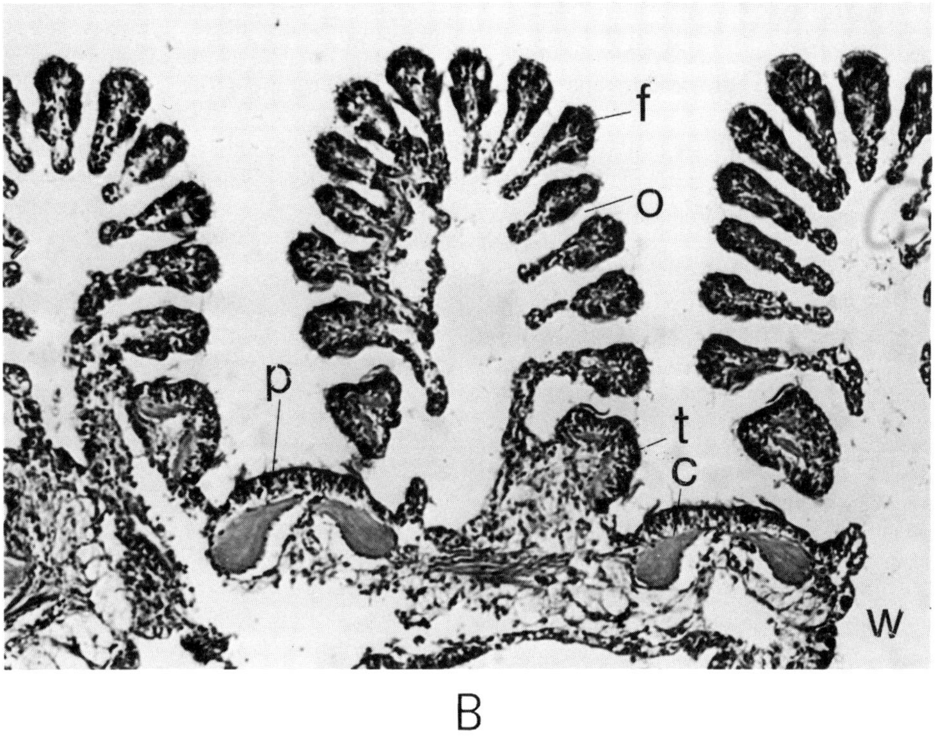

B

FIG. 12. A — cross section of part of a gill demibranch of a Pacific oyster. ×60. B — enlarged section of a single plica of A. ×15.0. *b* = blood vessel; *c* = chitinous rod; *f* = ordinary filament; *i* = interlamellar septum; *o* = ostium; *p* = principal filament; *pl* = plica; *t* = transitional filament; *w* = water tube.

partitions, called septa or inter-lamellar junctions, joining the two lamellae of each demibranch and forming what are known as water tubes. The junctions vary in height and may be seen by looking down into the gill system from above.

Individual lamellae are formed of single filaments which in turn are joined to each other by spaced strands of interfilamentary tissue, thus leaving spaces (ostia) which allow water to pass between filaments, above and below the interfilamentary junctions, into the branchial chambers. Further, filaments are joined together into folded groups called plicae, and in *C. gigas* there are between 11 and 17 filaments in a plica. There are three types of filaments, termed ordinary, transitional, and principal, in each plica; and their relationships are shown in Fig. 12. To support the filaments and various junctions, small rods of chitinous material are embedded in the tissue; these are flexible enough to allow changes in size and shape of the ostia by means of muscular activity of the gill tissue. In *C. gigas*, size of the ostia is sufficient to allow passage of ova, which are about 0.07 mm in diameter.

The whole gill system is abundantly provided with cilia, which are tiny hairs that bend or beat in unison, and it is the combined effort of the large numbers that provides motive power for the very considerable water currents that are produced and for movement of food particles. All filaments are amply provided with these cilia, each of which has a particular direction of beat and function, some being current producing, some concerned with cleansing, and others with transportation of food. Action of cilia on a gill may be observed by placing a small amount of finely ground coloured powder on the gill surface of a newly opened oyster. Cilia on the principal filaments all beat towards the gill base, to which material is consequently carried. On the other filaments, some cilia also beat in this same direction but they carry only the finest particles to the base of the gill while others beat toward the edge of the gills and carry larger particles.

Finally, at the bases of the gills and at their free edges are heavily ciliated tracts which transport food particles, carried to them by the gill cilia, toward the mouth. On each full gill, then, there are three grooves at the base and two on the free edges, making five tracts, or a total of 10 altogether, to carry collected food toward the mouth. However, in Pacific oysters the grooves at the free edge of the gills carry the major amount of material collected on them.

Closely associated with the gill system is a group of four palps which lie anterior to the gills and close to the mouth. These are fleshy, triangular-shaped bodies, with the free points directed downward and backwards (Fig. 9). The bases are attached dorsally to the body mass and covered anteriorly by the oral hood of the mantle. An outer pair of palps unite above the mouth in what might be termed the upper lip while an inner pair unite below the mouth as a lower lip. The inner opposed palp surfaces are deeply grooved, while the outer surfaces are smooth; but all surfaces are highly ciliated.

Although functions of the mantle and gills systems have already been indicated, these can now be described in more detail. Beating of the gill cilia is responsible for creating a flow of water between the mantles, which is controlled by the inner fold of the mantle edge. This flow, in an oyster of average size, may amount to between 20 and 30 L/h. A current enters the inhalant chamber along the ventral shell edge (Fig. 9) and then passes over the gill surfaces and through the gill ostia, into the water tubes, and up into the suprabranchial chambers. Thus, the gill system is a basket-like structure that acts as a sieve. Special cells on the gill surface secrete mucus which entangles food and other particles. Though the gill ostia are relatively large, the gill is able to prevent passage of very small particles such as bacteria and viral particles. Both single particles and groups enmeshed in mucus are then carried along the filaments, fine particles chiefly to the base of the gills, and larger particles to the free edge. Here ciliated grooves, both at the base and at the free edge, transport the strings of mucus, laden with food, toward the mouth. If the amount of material in the marginal grooves is too great it will fall out of the groove onto the mantle below it. Material that is carried forward in the grooves reaches the palps, where further size sorting takes place. In general, the grooved surfaces of the palps have ciliated tracts which direct fine particles forward to the mouth, while on the smooth surfaces the direction is backward toward the free apex of the palps, these tracts tending to carry heavier material which is to be discarded. This is dropped onto the mantle surface at a point near where gill lamellae meet the palps. This material is termed "pseudofaeces" for, although it is waste material, it has not passed through the digestive tract. Pseudofaeces rejected by the palps, as well as that dropped from the food grooves, is moved backward over the inner mantle surface, on which-ever mantle lobe is below, by means of other ciliated tracts, until it reaches a general area in the inhalant chamber just posterior to the point of mantle fusion (Fig. 9). With no ciliary mechanism for discharging the pseudofaecal material from the mantle cavity, this is done by sharp contractions of the adductor muscle which brings the valves together and forces water under pressure from the inhalant chamber. When oysters are just being exposed as the tide recedes, they may often be seen ejecting spurts of water; this is simply the action of the cleansing mechanism voiding pseudofaecal material.

Since gill filaments, interfilamentary and interlamellar junctions are supplied with blood vessels, passage of water through the gill ostia and water tubes allows for exchange of respiratory gases, so the gills act as a form of lung as well as a food-collecting organ.

After passage through gills and into the exhalant chamber, water, with its particulate matter and some removed oxygen, is discharged both behind the adductor muscle through the cloacal chamber, and in front of it, through the promyal chamber, and out between the dorsal mantle margins.

Adductor Muscle

The remaining visible portion of the anatomy of the oyster is the adductor muscle. As stated previously, this holds the two valves together and opposes action of the hinge ligament in springing the valves apart. Careful observation of the adductor muscle discloses a small posterior crescent-shaped section, pure white in colour, while a larger anterior area is light beige (Fig. 9). This larger section is composed of striated muscle fibres and is responsible for rapid contractions that can quickly close the valves when the mantle edge is stimulated, or in the "clapping" action which discharges pseudofaeces. Consequently, this part of the muscle is called the "quick" part, while the smaller white portion is termed the "catch" part and is composed of smooth muscle fibres. This part of the muscle is said to have a ratchet-like action and holds the valves in a set position for long periods while the "quick" portion is at rest. It has been shown for the Atlantic oyster that those oysters which live high in the tidal zone have a larger "catch" section than those living lower down, because the valves on these individuals are held closed for longer periods.

The adductor muscle of Atlantic oysters can withstand a pull of about 10 kg before the muscle tears in the middle, but the attachment of the muscle to the shell is not broken under this strain.

Digestive System

The digestive system proper begins with the mouth, which is a horizontal slit between upper and lower lips, formed by fusion of the anterior portions of the pairs of palps. It leads into a short tube, sometimes called the oesophagus, which in turn leads into the stomach proper. The form of a much grooved stomach is shown in Fig. 13 which is a photograph of a rubber cast of an oyster stomach. Entering the stomach from the side are a number of ducts leading from a dark mass of material that nearly surrounds the stomach (Fig. 14). The dark mass, sometimes brown (in winter) or green (in spring) is often called the "liver" but is properly named the digestive gland or digestive diverticulum. It is a system of fine tubules lined with special cells concerned with digestion; the colour is usually associated with the type of food being consumed.

Opposite the oesophagus is an entrance to the intestine (also called the mid-gut), a narrow tube which loops around the stomach to end in the anus, located in a cloacal chamber just above the adductor muscle. Close to the intestine entrance and associated

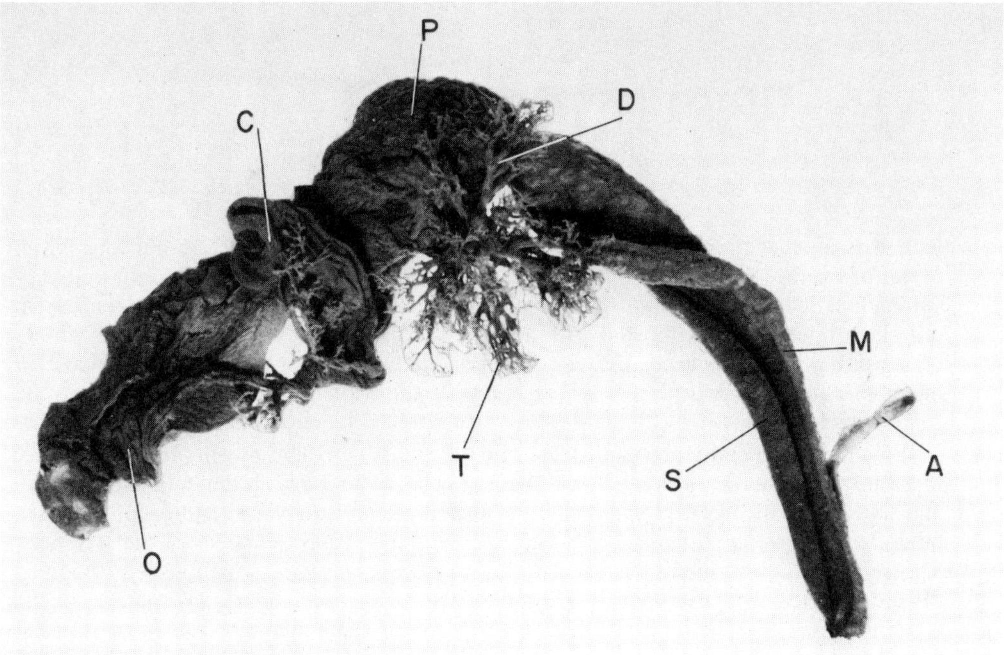

FIG. 13. A rubber latex cast on the stomach of the Pacific oyster. ×0.2. A = ascending limb of intestine; C = anterior appendage of caecum; D = caecal duct of digestive diverticula; M = mid-gut; O = oesophagus; P = posterior chamber of stomach; S = style sac; T = tubules of digestive diverticula.

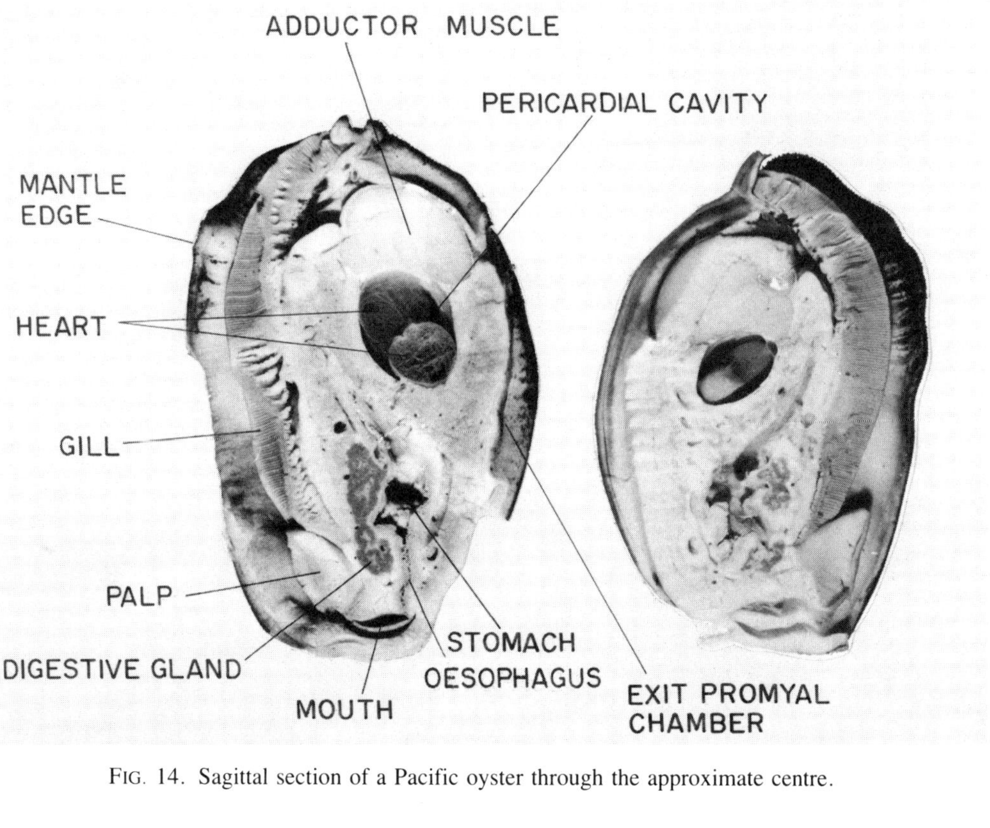

FIG. 14. Sagittal section of a Pacific oyster through the approximate centre.

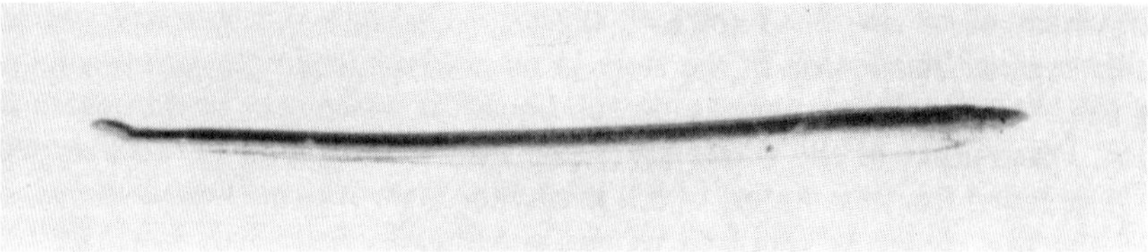

FIG. 15. The crystalline style of the Pacific oyster (*C. gigas*). ×2.

with it is another narrow tube which has a blind end and is about 2.5 cm long in an oyster of average size. This is the crystalline style sac which produces an extraordinary structure called the crystalline style, a stiff gelatinous rod (Fig. 15). It is golden brown or yellow in colour, flexible and with a tapered shape. Many mistakenly believe this to be a worm which infects the oyster. The crystalline style projects across the middle of the stomach to the opposite side where is rests against a hard transparent pad called the gastric shield.

While it has generally been considered that the style is rotated by cilia in the style sac, there are recent studies which indicate there may be, little if any, rotation in adult oysters and thin layers of digestive materials are moved along the style by action of style sac cilia. This brings about contact between style and the absorptive areas of the style sac. The style itself is formed of concentric layers of mucoprotein, which, when dissolved release digestive enzymes that convert starch into digestible sugars. Dissolution of the style in an actively feeding oyster occurs at the gastric shield where acidity of stomach liquids is lower than in the region of the style sac. Dissolution of the whole style also occurs soon after an actively feeding oyster is removed from the water, which is why the style is so seldom observed. However, if a live oyster is returned to the water and allowed to feed a style is reformed in about an hour. It is an integral part of the digestive system.

Contained in the crystalline style sac of the Pacific oyster (and other molluscs) are spirochaetes (*Cristispira*) (Fig. 16) which are elongated motile bacterialike organisms of microscopic size. Their function in the style area is not known but they are harmless to oysters and to man.

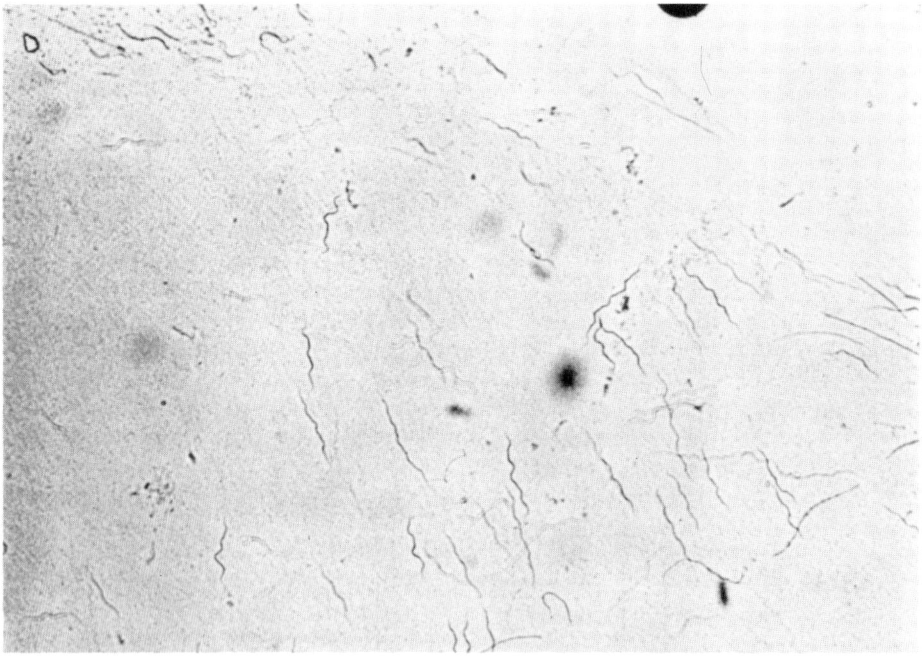

FIG. 16. Spirochaetes (*Cristispira*) from the crystalline style of the Pacifc oyster. ×226.

In addition to the style sac there are other shallow blind sacs or pouches in the stomach called caeca which are concerned with food sorting. The course of food and digestion in the oyster stomach is a very complicated process and the following is a simplified account. Mucus-enmeshed particles permitted to enter the oesophagus first meet the head of the crystalline style. Here food particles mix with enzymes which are released in this area. These enzymes effect what is called extracellular digestion of starches. By a combination of ciliary pathways in the sorting pouches and in the stomach itself, very small and partly digested particles are carried to the tubules of the digestive gland where intracellular digestion of fats and protein takes place.

In addition to these modes of digestion, some digestion is also carried on by blood cells which are able to migrate in and out of the stomach and, by an engulfing action similar to that used by an ameoba, are able to ingest individual particles of food. Once inside the blood cell, the particles undergo intracellular digestion. The stomach caeca, in addition to directing particles destined for the digestive diverticula, also direct unwanted material along a special path, called a typhlosole, to the opening of the intestine. Wastes from the digestive tubules are also carried to the intestine by this and other pathways. A function of the intestine is to compact waste materials into solid strings and to carry them, via the anus, to the cloaca whence they are carried outside the shell by exhalant water currents.

Circulatory System

In oysters, as in other animals, the function of the circulatory or blood system is to carry food, oxygen, and waste materials to various parts of the body. With exception of the heart and pericardial cavity, the circulatory system in oysters is difficult to trace, for the blood vessels are not well defined.

The heart, located above and close to the adductor muscle (Fig. 17), is enclosed in a chamber called the pericardial cavity. There are two auricles, one on each side of the single muscular ventricle. In a newly opened oyster, pulsations of the heart may be observed when the pericardium is opened. Oxygenated blood is received from the gills and pumped through the main arteries to smaller vessels. There are no fine vessels such as capillaries; instead the organs are simply bathed by blood. Deoxygenated blood is collected in veins and carried by them to the gills for reoxygenation or to the kidneys, sometimes called organs of Bojanus. The kidneys are a pair of tubes located under the adductor muscle, with an internal opening into the exhalant chamber, from where discharged wastes are carried away by the exhalant water current. There is also a pair of so-called accessory hearts which are slowly pulsating enlargements of radial blood vessels and are largely concerned with circulation of blood through the kidneys.

Nervous System

Since the oyster is sedentary and has relatively little muscle, it does not require a highly developed nervous system. Adult oysters have a pair of nerve cell aggregations near the mouth, called the cerebro-pleural ganglia. Another, larger pair of nerve cell aggregations, more closely united than the cerebro-pleural pair, occur under the adductor muscle and are the visceral ganglia. The latter are connected to cerebro-pleural ganglia from which nerves pass to the stomach, mouth and anterior areas generally. Nerves from the visceral ganglia are also connected to the adductor muscle and posterior areas. The only known sensory organ is located in the exhalant water chamber and is called the "abdominal sense-organ". Its function is unknown.

Reproductive System

This part of the anatomy can be readily observed only during the breeding season, for at this time reproductive organs may form at least 50% of the body volume. During winter, when there is no reproductive activity, the gonadal mass is replaced by a mass

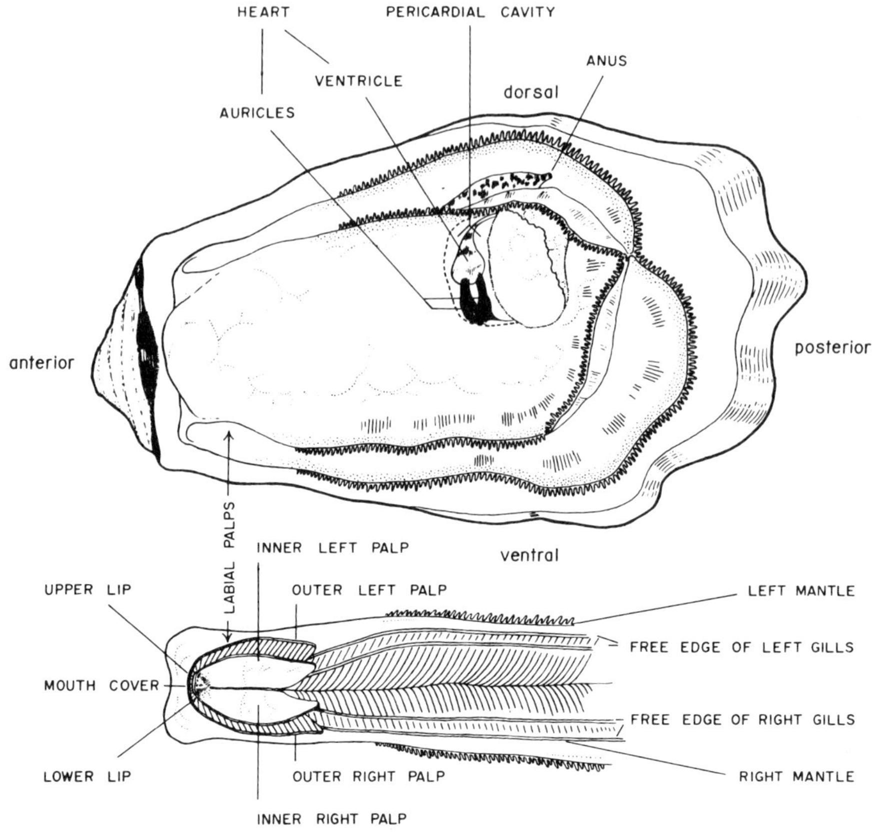

FIG. 17. (*Top*) Pacific oyster with the left valve removed and part of mantle cut away to expose the pericardial cavity. (*Bottom*) A ventral view of the mantle cavity to show the relationship between the palps and gills.

of connective tissue with vesicle cells containing fat and glyocogen. Embedded in this, close to the internal organs is a duplicate system, one on each side, of branching tubules, beginning at the anterior end of the body and finally uniting into a single tube, also one on each side, which ends in the genital pore. These enter the suprabranchial chamber at the posterior base of the adductor muscle, in close proximity to openings of the urinary ducts. In a fully ripe oyster, gonadal tubules, small in diameter at the anterior end of the oyster and thickening as they approach the genital opening, may be clearly seen on the surface of the body mass (Fig. 18). At this time, the two systems are almost completely inseparable, except at the genital pores.

In spring, glycogen-filled cells of connective tissue are gradually replaced by proliferating reproductive cells. Whether or not glycogen is actually used in gonad formation is not known, but disappearance of one coincides with appearance of the other.

Sexual maturity in Pacific oysters appears to be a function of size rather than age, and if there is early summer spatting with rapid growth, mature sex cells may be found by the end of the same summer. However, there is little possibility of these oysters spawning and the number of eggs or sperm would not be high. Spawning is reasonably certain to occur during the second summer but the quantity of gonadal products is not large. Under most British Columbia conditions, the first spawning of magnitude occurs during the third summer, when the oyster is 2 years old.

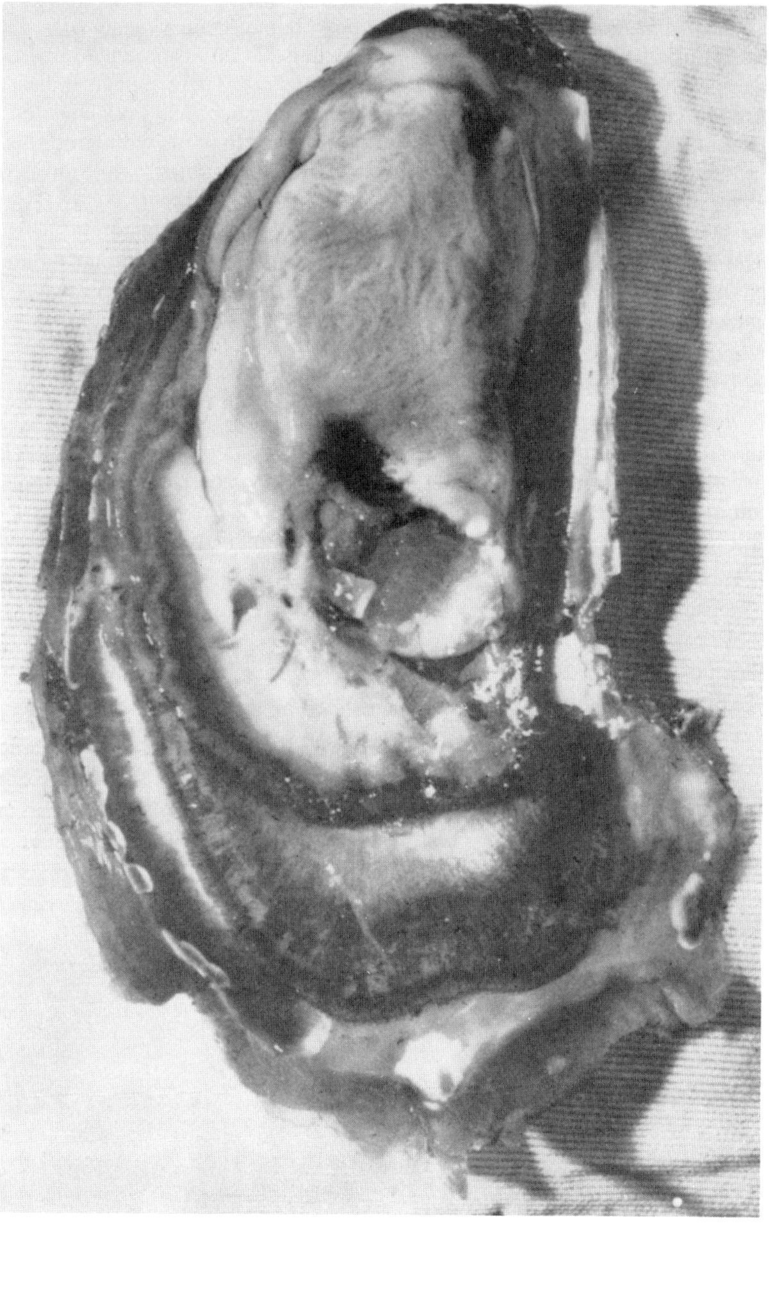

FIG. 18. Ripe Pacific oyster showing "spawn veins."

Biology

Sexuality

Sexes in the Pacific oyster are separate, that is, there are male and female oysters, though hermaphrodites occur occasionally (Fig. 20F). Sex can be determined only by examination of reproductive tissue. The sex may change from year to year, the change taking place during winter. There is a general belief that environmental conditions have a considerable influence on determination of sex in oysters. There appears to be a tendency for females to change to males where and when food supply is poor and for males to become females where and when supply of food is good. In areas with good food supply, the sex ratio in older oysters shows a preponderance of females, while in areas with a low food supply, the reverse is true.

In other species, such as the native oyster (*O. lurida*), an individual may spawn initially as a female, immediately change to a male and spawn as such within one summer. Also, the initial sexual phase in this oyster is always male. However, there always seems to be a satisfactory sex ratio to ensure successful breeding.

Gonad Development

Pacific oysters develop sexual products (egg or sperm), presumably by conversion of the winter store of glycogen, when water temperatures begin to rise in March. Under normal weather conditions, full ripeness is attained, in most British Columbia waters, by the end of June. Eggs of the female (Fig. 19) appear as tiny cream-coloured granules, just visible to the naked eye; while the sperm, extremely minute, form a pure white material which runs in thin streams.

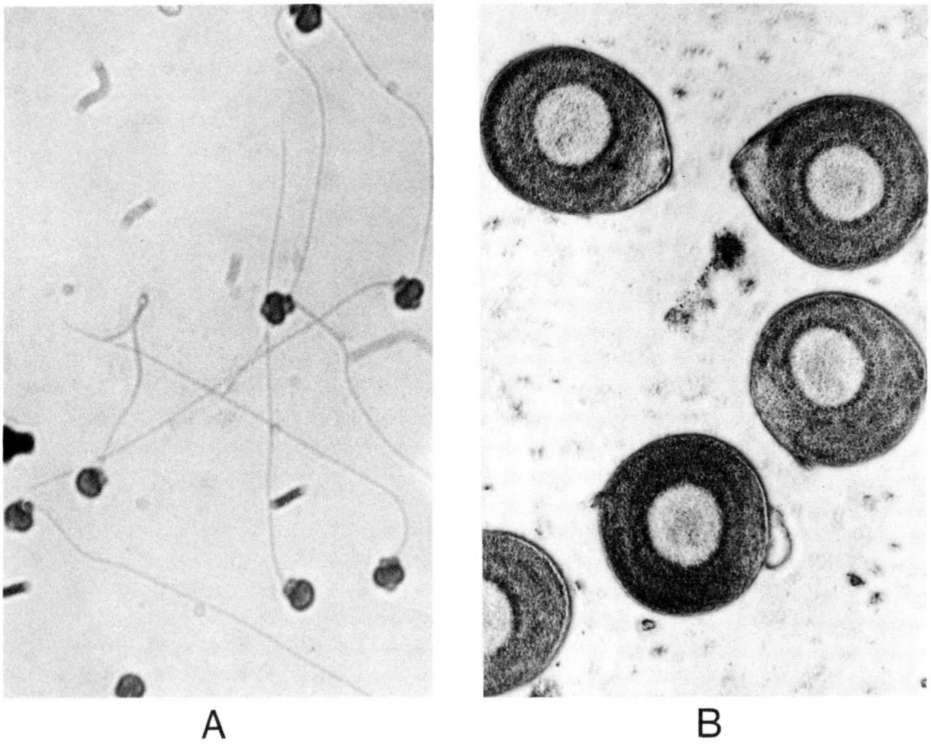

FIG. 19. Photomicrographs of sperm (A, ×1350) and egg (B, ×270) of the Pacific oyster.

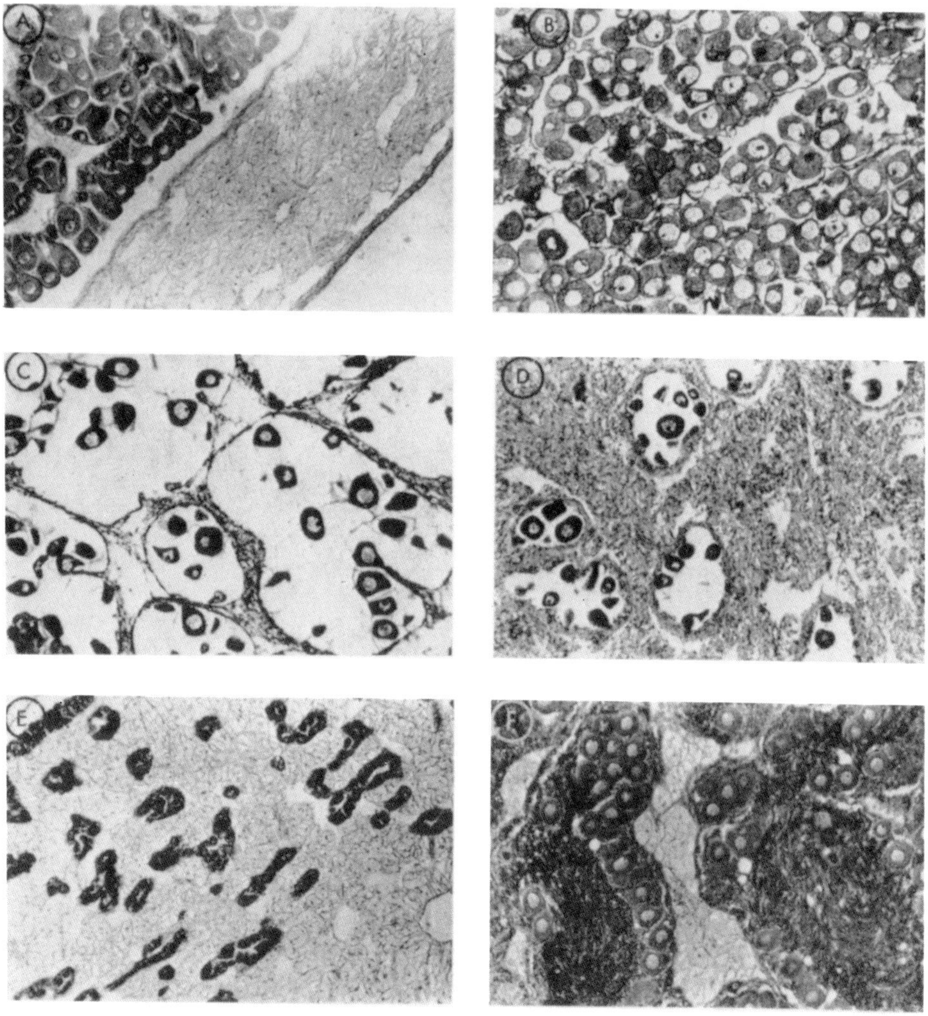

FIG. 20. Photomicrographs of sections of female Pacific oyster gonads to show seasonal changes. ×68. A = ripe female with gonad covered with a thin layer of glycogen-rich tissue; B = fully ripe female; C = spawned out female with a few relict eggs; D = female in fall condition with follicles closing in on the relict ova; E = female in spring condition with early-stage developing eggs on follicle margins; F = hermaphrodite with both eggs and sperm.

The number of eggs and sperm developed in a Pacific oyster is related to size of the oyster and to its state of nutrition. The number of eggs produced by an average market-sized oyster has been estimated at 50–100 million; the number of sperm very much greater.

During winter, the surface of a fat oyster is smooth, and even, but with onset of sexual maturity it becomes deeply veined (Fig. 18). This veined gonad covers both sides of the anterior end and forms a considerable proportion of the body weight. Fig 20A shows a thin section of a Pacific oyster gonad in which mature eggs are tightly packed in blind sacs called follicles, and tubules through which the eggs are discharged.

Details of the seasonal gonadal changes are as shown in Fig. 20 and 21. Spawning in Pacific oysters is usually total and may occur any time between late June and early September, but most often in late July and early August. After complete spawning the body is nearly transparent and the gonad follicles are collapsed and contain only a few relict gametes and tissue fragments. The condition is then at the lowest level of the year.

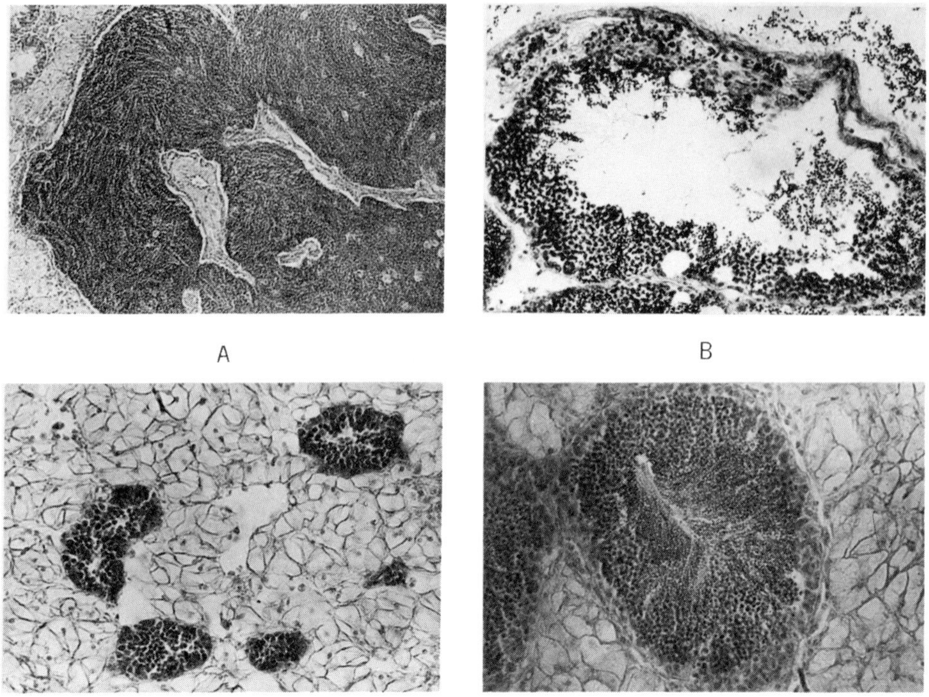

A B C D

FIG. 21. Photomicrographs of sections of male Pacific oyster gonads to show seasonal changes. ×69. A = ripe condition before spawning; B = spawned out condition with the folicle nearly empty; C = early development of male gonads; D = gonad approaching ripeness with developing cells in the outer portion of the follicle.

By November the level of winter condition has been established. Follicles have shrunk to small compact islands of germinal tissue scattered throughout the mass of vesicular connective tissue which has filled in interfollicular spaces as well as the area between the gonad proper and body epithelium. The relative amount of this connective tissue, particularly in the latter area, determines condition of the oyster. The main gonaduct separates the gonadal area from the outer connective tissue area where no germinal material occurs. At this stage some follicles that have not completely closed may contain a few relict eggs or sperm; in the few cases of partial spawning only outer follicles would be involved.

The gonad is generally undifferentiated as to sex at this time (November) and the condition exists throughout winter. It is not until April that early stages of gonad proliferation and differentiation may be noted. The maximum stage of development observed late in April occurs in females found with about 25% of gonad area occupied by follicular material. By mid-May gametogenesis is well under way and on average about 50% of the potential gonadal area is filled with expanding follicles.

By the end of June all animals are fully ripe with follicles tightly packed with eggs and sperm and only a very thin layer of vesicular connective tissue covers the gonad. This condition persists until spawning although an occasional partly spawned individual may show proliferation of connective tissues between spent follicles.

Spawning

In the spawning act, rather than releasing ova through the normal exhalant channel, a female discharges them into the suprabranchial chambers from which they are forced

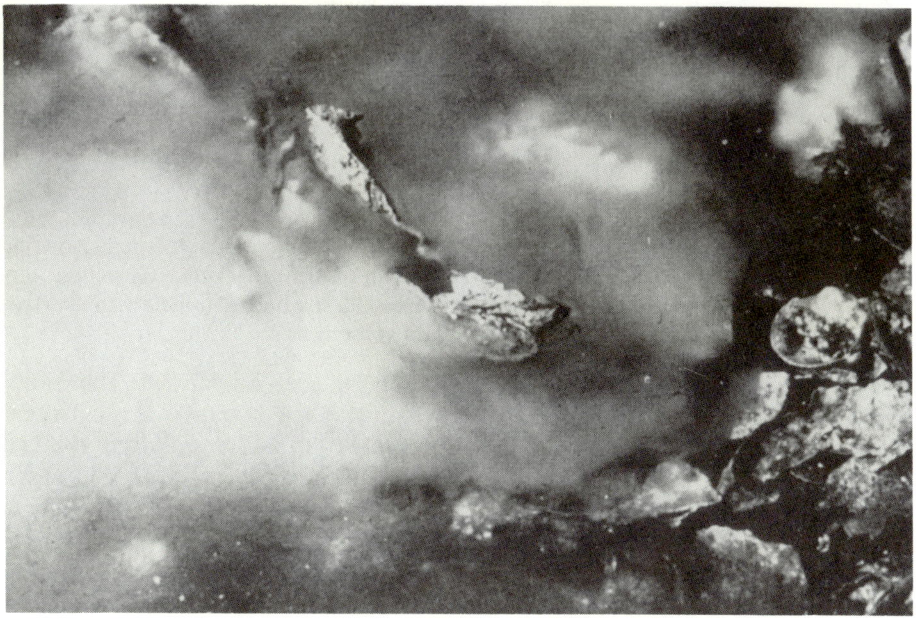

FIG. 22. Pacific oysters (*C. gigas*) in the act of spawning.

through gill ostia (apertures) into the mantle chamber, from which they are ejaculated in a small cloud (Fig. 22). This process is accomplished by adjustment of the mantle edges and by vigorous action of the adductor muscle. Discharge of eggs is intermittent, with a rate of 5–10 times per minute, eggs being forced distances of 30 cm or more from the oyster. The male, on the other hand, discharges its sperm in a thin steady stream, also into the suprabranchial chamber, but instead of passing through the gill apertures against the current as the eggs do, they are carried out in the normal exhalant stream of water. Thus eggs and sperm are discharged on opposite sides of the oyster.

Actual initiation of spawning may be brought about by temperature, chemical stimulation, or a combination of both. It is thought large mass spawnings that occur in molluscs such as the oyster, and indeed in a number of other marine animals, are necessary to provide concentrations of spawn needed to ensure fertilization when sexual products are discharged freely into open water. Presence of sexual products on which other oysters are feeding is enough to stimulate spawning. This fact makes it possible to force large quantities of oysters to spawn (Fig. 23).

FIG. 23. Spawn streak from Pacific oysters through the centre of the head of Ladysmith Harbour.

In Japan, the salinity range for development (breeding) of Paçific oysters is between 11‰ and 32‰; (‰ = parts per thousand symbol); the optimum is considered to be between 20‰ and 25‰. Development can take place within a temperature range 14–32°C, with the optimum at about 23°C. In British Columbia, salinity requirements are easily met, but optimum temperature occurs rarely.

Temperature is most important in initiation of spawning. Pacific oysters have been observed to spawn at temperatures as low at 15°C, but this may have been as a result of chemical stimulation.

In Pendrell Sound, the main breeding area in British Columbia, 18 recorded spawnings occurred at temperatures between 20 and 23°C, with an average of 21°C. These temperatures were recorded on a floating thermograph with the bulb 1 m below the surface. It is likely actual temperature at the spawning site would be 1–3°C higher.

Larval Development

Although eggs are pear-shaped when lying in the gonad (Fig. 20), on being spawned they assume a spherical shape, with a diameter of about 0.05 mm (200 eggs placed side by side would span a cm). Much smaller, swimming sperm seek out and fertilize the eggs, which are just able to float. Immediately after fertilization, which must occur within 10–15 h after spawning, rapid cell division and development takes place. In about 24 h, two tiny shells may be seen and various organ systems are being formed; within 48 hours the body is almost entirely enclosed by an enlarging shell. By means of cilia, the embryo is able to swim about 5 hours after spawning. Later, a definitive swimming organ, called the velum, is formed. This organ (Fig. 24) is a circular, ciliated platform which can be protruded from between the valves; its locomotive power enable the young oyster now called a veliger larva, to maintain position in the upper layers of water, although water currents remain the dominant force in moving larvae. In addition to its function as a swimming organ, the velum also acts as a food collecting apparatus. At this time, the larva is 80 μm in diameter, and the shell is haped like the letter "D" with the hinge on the straight side, and the larva is now said to be in the"straight-hinged" state.

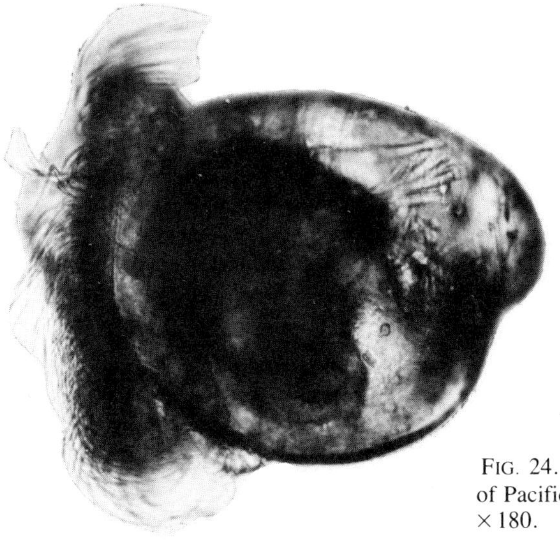

FIG. 24. Photomicrograph of an advanced stage of Pacific oyster larva with the velum extended. ×180.

In 3 or 4 days time, the umbones or beaks, which are protuberances of the shell near the hinge line, become visible, and the larva is now called "early umbo". Development continues rapidly and the umbones become more prominent and pointed (Fig. 25). In the meantime, a digestive system, complete with mouth, oesophagus, intestine, and digestive gland, is formed. In addition, there is a ciliated protrusible foot, which is a most active organ, and there are two adductor muscles; so at this stage the young oyster resembles

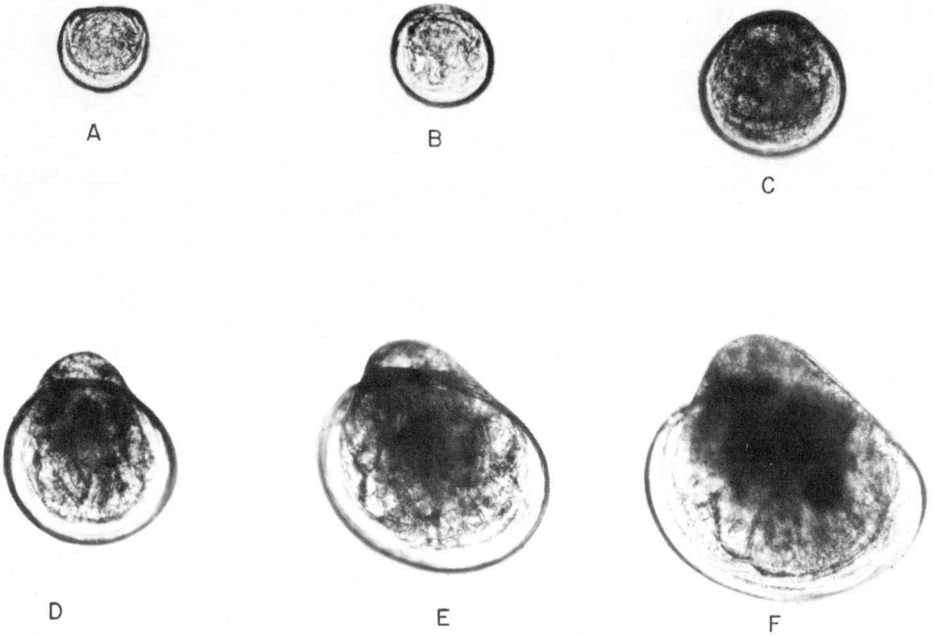

FIG. 25. Stages in the development of the larvae of the Pacific oyster. ×117. A = straight-hinged stage (90 μm); B = very early umbone (110 μm); C = early umbone (160 μm); D = umbone (200 μm); E = umbone (250 μm); F = advanced umbone (300 μm).

a clam more than it does an oyster. At the base of the foot is a cement gland whose function will be described later; there are also rudiments of gills, and a sensory organ called the statocyst.

Larval Growth

Not long before a larva reaches its maximum development, a black eye spot is formed on each side of the body. Growth of larvae seems to be directly associated with water temperature; the higher the temperature the faster the growth. Growth rate of Pacific oyster larvae, as determined in Pendrell Sound, is shown in Fig. 26; and in Fig. 27 length of the larval period at various temperatures, as determined from field observations is shown.

Setting

When a young oyster reaches a length of 0.30 mm, it has attained full development as a larva and has approached the end of its free–swimming existence. It is therefore ready to become attached and the process is called "setting" or "spatting". Indeed, when this stage has been reached, it must become attached to a solid object within a relatively short time, or perish.

The larva is carried by currents, with the velum active and the foot protruded, until it strikes a solid object. If it does so, it clings with the adhesive ciliated foot, by which it is also able to crawl. In this manner, with the shell upright above the crawling foot, a larva searches the surface; if the surface is unsuitable the larva will drop off, or swim away, until it strikes another. When this is located (the basic requirements are that it be clean and hard) a larva will carry on what appears to be an aimless search. When the right spot is found, and this is often where there is a slight uneveness or groove, a rocking motion of the shell forces out a small drop of cement from the gland in the foot. The

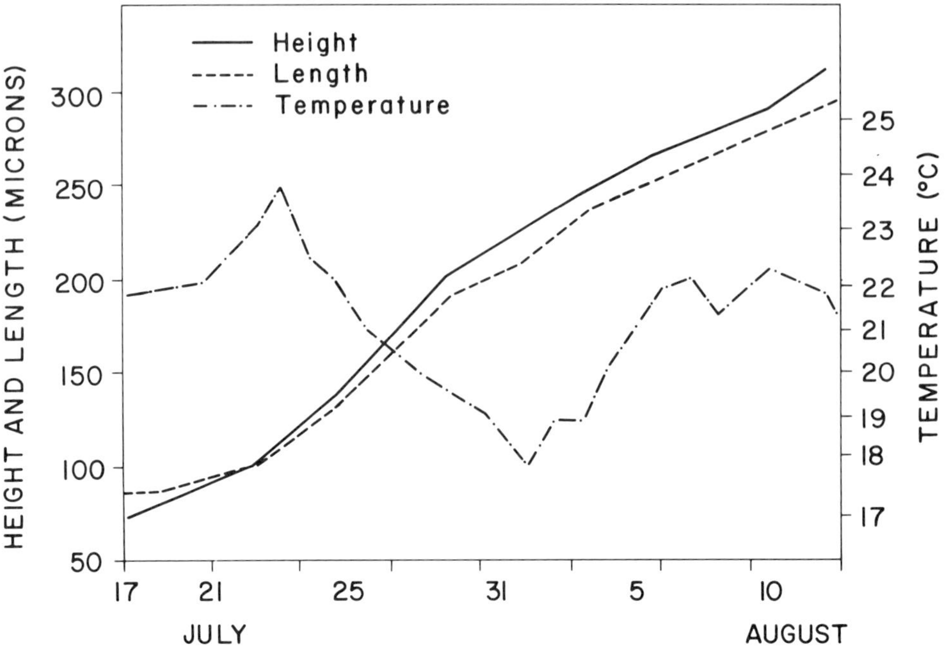

FIG. 26. Growth rate of the larvae of the Pacific oyster (*C. gigas*) in Pendrell Sound, August 1955.

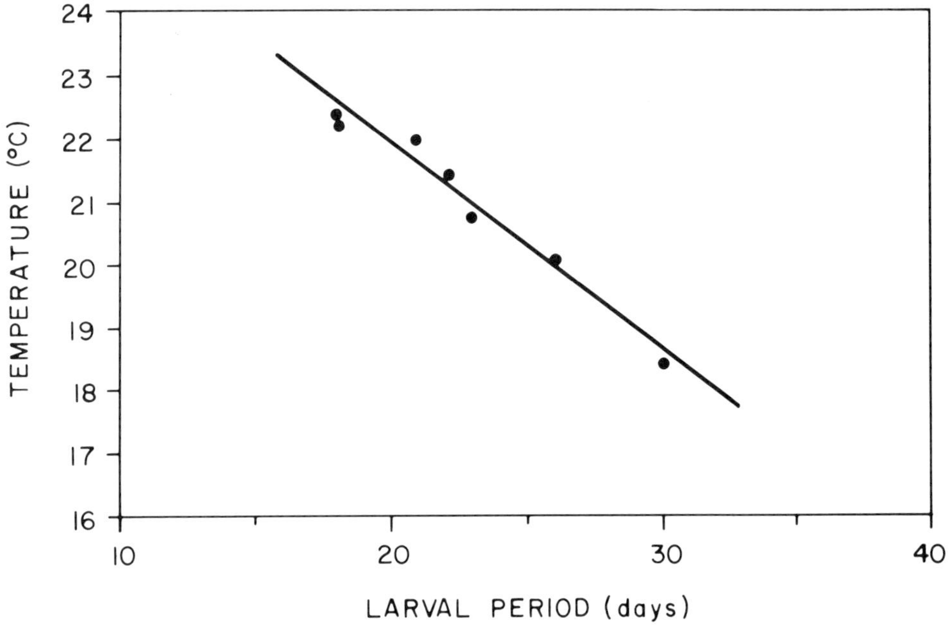

FIG. 27. Larval period of the Pacific oyster at various temperatures. Data from Pendrell Sound, 1950–55.

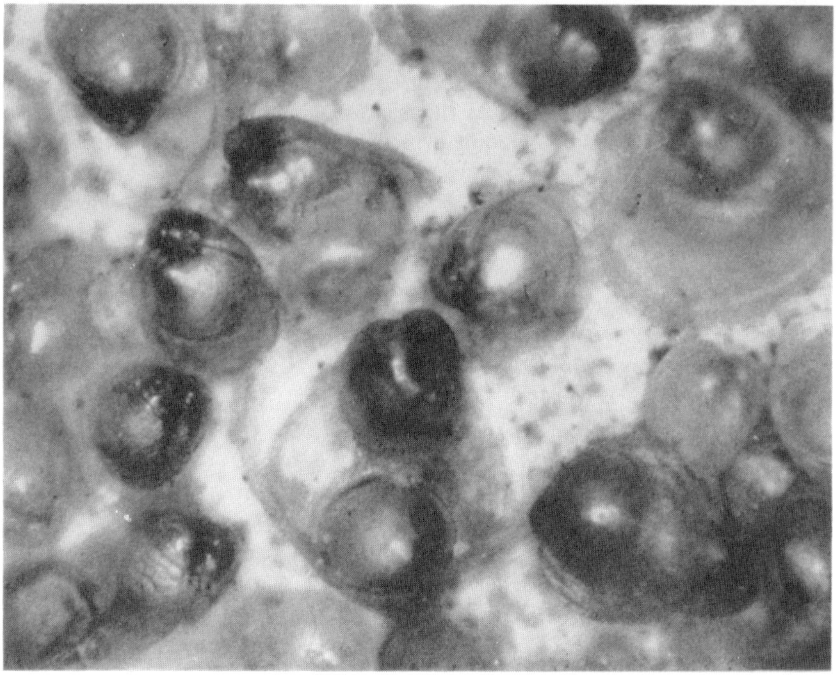

FIG. 28. Pacific oyster spat 1–4 days old. ×70.

left valve, never the right one, is then placed in the cement which sets in a matter of seconds, and the young oyster is attached for life (Fig. 28). Initial attachment is common to nearly all larval bivalves, but in the case of clams or mussels, attachment is by means of an elastic thread or byssus which is formed from the same type of gland in the foot.

Experiments with various chemicals have shown that larval oysters of *C. gigas* can be induced to settle and metamorphose by treatment with L-DOPA (L-dihydroxyphenylalanine). Larvae can be caused to metamorphose without settlement, thus producing single spat, by exposure to catechos, epinephrine, norepinophrine. Such control can have some benefits but the long term effects are still unknown.

Immediately after settlement, rather drastic anatomical changes occur in the oyster. The velum and foot are lost, as well as the anterior adductor muscle, while the posterior adductor takes a position closer to the centre of the shell. This change involves a rotation of the larval body in an anticlockwise direction, so the mouth is directed to the anterior rather than ventral as in the larva. The eye spot also disappears.

The gill system enlarges rapidly at this time and an associated increase in size of the mantle permits quite rapid growth of the calcareous adult-type shell. The larva shell, of different composition from that of the adult, called the prodissoconch shell, remains in a position at the umbone of the growing oyster and may persist for some time; indeed it has been found on fossil shells. It is only by means of the persisting larval shell that young Pacific oysters may be distinguished from young native oysters.

Breeding

Breeding History

Since the Pacific oyster is native to a country where relatively high summer water temperatures occur, the possibility of successful breeding in British Columbia appeared somewhat remote, at least on a regular basis. This has been shown to be largely true.

In the early years, between 1920 and 1930, when Japanese oysters were first introduced, the occurrence of spawning was of prime concern. The first recorded natural spawning of Pacific oysters in British Columbia occurred in Ladysmith Harbour, in 1926, when it was estimated that between 10 and 15 thousand oysters spawned and produced several hundred spat. The first major breeding which produced a commercial set occurred in Ladysmith Harbour in 1932.

Details of the history of Pacific oysters breeding in British Columbia are given in Table 2.

TABLE 2. Breeding of the Pacific oyster in British Columbia waters.

	Spatfall	
Year	Ladysmith	Pendrell Sound
1912–25	a few hundred (whole harbour)	—
1926	a few hundred (whole harbour)	—
1927	a few dozen (whole harbour)	—
1928	a few dozen (whole harbour)	—
1929	nil	—
1930	a few hundred (whole harbour)	—
1931	a few hundred (whole harbour)	—
1932	commercial (13 per shell)	—
1933	light noncommercial	—
1934	light noncommercial	—
1935	nil	—
1936	commercial (2 sets, 100 per shell)	—
1937	light noncommercial	—
1938	light noncommercial	—
1939	light noncommercial	—
1940	nil	—
1941	noncommercial	—
1942	commercial (50 per shell) (also in Pender Harbour and Baynes Sound)	—
1943	commercial	—
1944	commercial	—
1945	noncommercial (2 per shell)	—
1946	nil	—
1947	noncommercial	—
1948	noncommercial	commercial
1949	commercial (10 per shell)	commercial
1950	noncommercial (5 per shell)	commercial (44 per shell on shore cultch)
1951	noncommercial (5 per shell)	commercial (500 per shell)
1952	noncommercial (nil)	commercial (95 per shell)
1954	nil	noncommercial (6.5 per shell)
1955	nil	commercial (14 per shell, 10 per shell)
1956	commercial (25 per shell, 10 per shell)	commercial (3 sets, up to 1000 per shell)
1957	nil	commercial (21 per shell)
1958	commercial (several sets, up to 500 per shell)	commercial (several sets, 1000 or more per shell)
1959	nil	commercial (2 sets, 107 per shell, 100 per shell)
1960	nil	commercial (2 sets, 70 per shell, 33 per shell)

TABLE 2. (*Continued.*)

Year	Spatfall	
	Ladysmith	Pendrell Sound
1961	commercial (46 per shell, 30 per shell)	commercial (4 sets, 1041 per shell, 213 per shell, 261 per shell, 581 per shell)
1962	nil	commercial (2 sets, 1000 per shell, 100 per shell)
1963	nil	commercial (2 sets, 1000 per shell each)
1964	nil	commercial (75 per shell)
1965	light	commercial — very heavy — more than 1000 per shell
1966	light (15 per shell)	commercial (60 per shell)
1967	commercial (50 per shell)	commercial (38 per shell)
1968	light	commercial (2 sets, 40 per shell, 23 per shell)
1969	nil	commercial (25–50)
1970	—	commercial (17–69)
1971	commercial (30)	commercial (1000 +)
1972	nil	min. commercial (7–44)
1973	nil	noncommercial <10
1974	noncommercial (5)	commercial (40 +)
1975	noncommercial (< – 1)	commercial (>100)
1976	nil	nil
1977	noncommercial (1)	commercial (1000 +)
1978	noncommercial	commercial (100 +)
1979		commercial (50 +)
1980	nil	commercial (100 +)
1981	noncommercial	min. commercial (7–35)
1982	—	light commercial
1983	—	commercial
1984	—	commercial (100 +)
1985	—	commercial — later spat mortality
1986	—	nil

Year	Spatfall	
	Hotham	Pipestem
1973	commercial	—
1974	noncommercial (3–4)	—
1975	noncommercial (3–4 shell)	—
1976	nil	—
1977	commercial (30 +)	—
1978	commercial (30 +)	—
1979	—	—
1980	nil	—
1981	nil	commercial (100 +)
1982	—	—
1983	—	—
1984	—	commercial (25)
1985	—	commercial (25)
1986	—	nil

This successful 1932 breeding in Ladysmith Harbour was followed by another in 1936, part of which was the result of stimulated spawning of about 2 ha (5 acres) of oysters near the head of Ladysmith Harbour. Light setting occurred in succeeding years until 1942 when there was successful breeding in most oyster growing areas in the Strait of Georgia (Fig. 2). As a result oysters were distributed over a large part of this area and fortuitously, larvae were carried into Pendrell Sound in the north east corner of the Strait of Georgia, presumably from oyster growing areas in Pender Harbour. In the Strait of Georgia there is a general anticlockwise current movement.

Pendrell Sound, as a source of oyster seed was found in 1950 (Fig. 29). The potential had existed for several years previously but had not been recognized. Hotham Sound, a branch of Jervis Inlet (Fig. 30), also on the eastern shore of the Strait of Georgia, demonstrated its potential as a seed producer in 1973. This Sound is small, 6.4 km long and 1.6 km wide, but is very deep, up to 600 m (300 fathoms). It is well protected by mountains rising sharply from steep shores on which there is very little beach area.

In 1936 a cooperative venture between the Nanaimo Biological Station and the British Columbia Packers Limited planted Pacific oyster seed from Ladysmith Harbour at a number of sites along the west coast of Vancouver Island. More plantings were made later and resulted in wild populations in Barkley Sound and other inlets such as Sidney, Nootka and Kyuquot. Still more recently there is evidence that parts of Tlupana Inlet in

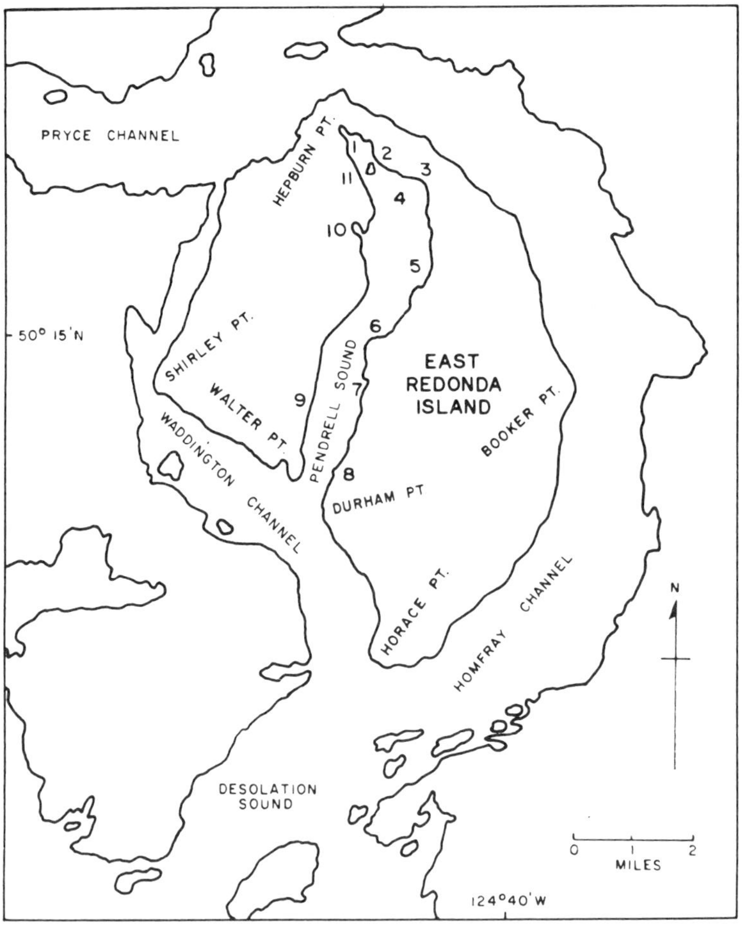

FIG. 29. East Redonda Island and Pendrell Sound showing location of sampling stations. Depth in metres.

FIG. 30. Hotham Sound oyster breeding area in Jervis Inlet.

Nootka Sound has potential as a breeding area for a commercial level spatfall was collected there in 1986.

Also in Barkley Sound a small oyster operation using Japanese oyster seed existed in Useless Inlet in the late 1930's supplementing the 1936 plantings and there was light breeding there, probably beginning in the 1940's. By the mid 1950s further breeding had distributed oysters throughout the inner Barkley Sound, including Pipestem Inlet (Fig. 31). This is a narrow inlet about 0.8 km wide (½ mile) and 8 km long (5 miles) in the north east corner of Barkley Sound. Its depth ranges from 30 to 60 m (10 to 20 fathoms) with a semi-sill about 15 metres deep at the entrance. Pipestem Inlet began to develop as a seed producer in 1981.

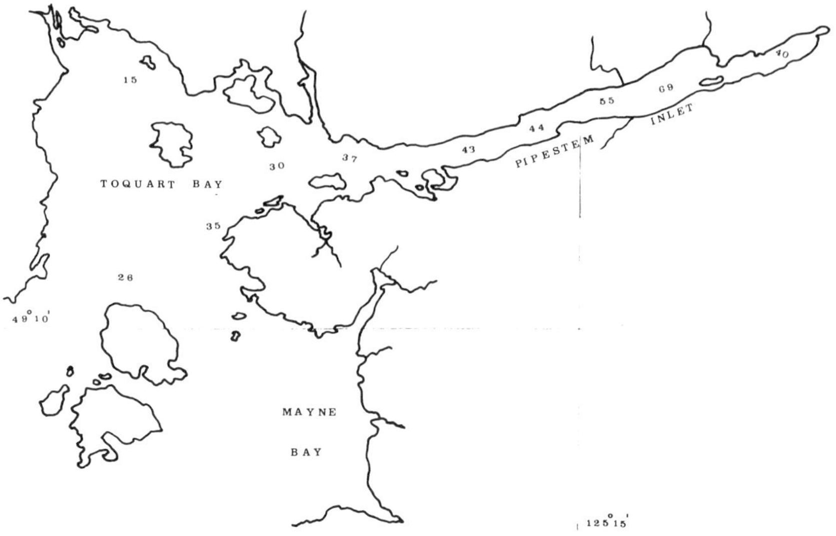

FIG. 31. Pipestem Inlet oyster breeding area in Barkley Sound on the West Coast of Vancouver Island. Depth in metres.

Stimulated Spawning

Initiation of spawning, notwithstanding the 1932 success in Ladysmith Harbour continued to be a problem, for when spawning did occur it was usually late in the summer and the larvae were unable to complete development owing to the rapid onset of cool weather. Also, delayed spawning or nonspawning caused adult oysters to enter winter in poor condition with subsequent marketing complications. This problem was partially solved by the discovery that spawning could be stimulated on a large scale on the beds. By stimulating spawning in early summer, advantage was taken of better water temperatures for larval development, and of a longer fattening period for spawned-out oysters.

Stimulated spawning on a large scale may be accomplished by loading a scow with several tons of oysters on the low tide of the day on which stimulation is to take place. Shortly before high water slack, usually in late afternoon or evening, the oysters are transferred to a sink float or live box. After the oysters have begun to feed, several dozen oysters are opened and macerated thus freeing spawn which is added to water in the sink float. Feeding on a suspension of eggs and/or sperm stimulates ripe oysters to spawn. Should this fail, it was discovered, strangely enough, if the sink float was towed at a speed of 2 or 3 knots for 5 or 10 min so the water rushed over the oysters, spawning was almost certain to begin. When spawning is well underway the oysters are dropped to the oyster bed below. The sink float should be so anchored that the ebbing tide will carry the spawn over the main oyster bed and timing of the operation should be such that the slow current, just at the turn of tide from flood to ebb, permits concentration of spawn among oysters adjacent to the spawners from the sink float. Gradually, more and more oysters on the bed are stimulated by spawn from their neighbours, and, within an hour, most oysters may be spawning.

Beds of up to 1 ha in area have been stimulated to spawn in this manner in Ladysmith Harbour. In 1936, a commercial set resulted from a series of stimulated spawnings. However, the practice of stimulated spawning has been little used by the industry.

It appeared that, by the mid-1940's, initial indifferent success in natural spawning had been overcome, although the reason is not clear; now, only in the coolest of summers do oysters in most areas fail to spawn. However, successful spawning does not necessarily mean successful breeding. As described elsewhere, after spawning and fertilization there is the free-swimming larval period and mortality may occur in larval broods, particularly during the first week or 10 days.

Time of Spawning

Some consideration has been given to the possibility of a connection between phases of the moon and spawning, and this is believed to exist in the European flat oyster (*O. edulis*), where spawning tends to reach a maximum at both spring tides, about 10 days after full or new moon.

In British Columbia, of 30 Pacific oyster spawnings when times of spawning were accurately recorded, 12 in Ladysmith Harbour and 18 in Pendrell Sound, all occurred on or within 3 days of a specific phase of the moon, 6 on the new moon, 10 the first quarter, 7 on the full, and 7 on the last quarter.

There appears to be a tendency for spawning in Ladysmith Harbour and Pendrell Sound to occur at about the same time. During the period, 1951 to 1956, 14 spawnings were recorded in Pendrell Sound and 9 in Ladysmith Harbour; 7 took place on the same day as, or within 1 day of, spawnings in Pendrell Sound. This is probably a reflection of weather and tidal conditions operating together to create similar water temperature regimes in the two areas. Sunny, warm weather in conjunction with long tidal run-outs generally causes a sharp upward trend in surface water temperatures and the oysters themselves are subject to long periods of high air temperatures.

Spawning of Pacific oysters does not occur throughout British Columbia every year, but in the recognized breeding areas, at least a proportion will spawn each year. As a general rule, once a Pacific oyster begins to spawn, all the gonad products are discharged at that time, and, on an oyster bed that has spawned, few oysters can be found with significant amounts of gonad material remaining. There may be several distinct spawnings occurring at different times in different parts of any one general breeding area in one year.

Temperature

Water temperature appears to be the main limiting factor in breeding success. It may not have a direct effect on larvae but along with sunlight it may have an effect on the production of larval food. The direct effect of temperature on larval survival is shown best in Ladysmith Harbour, which is a long, narrow, shallow estuarine-type bay. During calm, sunny weather in summer, water temperature can build up rapidly, partly because of heat transfer from the extensive tidal flats which are exposed to the sun's rays for considerable periods during spring tides. But because it is so shallow, a continuous breeze of any duration either from the southeast or northwest can cause vertical mixing and a rapid decline in water temperature. If the temperature fall is rapid enough or extensive enough, it can affect larval survival directly and immediately. If the temperature change is more gradual, the effect may be more indirect, probably acting on the development of suitable food.

In comparison, as described below, Pendrell Sound is able to maintain high water temperatures in spite of adverse weather conditions, unless these continue for extended periods.

In 1942, a particularly warm summer produced abnormally high water temperatures (Fig. 32) and breeding of Pacific oysters occurred in all oyster producing areas in British Columbia. As a result, larvae, which subsequently spatted, were distributed over the whole northern half of the Strait of Georgia and into long mainland inlets and sounds such as Jervis, Hotham, Malaspina, and Pendrell. The progeny from this spawning was so abundant it served to maintain a supply of oysters to the industry during the wartime period when Japanese seed was not available. Indeed, there was a sharp rise in oyster production during the latter war years, due largely to these ''wild'' oysters, as they were called, from the 1942 spatfall, and this was repeated again after the 1958 general breeding.

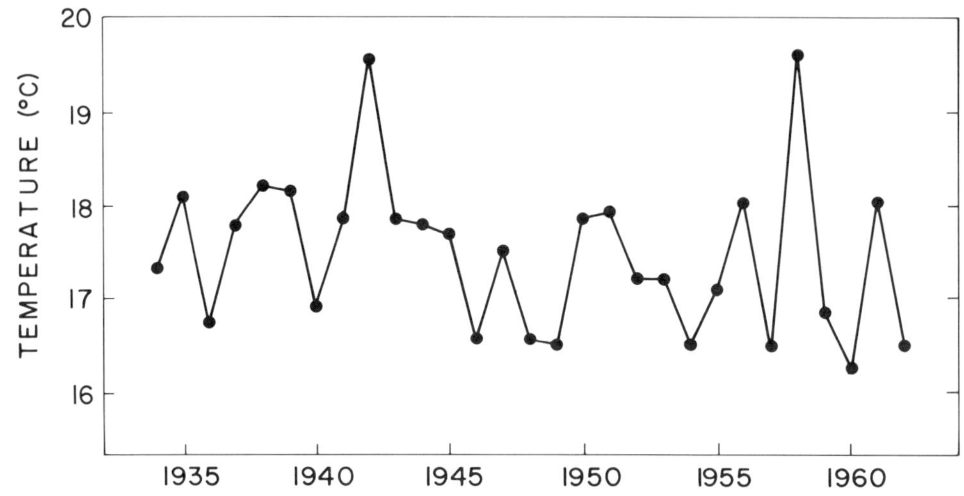

FIG. 32. Mean July–August seawater surface temperature at Departure Bay, B.C.

In addition, these oysters were abundant enough in many areas to form the nucleus of breeding populations. In Pendrell Sound, in particular, and, to a lesser extent, in Hotham Sound, conditions occur suitable for propagation of Japanese oysters.

Larval Distribution

Development of larvae after spawning has already been described, but little said about larval behaviour. Since Pacific oyster larvae have a free-swimming period of up to 4 wk and movement is largely dependent on currents, they may be carried considerable distances from the spawning area. An excellent opportunity for direct information on this point was obtained from the 1932 and 1936 spawnings in Ladysmith Harbour (Fig. 33, 34). As a result of these breedings, oysters were found spatted on rocks many kilometres from

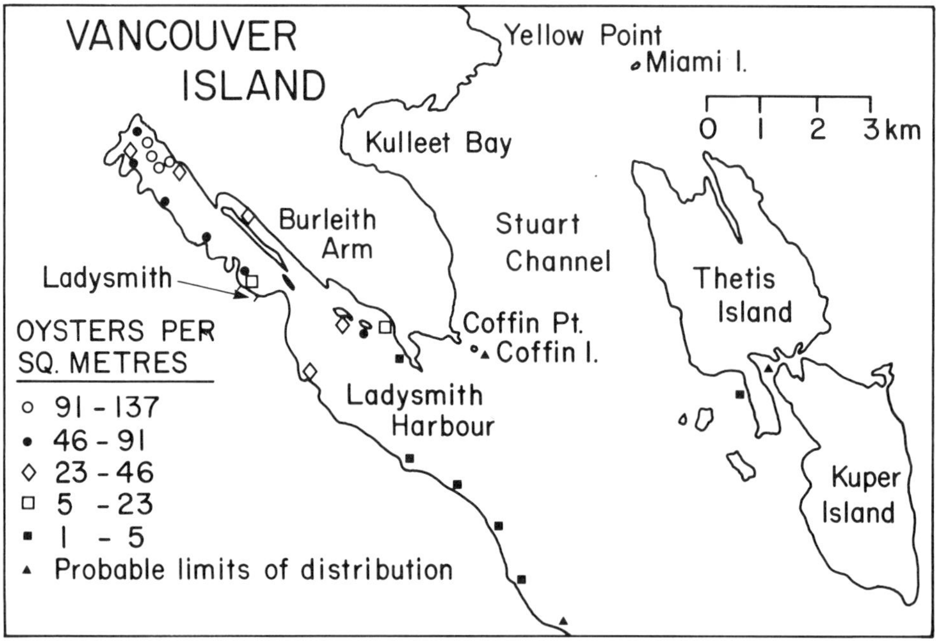

FIG. 33. Distribution of the 1932 Pacific oyster spatfall.

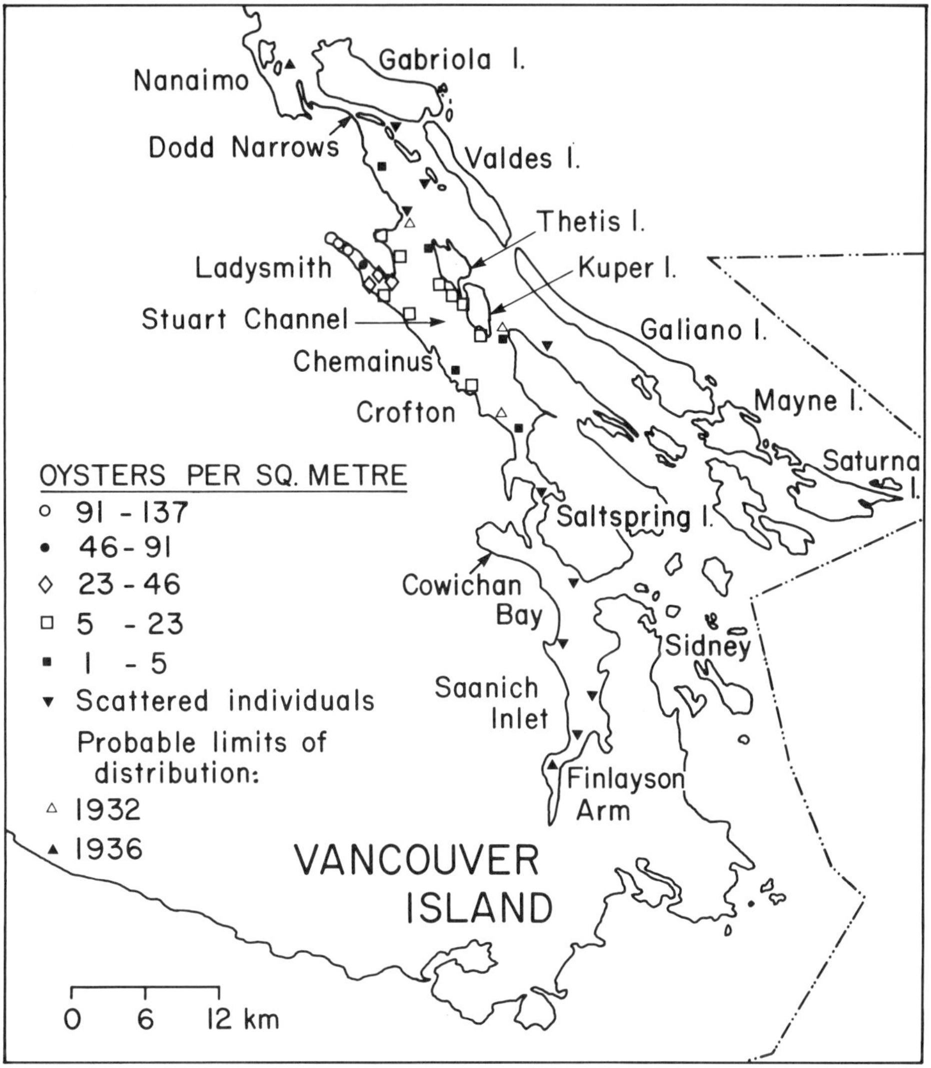

FIG. 34. Distribution of the 1936 Pacific oyster spatfall.

the head of Ladysmith Harbour, which at that time held the only stock of oysters from which these larvae could have originated. The greatest intensity of spatting in these two years occurred near the head of the harbour, where oyster beds were located, and numbers of spat per unit area decreased with distance from these beds. The most northerly point at which spat were found was just south of Nanaimo at False Narrows, indicating the larvae had been carried about 32 km in that direction. The most southerly point of spatting was at the head of Saanich Inlet, about 56 km from the head of Ladysmith Harbour (Fig. 34).

Another notable instance of movement of larvae from known stocks occurred in 1942, when larvae were presumed to move from Pender Harbour into areas such as Pendrell Sound.

The horizontal distribution of oyster larvae has also been studied in some detail in Pendrell Sound. There larvae become spread from the spawning area throughout the Sound within a week. Sampling indicates an aggregated rather than a random, or even, distribution.

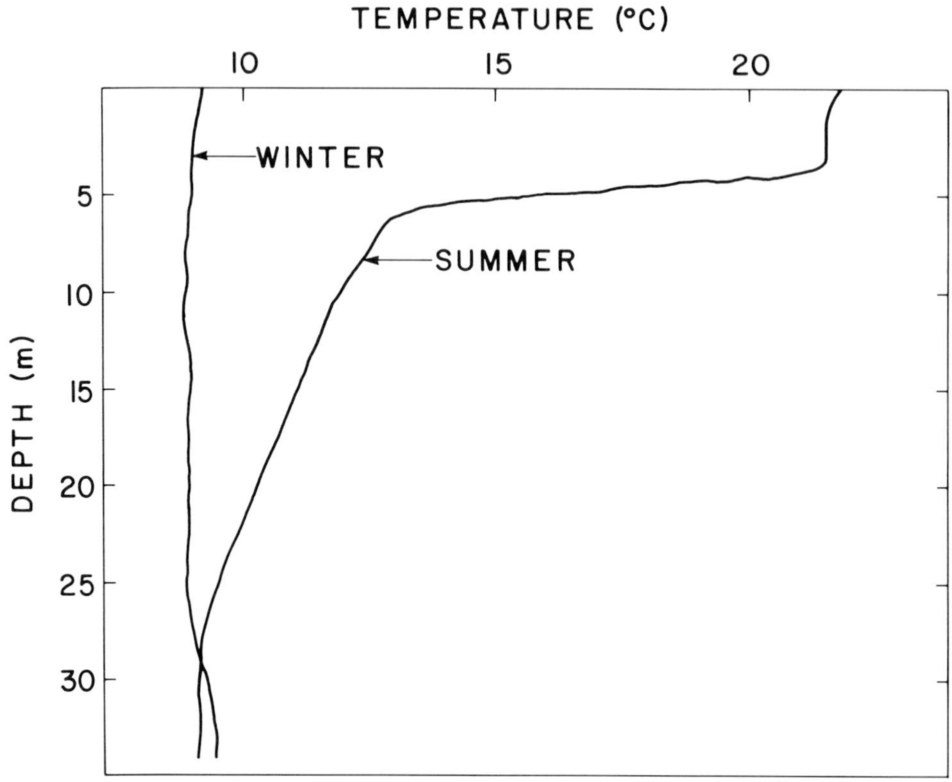

FIG. 35. Winter (March 26, 1956) and summer (July 30, 1956) temperature structure, station 4, Pendrell Sound.

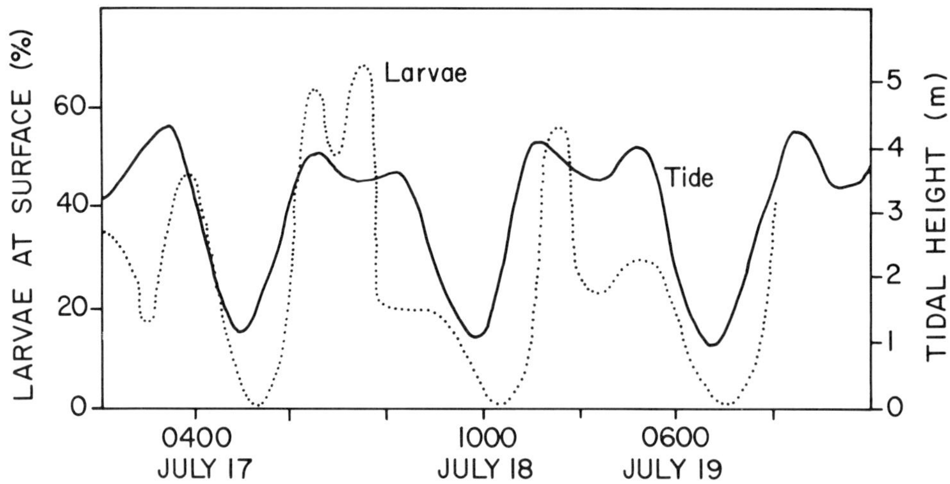

FIG. 36. Relationship between the tidal cycle and percentage of straight-hinged Pacific oyster larvae occurring at the surface. Percentage is the surface fraction of larvae in six samples taken between the surface and a depth of 5.5 m at 3-h intervals. The solid bars along the baseline indicate the periods of darkness. Station 2, Pendrell Sound, July 16–19, 1956.

It would be expected there might be a direct relationship between numbers of advanced larvae at any spot and density of spatfall and undoubtedly there is. However, repeated attempts to correlate the two have shown only slight indications. This is partly because methods of sampling larvae are inadequate. The quantity of larvae-bearing water passing cultch per unit of time, rather than the absolute number of larvae per unit volume of water at a specific time, may be the critical factor.

The vertical distribution of larvae in Pendrell Sound has also been studied by sampling at depths 0, 1, 2, 3, 4, 6, and 8 metres below the surface, sometimes at 3–h intervals for several days. It seems the majority of larvae do not descend below a depth of about 8 m, for only small numbers are found below this depth. Just above this (4–6 m) is an area of temperature change (Fig. 35) called the thermocline. The water above has a temperature of 21°C or more, while below that depth, the temperature decreases very rapidly. There is also a salinity change called the halocline, at approximately the same level. These water characteristics no doubt prevent the occurrence of significant numbers of larvae below the thermocline.

There may be some vertical movement of larvae as shown in Fig. 36. In a series of samples at various depths, the proportion of larvae that occur at the surface shows a fairly regular fluctuation that follows the tidal cycle. However, high tides occur in the evening or at night and it may be that, like many other marine invertebrate organisms, there is a tendency for oyster larvae to move near the surface at night and to retreat to deeper levels during daylights hours. If such movements are real or consistent, they must be taken into account when sampling oyster larvae for spatfall forecast purposes.

Larval Mortality

Because it is virtually impossible to estimate the number of adults participating in a spawning, it is difficult to assess larval mortality. It is thought, however to be of considerable magnitude. In Pendrell Sound, when initial numbers of straight-hinged larvae were compared with numbers of advanced-stage larvae just at setting time, the survival rate was found to vary between 0.5 and 9%. In some years, however, there has been total larval mortality.

If water temperatures are satisfactory, mortality is apparently due mainly to dispersal and predation. Water currents can disperse larvae and carry them away from spawning sites into unfavourable water conditions where they may perish. Among predators of oyster larvae are adult oysters which, in the feeding process, filter out all planktonic organisms, including oyster larvae, that cannot escape by active movement. Oyster larvae may be found in faeces and pseudofaeces, very often alive, but so enmeshed in mucus they are unable to survive. When it is considered that each of many millions of adult oysters in Pendrell Sound may filter up to 80 litres of water per day and that this water may contain many larvae per litre, mortality from this cause alone may be of some significance during a larval life of approximately 3 wk. In addition to oysters, there are other filter feeders such as clams and barnacles.

Still another predator is the sea gooseberry or comb jelly, the ctenophore (*Pleurobrachia pileus*), which captures oyster larvae by means of its two long tentacles (Fig. 37, 38). As may be seen from this photograph, a single ctenophore can account for about 20 oyster larvae at one time. Schools of anchovies frequent Pendrell Sound during summer and the mode of filter feeding can label them as larval predators.

Eventually, even though it is thought that many larvae can postpone metamorphosis for a time, oyster larvae must find an object to become attached to, or otherwise perish. In Pendrell Sound, the only setting area is the shoreline or floating cultch. With the small tidal action, it is likely that a considerable proportion of larvae fail to make contact and so perish. This must be a significant source of larval loss.

In spite or mortality by these causes, enough larvae survive to give spatfalls of high intensity near the spawning source and some spatting at considerable distances away.

FIG. 37. Ctenophore or comb jelly (*Pleurobrachia pileus*) with Pacific oyster larvae in the digestive tract, Pendrell Sound, July 1965. ×10.

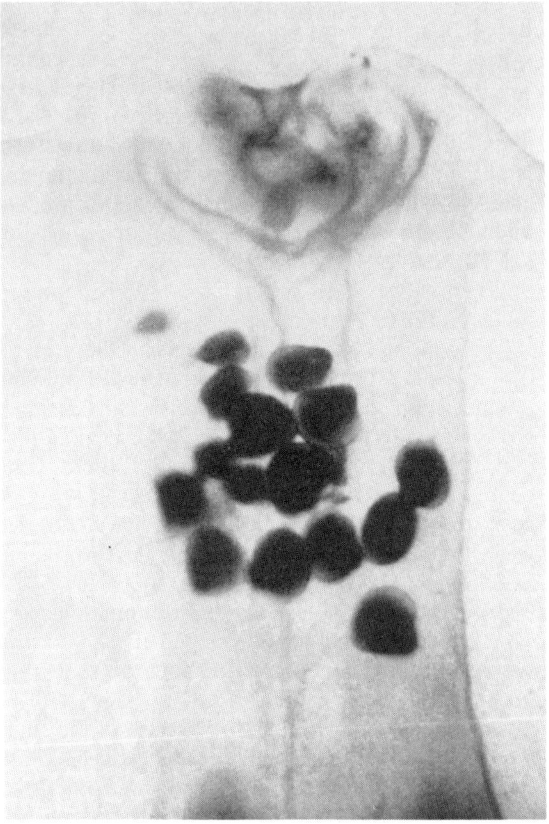

FIG. 38. Enlargement of Fig. 37 to show Pacific oyster larvae in the digestive tract of the ctenophore or comb jelly (*P. pileus*). ×33.

Hatchery Seed

The idea of artificial production of oyster seed has been in existence for a long time but it was not until the 1940's that the possibility was fully realized. The critical factor was shown to be need for suitable larval food. Through the years many attempts were made by a general enrichment of water in which larvae were held and some were successful if it happened the right food organisms were present, but there was no consistency. Such efforts were conducted in Ladysmith Harbour in the 1930's, and large cement breeding tanks were built at the Biological Station at Nanaimo in the early 1940's, but success was limited.

In about 1940, experiments in Great Britain demonstrated larvae required specific types of organisms of a certain size and two of these were isolated and named as *Isochrysis galbana* and *Monochrysis lutheri*. More species have since been added to the list of larval foods.

The quality of water for molluscan hatcheries is critical. It should be initially sand filtered and then sterilized with ultraviolet light. To produce larvae, oysters of both sexes are conditioned to ripen if they are not already in this condition, by holding and feeding them at breeding temperatures. For Pacific oysters, this is between 18 and 20°C and the process may take up to a month.

Spawning of ripe oysters is induced by rapid raising or lowering the water temperature or by feeding macerated sex products from other oysters. Certain chemicals may also induce spawning. Fertilization is carried out by mixing a small amount (several mL) of sperm to a suspension containing up to a million eggs. If necessary, a screen with a mesh of 80 microns filters out debris if necessary. The larvae are reared in concentrations of about 10 000/L at a temperature of 25°C and a salinity not beyond a range of 20 to 30‰.

Initially the larvae are fed daily to provide a concentration of approximately 25 000 algal cells/mL. As the larvae grow the algal concentration is increased three fold by degrees and the number of feedings increased. The water is changed at intervals by draining the tanks through filters of the appropriate size. The survival and growth rate of the larvae is monitored regularly to detect any deviations from the normal course of events. Oyster larvae are subject to disease as well as the adults, and these are of serious concern to hatchery operators. Commercial hatcheries produce enough larvae to provide shipments up to 5 million to individual clients for remote setting during the period between April and August.

When larvae reach setting size of 0.3 mm after 2–3 wk the hatchery may take three routes to produce spat.

The first method is to introduce into the growing tanks, the cultch of choice, such as oyster shell. This presents two logistical problems: first the number of tanks required to produce appreciable numbers of spat, considering the bulk of shell cultch. Secondly, the grower usually wishes to receive spat of an appreciable size so this means feeding spat until they reach an appropriate size. Also growers like to receive spat just before growth begins in spring when difficulties caused by winter weather are over. This holding is costly in terms of food and in tank space. For a short period of the year when outside water temperatures are nearly equivalent to the culture temperature they may be placed in the sea.

To overcome problems of space limitation, a system of producing single–sometimes called cultchless-spat was developed. This is accomplished by using small pieces of cultch such as marble chips or by brushing the spat from pliofilm soon after they have set on it. After these spat have grown to a diameter of 2–3 mm they could be sent to growers in small containers (Fig. 39a). Because of the small size, spat must be grown initially in a protected environment such as trays (Fig. 39b, c). At first most trays require fine mesh liners. The trays should be held in an area of still water, otherwise the spat will be washed and piled in one corner, requiring constant redistribution. In still water, they also require attention, for there is a tendency for them to attach to each other and to the tray. For appreciable production a considerable number of trays are required. As a result of these problems, production of single spat has been virtually abandoned. The problem of timing still persisted.

A third method now in general use is remote setting of eyed larvae. In this case the hatchery grows larvae to a stage when the eye spot develops which is just before attainment of setting size. These larvae are collected and shipped to growers wrapped in a damp cloth and other water absorbent materials in a cooler and held at a temperature of about 5°C. About two million larvae can be contained in a match box. In this condition the larvae can be held for a period of about a week but preferrably as short a period as possible. On receipt of the larvae, the grower places them in tanks of heated water at a temperature of about 25°C and in which the cultch of choice has already been placed. Some water stirring device such as air bubbled from the tank bottom distributes larvae among the pieces of cultch. Mechanical stirring is preferable. Under normal conditions approximately 20% of the larvae are spatted.

Spat are held in the tank at the same temperature or it may be lowered slowly to approximate the outside sea temperature when the spat may be removed from the tank. During this period the spat may be fed with appropriate organisms such as *Phaeodactylum, Dunaliella, Chaetoceros, Cyclotella*, or *Isochrysis* either from cultures or in paste. The tanks may be of various sizes, but larger tanks (circa 22 000–44 000 L, 5 000–10 000 gal) appear preferable and materials such as plywood, fiberglass, cement should be insulated. They may be heated electrically, by steam or hot water heat. The tank should be constructed to permit adequate cleaning and with drainage valves. Alternatively tanks may be refilled with raw sea water to provide food if the temperature differential between tank and sea is not too great. Depending on the area and time of year (organisms in the sea) it may be necessary to filter the tank water. If it is necessary to change the water in the tank while the larvae are still present it can also be drained out through a sieve

FIG. 39. Hatchery bred single spat. a. shipping container. b. trayed single spat 15 mm in diameter.

called a "bongo" filter which is usually a 15 cm diameter plastic pipe holding special nylon screens of various pore sizes. The filters are available from the larvae suppliers.

It is considered 100 larvae per averaged size Pacific oyster shell cultch piece will produce in an average of about 20 spat per shell with a range of about 5 to 100. This is equivalent to a density of about 100 larvae per litre (approximately 500 per gallon). Alternatively, about 100 larvae per piece of shell cultch is a suitable density. With 2-m tubes commercial operators aim for 500-600 spat per tube from a provision of 3 000 larvae per tube.

In most B.C. areas, summer water temperatures seldom approach the tank temperatures required for setting eyed larvae so to minimize water heating fuel costs and long time retention in the tanks, this method is essentially a summer operation. The size of spat when transferred to the sea is of less importance than the temperature differential and is a function of subsequent handling of the cultch. It may be suspended for continuous immersion but preferably stored intertidally for conditioning (hardening) to improve subsequent survival. Initial intertidal storage should be at a low level and preferably during a neap tide series so the initial air exposure, is minimal, gradually increasing as the neap series changes into a spring tidal period.

Genetics

There is a developing interest and activity in the field of oyster genetics mainly because breeding control is now possible with the advent of hatchery technology. Maintenance of broodstock improvements are only possible through continued hatchery production.

It has been shown there are genetic differences between oyster populations of the same species and by genetic manipulation selective breeding is possible. The potential for this is considerable but the progress is and will continue to be slow and in British Columbia there is no activity in this field at the present time.

Of some concern is the possibility of deleterious effects from inbreeding in the Pendrell Sound oyster population which is probably a pure line (gene pool) from the original 1942 population. Any impact from the outside, which was possible from the extensive 1958 breeding throughout Georgia Strait, would likely be suppressed by the large and concentrated Pendrell Sound population.

Triploidy

There is at present considerable interest in producing oysters with three sets of chromosomes (triploidy) instead of the normal with two sets (diploid). This is accomplished in molluscs by treating newly fertilized eggs 10 min after fertilization with 0.05 mg/L of the chemical cytochalasin B for about 20 min. This interferes with the normal chromosome sorting after fertilization, resulting in three sets instead of the normal two. The proportion of triploids developed is variable, depending on the exact timing of the operation. The triploid animals are undetectable until sacrificed. The altered chromosome number inhibits normal gonad development in the adult apparently making the oyster more palatable during the summer months when normally this is at a low level. Survival and growth in triploid oysters is considered to be as good as in the normal diploids. Triploidy, of course, can be initiated only in a hatchery setting. The manipulation is still in the early stages of development as well as its applicability to large-scale operations.

Spatfall Forecasting

Necessity for Forecasting

In British Columbia the Pacific oyster is under a decided handicap with respect to breeding, owing to low water temperatures. The exception is Pendrell Sound, a small area in the northeast Strait of Georgia (Fig. 29) where summer water temperatures often exceed 20°C for some time. Other breeding area temperatures are usually reached but are not often maintained for a period long enough for larvae to survive to setting.

Since spatting may be erratic and timing variable, it is important to avoid fouling of the cultch by marine organisms, so "blind" cultching is both impractical and uneconomical for suspended cultch. These problems may be partially eliminated by greater use of cultch placed on the shore where it may be left *in situ* for the next year if there are problems of failure or partial spatfall failure. For floated cultch, prediction of the time and intensity of spatfalls is desirable and collection of seed in B.C. has come to be based almost entirely on spatfall forecasting. This was originally carried out by the Provincial Shellfish Laboratory at Ladysmith and forecasts were relayed individually to firms interested in seed collection or broadcast over the then CBC *Fishermen's Program* and published in oyster spatfall bulletins. In 1959 the Pacific Biological Station at Nanaimo took over. In 1981, the Marine Resources Branch of the Provincial Government assumed responsibility for the forecasts which it did for a time, but now contractors provide the forecasts.

Originally forecasts were made only for Pendrell Sound and Ladysmith Harbour, with the later addition of Hotham Sound. More recently Pipestem Inlet is included in the forecasts.

Technique

Oyster spawning in British Columbia is usually sudden and complete for any single oyster bed, and is often observed visually (Fig. 40). Therefore, individual larval broods may be followed closely with respect to number and distribution of larvae. This is done by means of plankton sampling.

In Pendrell Sound, which has an area of about 12 square km, plankton samples are taken at a number of stations. A known volume of water, as measured by a water meter in the line, is pumped from a standard depth of one metre through a fine mesh net where the larvae are trapped (Fig. 41). For adequate prediction, samples should be taken at intervals of not more than 3 days, but this is not always possible. Larvae in these samples are counted and the results show trends in the number of larvae well enough for prediction. There are many difficulties in sampling a microscopic organism such as an oyster larva which can carry out some swimming activity but whose distribution is largely the result of current action. A recent development is a plankton sampler which collects evenly from all depths down to 3 m and which can also be operated on a moving boat, so horizontal

FIG. 40. Spawning of a group of Pacific oysters in Pendrell Sound.

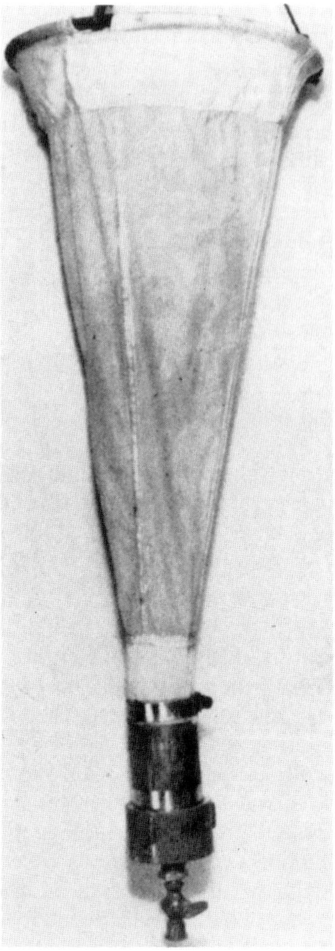

FIG. 41. Plankton net used in sampling oyster larvae. ×0.2.

as well as vertical sampling can be carried out simultaneously, thus increasing the accuracy of the determination of larval abundance (Fig. 42, 43).

Although the number of larvae per unit volume of water about a week before setting is the ultimate basis for prediction, rate of reduction of numbers is also important for this usually indicates a continuing trend. Along with this, trends in water temperature and weather conditions which may affect larval survival must be considered. Declining water temperatures, caused by rain or cloud cover, frequently coincide with diminution of the size of larval broods, or even their decimation. Gale force winds, particularly in an area like Ladysmith Harbour, reduce water temperatures so rapidly and drastically that larval broods may be eradicated within a day or two. Pacific oyster larvae appear to be able to withstand large, slow drops in temperature better than a small rapid decrease.

Spawnings are reported to the industry as soon as straight-hinged larvae are observed in the plankton. The number of larvae in the immediate vicinity of the spawning is used to make a preliminary estimate of the probability of a commercial set. Reports are provided at approximately 5-day intervals.

Knowing the rate of growth of larvae and length of larval life at different water temperatures (Fig. 27) it is possible to predict or to modify predictions of the date of the start of setting. About a week before the expected date of the set a final decision on time and approximate intensity of the set is given. Intensity is usually given in terms of whether

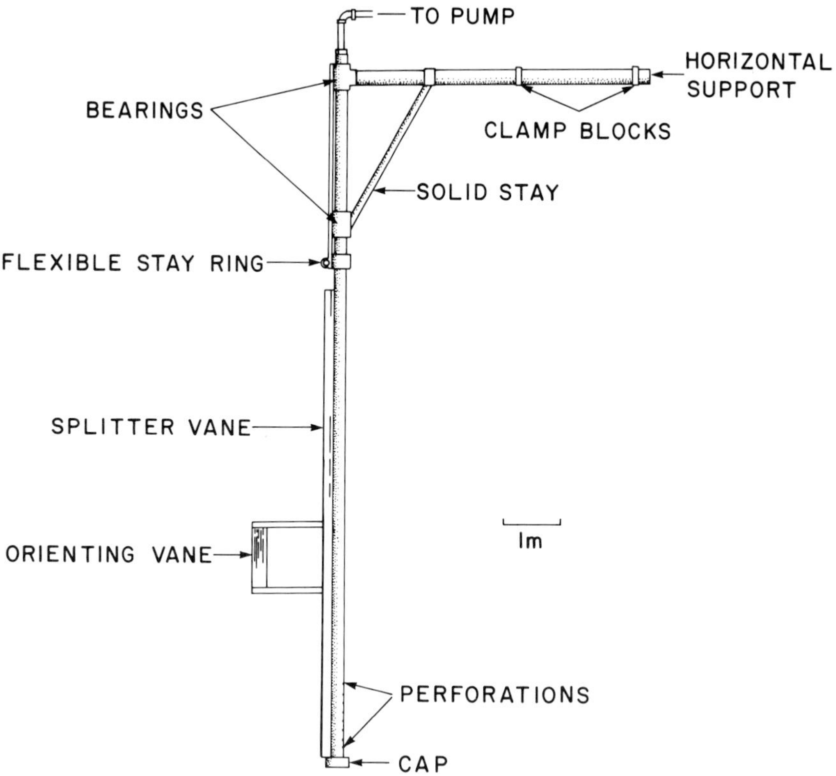

FIG. 42. Pipe sampler for oyster larvae.

FIG. 43. Pipe sampler for oyster larvae mounted on the starboard bow of a launch. The pipe leads back to a water meter and to the motor-powered pump; the long marked hose is used for taking spot samples from various depths.

or not the set will be commercial. An average of 10 spat per piece of shell or more has come to be the accepted criterion of a commercial spatfall. However, seed collected for sale usually requires 25 spat per shell, while seed collected for private use is often quite acceptable with five spat per shell, so that, in reality, there are several criteria.

It appears that an average initial concentration of 3–6 straight-hinged larvae per litre of sea water is required to produce a minimum commercial set under optimum conditions. This number of early stage larvae is usually reduced to about one advanced or setting stage larva per 4.5 litres at the end of the usual 18–21 day larval period.

In Ladysmith Harbour, commercial spatfalls (an average of not less than 10 spat per shell) have occurred on floating cultch in 11 of the last 53 years. In Pendrell Sound there have been commercial sets in 30 of the last 35 years.

In general, spatfall predictions have been remarkably accurate. In Pendrell Sound, results have been close to the forecast. In Ladysmith Harbour, the record is not quite so good, but at no time has cultch been lost as a result of inaccurate predictions. Accuracy depends to a considerable extent on the effort put into larval sampling; more frequent sampling results in more accurate forecasting. An indication of the usefulness of forecasts is that growers and seed collectors have come to rely on them and do not expose cultch unless the prediction indication is positive.

Pendrell Sound Seed

The Pacific oyster industry in British Columbia is based on the principle of bringing seed oysters from external sources to growing areas. The quantity of oysters harvested is then directly related to the quantity of seed planted, provided normal cultural conditions are met.

The west coast of North America has been fortunate in the availability of Pacific oyster seed through the years. Initially the main source was Japan, but later local seed became available as a supplementary source from Pendrell Sound and other areas in British Columbia, and from Dabob Bay and Willapa Harbour in the State of Washington. At about the time Japanese seed became quite costly and in reduced supply in the early 1970's, the development of hatcheries acted as an alternative source. Local seed and hatchery seed should be complementary to each other rather than reliance on either as the sole source and this is the existing situation.

Description

The history of breeding in Pendrell Sound has been followed in some detail since 1950 and there has been consistent spatting since then except for 2 or 3 years and when spatfall intensity was near or a little below the minimum commercial level. Consequently, from this area at least, a reasonably dependable supply of seed becomes available and Pendrell Sound has been a significant oyster seed producing area. Productivity is limited only by the amount of cultch exposed and there are at least 5–8 square km of area for this purpose. In some years the value of seed produced in Pendrell Sound has been equivalent to that of marketed oysters in all British Columbia.

As soon as the potential of Pendrell Sound was realized in 1950, the British Columbia government set aside the Sound as an area for use and enjoyment of the public. This means that anyone, public or oyster growers, may collect seed or oysters there; leases are not granted. The official reserve notice is shown in Appendix A.

Pendrell Sound penetrates deep into East Redonda Island in the northeast corner of the Strait of Georgia and is a deep fiordlike inlet about 10 km long and 1.6 km wide. At the head, a neck of low land less than 1 km wide separates it from Pryce Channel (Fig. 29). Hills, rising to 1 800 m high within 1 km or so of the shores, surround most of the inlet and no doubt serve as a wind break. On at least one peak, some snow may remain throughout the summer. The shores now, largely covered with oysters (Fig. 44), are steep

FIG. 44. Oyster reef in Pendrell Sound, July 1955, showing abundance of the spawning stock.

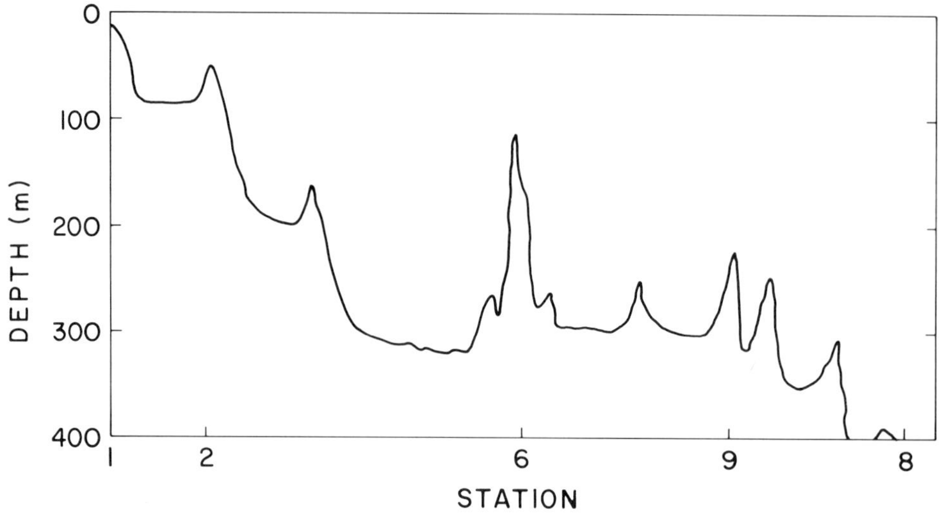

FIG. 45. Longitudinal depth profile through the centre of Pendrell Sound from the head at station 1 to the mouth at Durham Point (station 8).

and there is little beach area. Except for seepage-type runoff there is no significant freshwater drainage into the inlet. The depth profile from head to mouth is shown in Fig. 45.

Prevailing winds during summer in the Strait of Georgia are westerly and northwesterly, and, in winter, southeasterly. Because of the protective nature of the high hills surrounding Pendrell Sound, wind forces less than about 35 km/h, as recorded in the Strait of Georgia, have little effect in Pendrell Sound.

Westerly winds strong enough to have an effect result in breezes from the south, up the outer and into the middle sound. Under these conditions a calm area occurs in the middle sound area in the vicinity of the lagoon. Only near-gale-force southeast winds in the Strait of Georgia cause south wind effects to reach the head of the sound. Local westerly winds of considerable force may occur occasionally at night in the upper sound in July and August.

Temperature

One significant factor in Pendrell Sound, relative to oyster breeding, is maintenance of high water temperatures at the surface during July and August (Fig. 46). Surface water temperatures as high as 28°C have been recorded. The reason for this may be associated with the fact that heat resulting from solar radiation is not dissipated to any extent by vertical or horizontal mixing. Since the inlet is so deep and without a significant sill (shallow bar) at the entrance, there is little horizontal tidal movement. Absence of shallows and surface drainage inhibits vertical mixing. Also a layer of water, 3–6 m deep, with relatively low salinity, established during early summer, no doubt acts as an insulating layer which prevents vertical movements of deeper cold water. As early in the year as May 30 (1956), a temperature of 21°C was recorded at a depth of 1 m, although usually the temperature only begins to rise significantly toward the end of June. By mid-July, temperatures of 20°C and above are usually quite firmly established. Vertically, the 21°C level lies within 0.3–0.6 m of the 3-m depth (Fig. 35). Of frequent occurrence (4 years of the last 5) is a midsummer drop to below the 20°C level. This is usually associated with southeast winds and cloudy or rainy weather. This water temperature depression causes an hiatus in spawning activity and may affect survival of larval broods already present in the plankton.

Surface water temperatures are quite uniform in the middle and upper sound. In the outer sound temperatures are slightly lower.

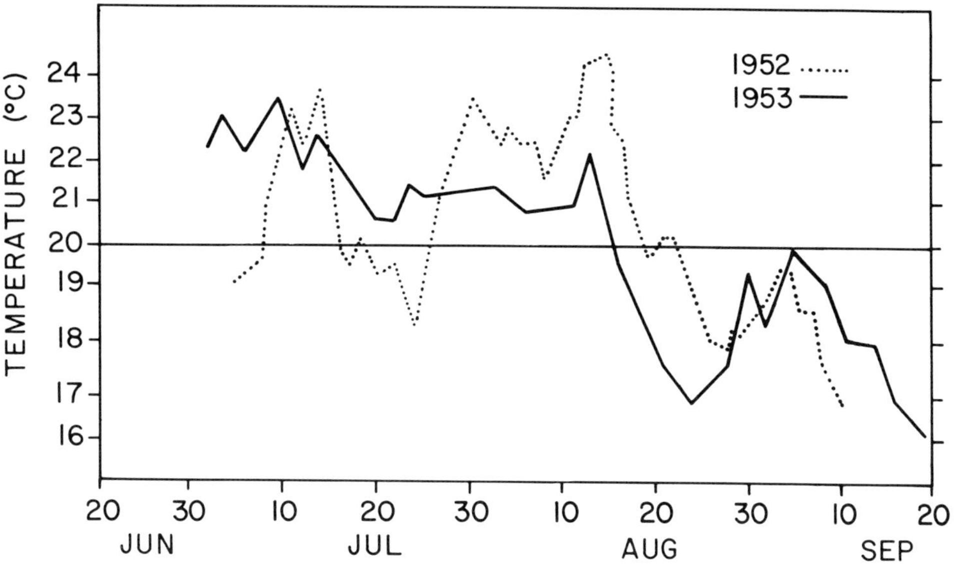

FIG. 46. Seawater temperature at a depth of 1 m at station 2, Pendrell Sound, June to September, 1952 and 1953.

Salinity

Salinity of surface waters in Pendrell Sound undergoes a pronounced seasonal cycle. This is shown in Fig. 47. During winter, salinity, down to at least the 25-m level, is quite uniform at 28–29‰. In May, an intrusion of low salinity water occurs along the surface. This water originates from Toba Inlet into which several large rivers flow. At the entrance to Toba Inlet, surface waters may have salinities of less than 10‰. The relationship between river runoff into Toba Inlet and salinity in Pendrell Sound is shown in Fig. 48. The salinity profile of waters surrounding Pendrell Sound is shown in Fig. 49. Salinity drops slowly until the beginning of August, when it begins to rise. This rise

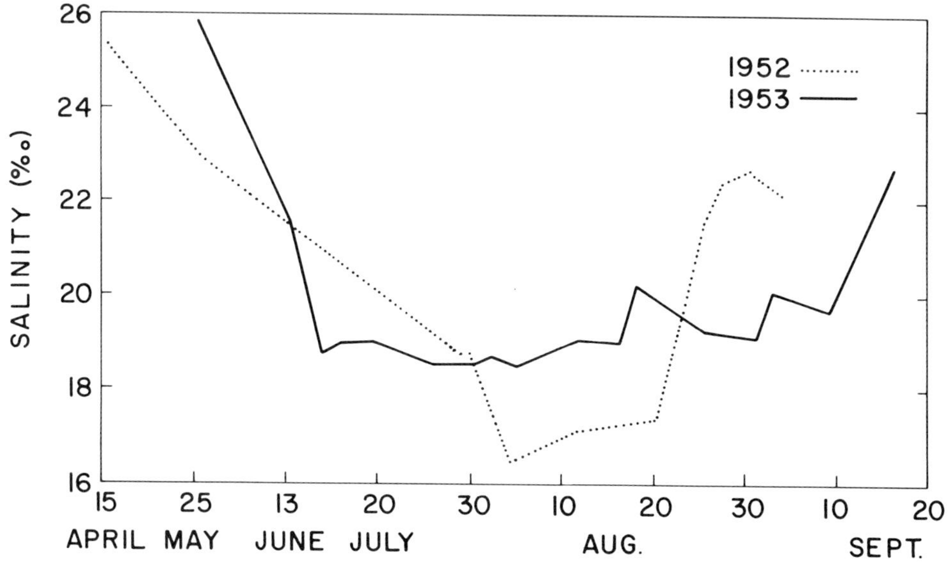

FIG. 47. Seawater salinity at a depth of 1 m at station 2, Pendrell Sound, April–September, 1952 and 1953.

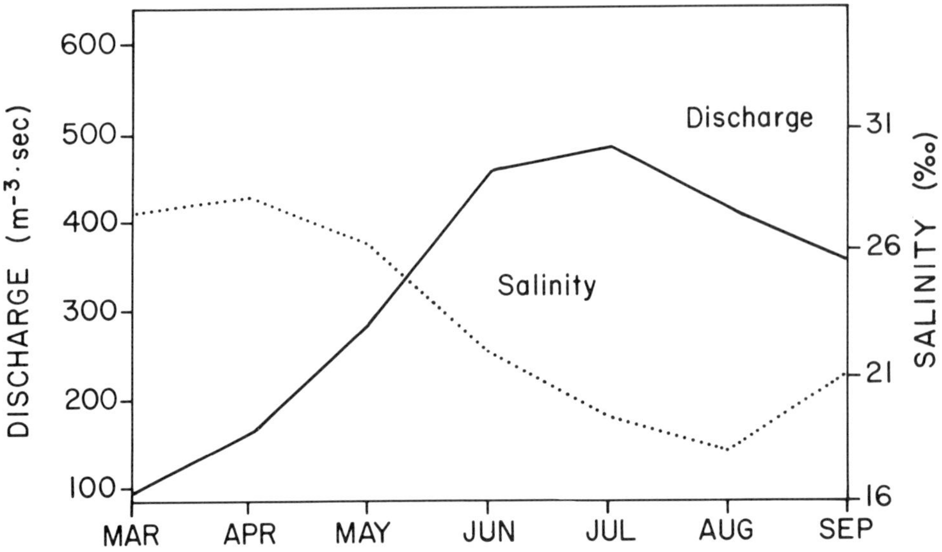

FIG. 48. Relationship between the mean monthly discharge of the Tahumming River, Toba Inlet, for the years 1923–31 and the typical salinity pattern at station 2, Pendrell Sound.

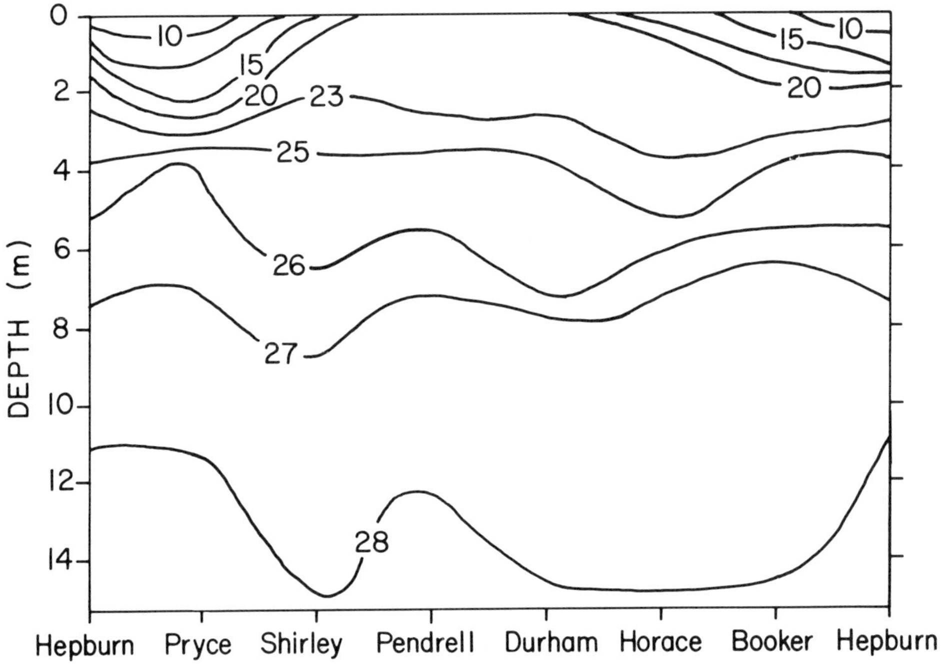

FIG. 49. Salinity-depth profile in the waters surrounding East Redonda Island and Pendrell Sound, July 16, 1956.

is accelerated by the advent of autumnal storms which cause vertical mixing and, by November, salinities are restored to normal winter levels. As with vertical temperature distribution, the halocline (area of sharp salinity change), in summer, occurs at about the 3-m level (Fig. 50).

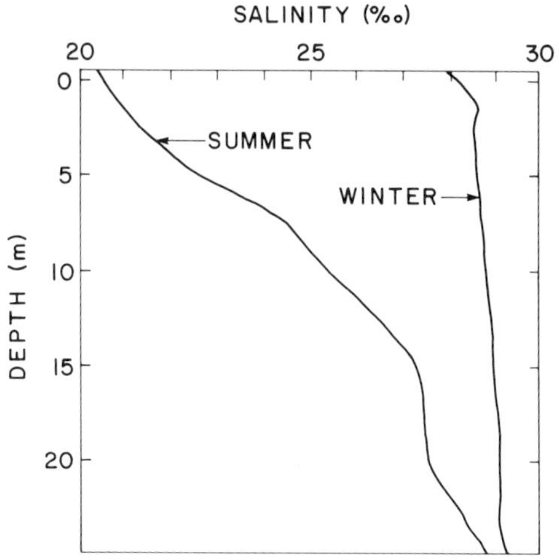

FIG. 50. Winter (January 31, 1956) and summer (June 20, 1956) salinity structure. Station 2, Pendrell Sound.

The relatively low salinity in Pendrell Sound, compared with Ladysmith Harbour, may be one of the factors involved in the high level of breeding success in Pendrell Sound. However, in years of exceptionally heavy river run-off as in 1967, the salinity in the upper sound may fall as low as 10‰, at which point larval survival is affected.

Currents

In a Pendrell Sound type of inlet, times of slack water at the head and at the entrance are almost the same, which results from the whole surface of the inlet rising and falling according to the tidal impulse operating at the entrance. Consequently, surface currents are weak. These have been studied in Pendrell Sound by means of free-floating drift poles, 4–5 cm, 1 m, and 3 m in length.

The conclusion from these studies is that surface currents in Pendrell Sound, particularly in the middle and upper sound, are weak, variable and generally of the order of 2 or 3 km/day, with a maximum of less than 1.8 km/h. Movements of surface floats are greatly influenced by wind. When there is no wind in the upper sound, movement of surface floats is inward on the ebb and outward on the flood. In the middle sound there is a tendency for surface floats to move inward. Free-floating debris has been observed to circulate about the middle sound within a radius of 2 km for periods up to 6 weeks. Consequently, currents are such that the slow movements of oyster larvae in the sound allow only a relatively small portion of the population to move out of the sound. Only on the south side between stations 5 and 6 do fairly rapid outgoing surface currents develop on large flood tides.

Spatting

To determine the location of the best spat collecting areas in Pendrell Sound, studies were made of the horizontal distribution of spatfall on floating cultch; the data are shown in Table 3, with position of the stations indicated on Fig. 29. Stations in the central sound area (3, 3A, 5, 6) collected somewhat more spat than those in the upper sound (1 and 2) which in turn collected more than those in the outer sound.

To date, virtually all commercial collecting has been done in the upper sound, between stations 1 and 2, largely because facilities for anchoring rafts are better there than elsewhere and setting has usually been adequate. It is, however, quite feasible to hold rafts in other parts of the sound, at least during July and August, but thereafter they should be moved to safer waters. With advent of long lines, increased use has been made of the areas adjacent to stations 3, 5, and 6.

TABLE 3. Horizontal distribution of spatfall of Pacific oysters at standard stations on floating shell cultch, Pendrell Sound, 1952–56.

	No. spat per shell							
	1952					1956		
Station	Set A	Set B	1953	1954	1955	Set A	Set B	Mean
1	30	12	36	6	9	16	489	85
2	14	33	55	6	11	267	405	114
3	31	46	91	5	12	235	430	121
3A	27	35	—	9	19	318	—	84
5	37	36	440	13	57	255	—	156
6	65	65	124	12	15	434	335	121
8	2	7	35	1	2	140	150	48
9	11	48	18	6	3	475	—	94
10	15	9	16	8	8	173	245	68
11	—	—	40	2	10	154	229	87

Vertical Distribution of Spatfall

Of interest and importance also is the vertical distribution of spatfall (Fig. 51) and the two phases of this are relevant to positioning of both shore and floating cultch. This is partly related to the vertical distribution of larvae (Fig. 52).

Shore Cultch

As previously indicated, shore cultch catches less spat than does cultch suspended from rafts; but the difference decreases with increases in the spatfall intensity.

In years of average (100 spat per shell) to excellent (1 000 spat per shell) spatting, shore cultch collects an adequate number of spat. A comparison of shore and floating cultch is shown for 1 year in Table 4.

Table 5 shows the vertical distribution of spatfall on long strings of cultch hung against a vertical cliff near station 2 in Pendrell Sound. It may be seen that actual spatting occurs to the level of the very highest tides. However, mortality of spat at higher levels occurs

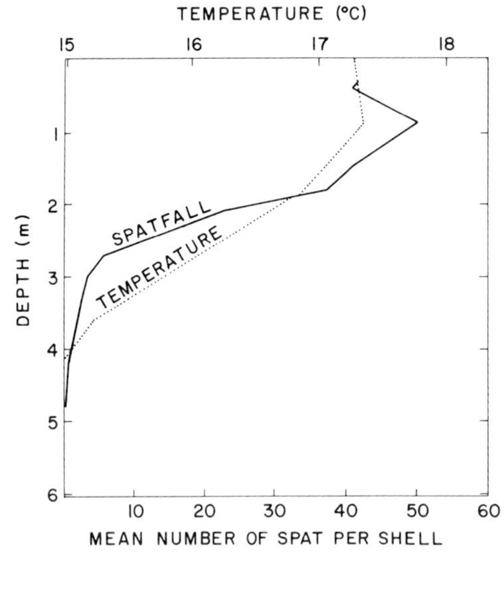

FIG. 51. Vertical distribution of spatfall on floating cultch, Pendrell Sound, 1953, as related to vertical temperature changes.

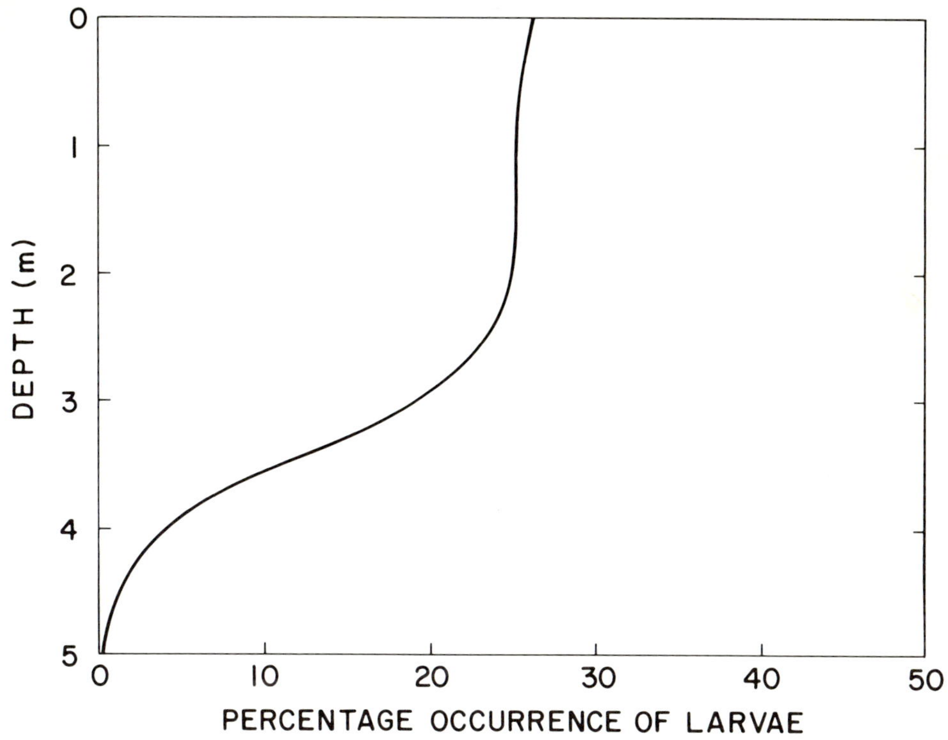

FIG. 52. Percentage occurrence of Pacific oyster larvae with depth, based on a total of 288 samples (day and night combined). Station 2, Pendrell Sound, July 25–28, 1956.

almost immediately owing to desiccation. At any rate spatting occurs at tide levels considerably higher than those at which significant survival occurs. At the other end of the scale spatting falls off very rapidly below zero tide level and for practical purposes the numbers spatted are inconsequential. Where no specific cultch has been exposed, spatfall below the zero tide mark has not been found. In Pendrell Sound, natural spatting and survival on the shore occurs between zero and a tide level of 4 m. In Ladysmith Harbour, the large majority of spat are found consistently between the levels of 1 and 3 m and much the same situation exists in Puget Sound. In Willapa Harbor, Washington, however, where there is little evidence of temperature and salinity stratification, maximum spatting occurs nearer low tidal areas than in Puget Sound or Ladysmith Harbour.

TABLE 4. Comparison of spatfall collected by shore and floating cultch, Pendrell Sound, 1953.

Station	Floating cultch (mean spatfall per shell)	Shore cultch (mean spatfall per shell)
1	36	10
3	91	5
6	124	50
8	35	9
9	18	5
19	16	2
11	40	11
Total	360	92

TABLE 5. Vertical distribution of Pacific oyster spatfall on shore cultch, Pendrell Sound.

Tide level (m)	Mean spatfall per shell per 30 cm 1952	Mean spatfall per shell per 30 cm 1953	(m)	Mean spatfall per shell per 30 cm 1952	Mean spatfall per shell per 30 cm 1953
4.8	0.1		0.3	22.5	6.7
4.5	1.2		0.6	13.0	3.2
4.2	5.5		0.9	6.0	0.6
3.9	7.9		1.2	2.0	—
3.6	44.4	1.4	1.5	2.9	0.1
3.3	76.3	2.0	1.8	2.1	0.4
3.0	90.9	10.0	2.1	1.3	0.5
2.7	97.0	32.2	2.4	1.0*	—
2.4	—	—	2.7	1.0*	0
2.1	90.0	58.6	3.0	4.0*	
1.8	63.3	72.3	3.3	1.0*	
1.5	54.3	70.6	4.5	1.0*	
1.2	42.9	—	5.7	1.0*	
0.9	39.5	101.1	6.0	1.0*	
0.6	36.4	56.6	6.6	3.0*	
0.3	35.5	46.1	8.4	1.0*	
0	24.4	—			

* Total spatfall per 30 cm.

Lack of spatting below the zero tide level in Pendrell Sound may be due to several factors. First, in waters below zero tide levels, silting is such that there is probably little or no clean, suitable setting area. Also, larvae occur in the upper 3 to 4 m of water where suitable temperatures occur. At low tide, the shore below zero tide level is certainly exposed to larvae-bearing water but at high tide this area is in a zone where water temperature is not conducive to either larval or spat survival. Also, experiments have indicated that maximum spatting occurs near the time of high water when tidal action is minimal and there would be few larvae in the layer of water covering the subtidal zone.

The conclusion is that shore cultch in Pendrell Sound may be exposed between the tide levels of zero and 3 m. An advantage of shore cultch is that best hardening conditions occur between the 1 and 3 m levels; also, the danger of attack by intertidal sea stars which, particularly in summer, are concentrated closer to the zero tide level than in winter, is removed.

Floating Cultch

Long shell strings were suspended from a raft at station 2, and the results of these experiments are shown in Table 6. In 1951, there was significant setting down to about 6 m below the surface but, in 1952 and 1953, 3.0 and 3.3 m marked the lowest level of main spatting. In 1956, spatfall was so intense (over 500 per shell) that counting was impossible. Setting was heavy from the surface to a depth of 4 m, where a sharp drop in numbers occurred. From the 4-m level down to 5 m, there were relatively few spat, and below that there were none.

The standard length of a string of shell cultch is 2 m or less for bags,so, with floating cultch, the shell lies within the zone of heaviest setting (Fig. 51), correlating with high concentration of larvae (Fig. 52). Somewhat longer strings could be used and still be

TABLE 6. Vertical distribution of spatfall on suspended cultch, Pendrell Sound.

Depth (m)	Mean spatfall per shell per 30 cm			
	1951	1952	1953	1956
0.3	46.1	160	42	500+
0.6	27.7	302	40	500+
1.0	36.9	232	50	500+
1.3	27.8	218	—	500+
1.6	34.6	129	40	500+
2.0	41.7	116	37	500+
2.3	42.4	97	22	500+
2.6	28.1	69	—	500+
3.0	24.2	53	6	500+
3.3	15.8	15	3	500+
3.6	16.1	26	3	500+
4.0	30.8	19	—	500+
4.3	30.8	14	2	
4.6	26.9	7	0.5	
5.0	22.3	7	0.2	
5.3	25.4	3	—	
5.6	17.7	2	0.2	
6.0	19.7	2	0.08	
6.3	7.7	—	0	
6.6		18*		
7.0		12*		
7.3		14*		
7.6		19*		
8.0		14*		
8.3		5*		
8.6		14*		
9.0		4*		
9.3		—*		
9.6		1*		
10.0		5*		

* Total spatfall per 30 cm.

within the zone of optimum spatting, but experience has shown that strings 2 m long are about the maximum length for efficient handling with present methods.

Although knowledge of the specific time of day when spatting occurs has little practical application, it has been investigated briefly, and it appears that most intense spatting takes place during the period of minimal tidal action, which is usually associated with the high tide of the day.

Another problem related to spat collection is the relative amount of spatfall on the upper and lower surfaces of shell cultch. Strings are made up with the rough sides of all valves uppermost so that the cups of all valves are on the underside. They are hung in this way so that silt will not collect in the cups of the shells. Table 7 gives the catch on upper and lower surface of valves and the data show the upper surface collects more spat than the lower. Shells in bags are randomly arranged so the position is not of concern.

In this same connection, a study was made of setting intensity relative to the angle at which cultch is held by placing cement coated laths at various angles. The data are shown in Table 8 and Fig. 53.

TABLE 7. Pacific oyster spatfall on upper (rough or outer) and lower (smooth or inner) surfaces of shell cultch, Pendrell Sound, 1951–55.

	Percentage of total catch	
Year	Upper surface	Lower surface
1951	36	64
1952	65	35
1953	73	27
1954	54	46
1955	66	34
Mean	59	41

TABLE 8. Angle of setting surface and setting intensity of Pacific oysters, Pendrell Sound, on double laths 2.5 cm apart.

Angle	Upper lath	Lower lath	Mean of both laths
0° (under horizontal)	42.5	46.5	44.5
45° (lower diagonal)	14.0	26.0	20.0
90° (vertical)	1.0	0.0	0.5
135° (upper diagonal)	3.0	38.5	20.7
180° (upper horizontal)	1.0	100.0	50.5

The upper lath caught little or no spat on the vertical and upper surfaces and the lower horizontal caught three times as many as the lower diagonal. This is close to the typical pattern found for both native and Pacific oysters in most other Pacific coast areas. The lower lath caught twice as many spat on the upper horizontal as on the lower horizontal

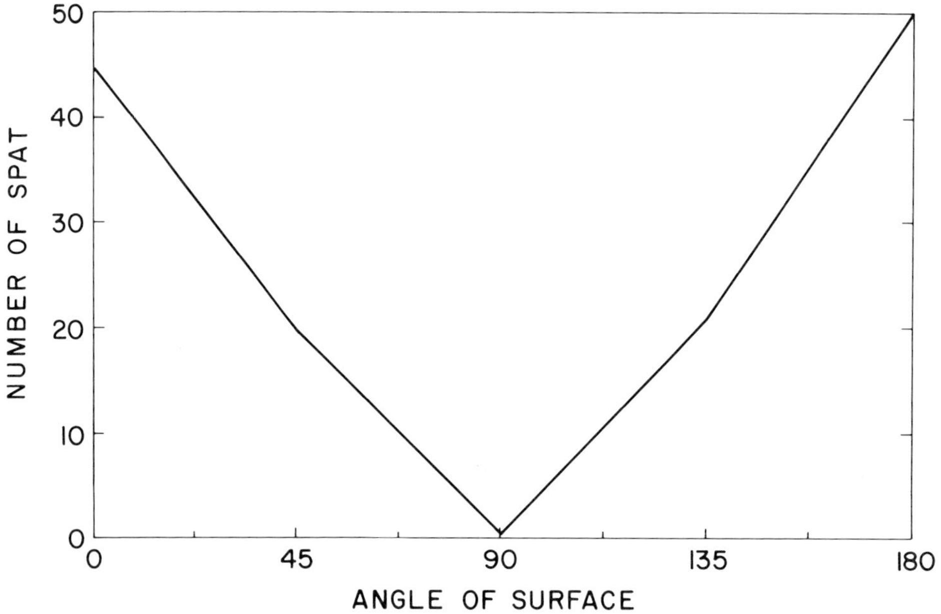

FIG. 53. Relationship between the angle of setting surface and spatfall, Pendrell Sound. (0° = lower horizontal; 180° = upper horizontal; 45° = lower diagonal; 135° = upper diagonal; 90° = vertical).

with none on the vertical and nearly equal amounts on the upper and lower diagonals. This approximates the pattern of setting on shells in Pendrell Sound. There was virtually no silting during this experiment so that this factor, which is often an important one, may be disregarded in this experiment. The position of collector laths for stick culture must be considered in the light of these results.

Oyster Cultch

Cultch is material used for collecting oyster spat, in other words, the substratum provided for attachment of larval oysters.

To collect oyster seed, cultch is placed in the water at an appropriate time and place. After settlement on the cultch, spat are usually allowed to grow for a time before removal to growing areas.

Considering seed collection in detail, first comes selection of the type of cultch. There is a multitude of materials to which oyster larvae will attach but few are suitable as cultching material from a practical and economic of view.

Ideal Cultch

So far an ideal universal cultch has not been found, for there are so many types of culture each requiring cultch with different characteristics. Eventually, no doubt, a cultch may be developed which will have the basic attributes common to all good cultches, but which can be modified to suit a particular type of culture. The following would be some of the characteristics of such an ideal cultch:

(i) Low cost.

(ii) Larvae require a clean solid object on which to settle. Waxy or slick-surfaced ones are not acceptable; colour appears to be immaterial.

(iii) The material should be light in weight although it must be sinkable. Weight is important because of handling, shipping, and buoyancy problems.

(iv) A maximum surface per unit of volume is required, for this will affect packaging cultch into larger units for exposure and transport. Stringing individual shells on a wire is a form of packaging; other forms of packages are wooden or wire crates and wire or net bags and bundles of tubes or laths. Satisfactory and economical packaging or a cultch are prime requisites.

(v) It is necessary for cultch to permit a free flow of water so larvae may reach all parts of the collecting surface.

(vi) If cultch could disintegrate of its own accord after a specified length of time, the cost of separating clusters would be reduced. The time when cultch should disintegrate is related to growth rate of the spat and pore diameter of the cultch, if it is of the lattice type or bundles. Spat must be large enough to survive a modicum of silting, and separation should take place before the spat have grown together in clusters. A fine balance between strength to withstand handling and fragility for ease of separation is required.

(vii) The pore diameter (circa 2 cm) should be large enough to permit adequate circulation, either in the unit or package, and of a shape to prevent collection of silt. The pore should also be large enough to allow spat to grow to a size at which it will be separated.

The silting problem, discussed elsewhere, varies with circumstances. Silting of suspended cultch is not a serious problem in Pendrell Sound but it may be elsewhere. When cultch is placed on the bed the spat should be nearly 2.5 cm in diameter (although it is usually less in practice) or silting mortality will be significant on most ground. This is only approximate, for the size will vary with rate of growth, density of planning, wave action, and degree of silting.

Shell Cultch

As already mentioned, Japanese oyster seed was collected on oyster shell cultch, and this is still the most frequently used cultch in North America, on both the Atlantic and Pacific coasts. The main advantages of shell cultch are low cost, and the availability. Disadvantages are heavy weight, hardness, and strength. The great weight requires considerable flotation power if cultch is to be suspended and quite heavy lifting gear if handled in quantity, both before and after exposure. Hard, strong shell, of large size, may create problems in separating clusters. The softer and smaller shells from raft culture are generally preferred to bottom grown shell cultch. Through the years in British Columbia shell cultch has seldom attained its potential efficiency for growers appear loath to expend the effort necessary to ensure its cleanliness which is absolutely necessary.

Shells, which are clean, weathered, and aged (making them softer), are punched through the centre with a hammer to which a 5 mm spike has been welded (Fig. 54), with a punching machine (Fig. 55), or an electric drill. These punched shells are strung like beads on 2 m lengths of number 14 gauge galvanized wire (Fig. 56a). Sometimes, double strings are used, and, in this case, each end of a 4 m-long wire is strung with 2 m of shell leaving a blank of 1 m between. This type of string can be draped over a float log or long line and is not so easily lost by accident as a single string. This type of cultch is used for suspended string culture where the shells are held apart by separators.

Shell may also be packed in bags made of chicken wire, discarded fish net, or vexar tubing (Fig. 56c). The bags normally 1 m in length, must be rather small to permit adequate water circulation and the diameter, when filled, should not be greater than 20–25 cm. Wooden frames covered with wire netting have also been used, as have many other methods. This non punched shell is strung for suspended string culture by insertion between the strands of double lay rope.

Other Cultches

Another type of cultch developed in British Columbia is a cement-dipped, wood veneer ring. Veneer strips, 60 cm long by 8 cm wide by 2 mm thick, are obtainable from specialty sawmills. These strips are made into 15 cm diameter circles with an ordinary desk stapler (Fig. 57). They are then dipped in a slurry of cement, about the consistency of thick soup, after which they are dried slowly. Fast drying causes the cement to flake off. Each circle provides a setting area of slightly more than 900 square cm.

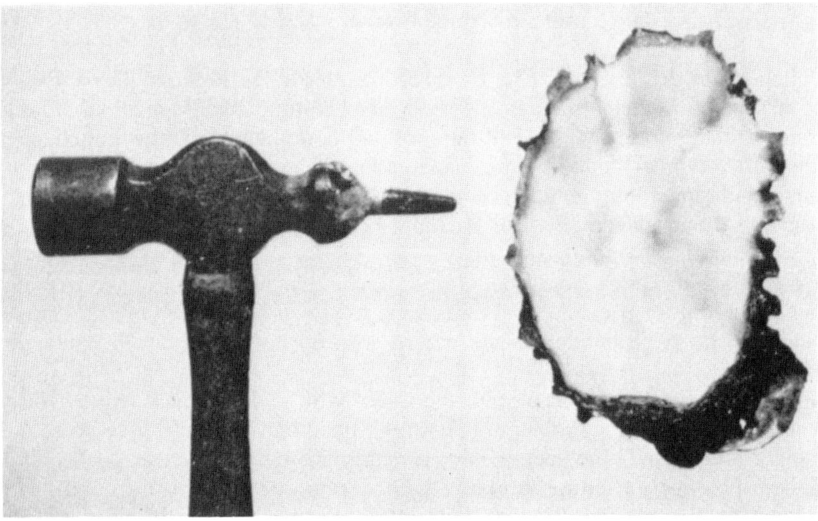

FIG. 54. Shell-punching hammer.

FIG. 55. Machine for punching oyster-shell cultch.

The area of one bag of shell cultch is roughly equivalent to the area of 20 circles, which may be simply and rapidly strung on galvanized wire or polypropylene rope. A string of 30 makes a handy bundle about 2 kg in weight compared to an equivalent area of shell surface which would weigh nearly 20 kg. Also there is good circulation through a bundle of circles, for there is sufficient variation in size and shape to prevent locking. They have been used extensively in eastern Canada.

Other types of cultch are in use. In Japan, considerable use is made of scallop shells. In Europe, limed roofing tile is a common cultch used year after year, for the coating of lime with attached spat is scraped off when the spat become about 1 cm in diameter. Cement dipped building laths are used in the USA as are cement coated cardboard egg separators, particularly for Olympia (native) oysters. Brush and discarded fish net dipped in cement are also satisfactory collectors.

French oyster growers have been assiduous in cultch development in recent years. In addition to traditional limed roofing tiles, especially for the flat oyster (*Ostrea edulis*) there are P.V.C. tubes described elsewhere (Fig. 56b) and the Chinese hat (Fig. 58) made of plastic mesh which is coated with a lime mixture as well as smaller plate-like plastic discs. The recipes for lime or cement coatings are many and specific for tiles, fibrocement (requiring sawdust) or plastic. Various types of lime, sand and mud are used in these recipes.

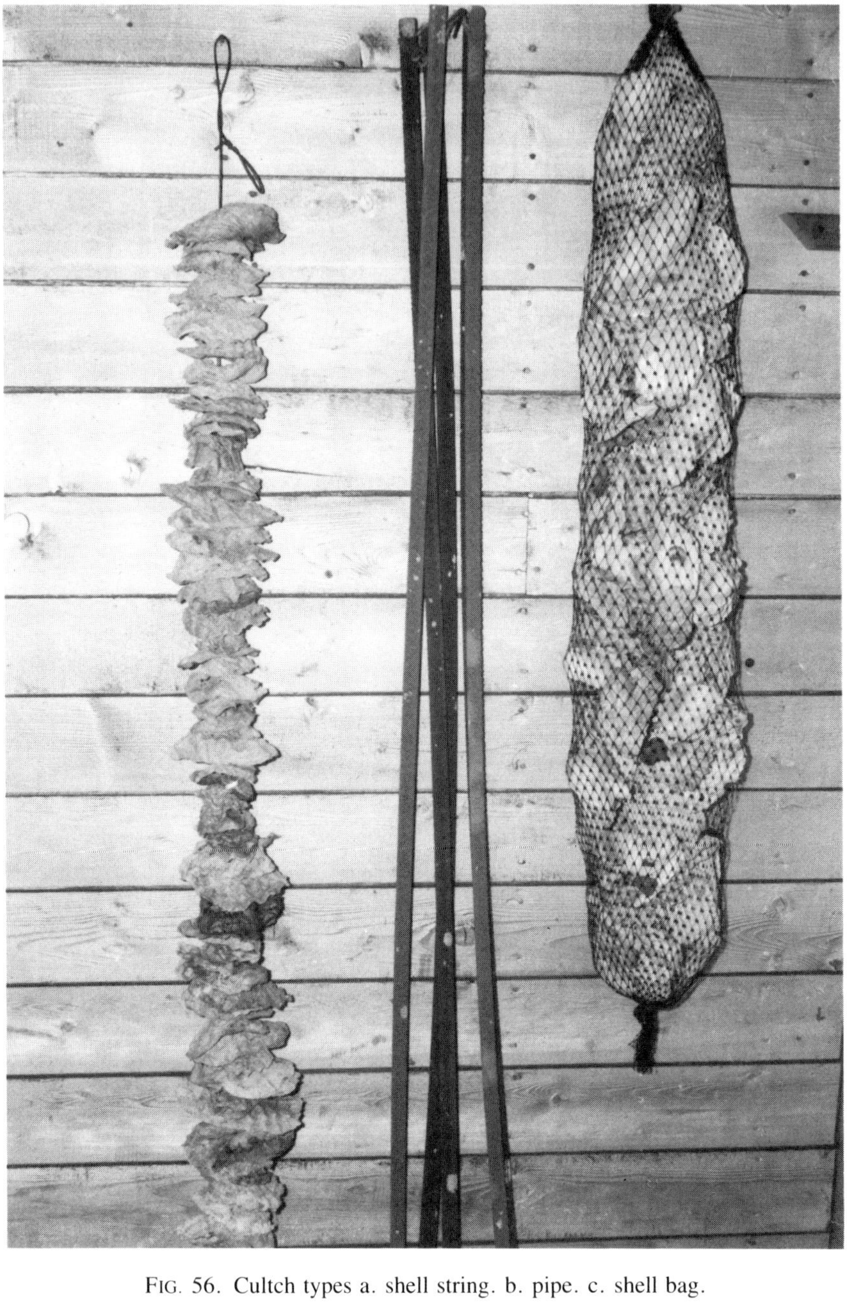

FIG. 56. Cultch types a. shell string. b. pipe. c. shell bag.

FIG. 57. Wood-veneer cement-dipped circles used for oyster cultch. ×0.3.

FIG. 58. "Chinese hat" oyster spat collectors.

Diamond chip cultch of cement is shown in Fig. 59a, b.

In B.C. an effort to develop a self disintegrating cultch resulted in what are called cultchettes (Fig. 60). This was a cultch which began with providing the basic principles of an ideal cultch to a research organization with chemical and mechanical engineering facilities. Some materials and combinations of them were processed by castings, hot moulding, hot pressing, cold pressing, spraying and dipping and extrusion. Cold pressing at pressures from 900 to 3600 kg per 6.5 square cm formed discs which were tested in the field.

FIG. 59. a. diamond chip cultch. b. diamond chip forms in suspension.

FIG. 60. Cultchettes, artificial self-disintegrating oyster cultch. × 2/3.

Form of the cultch was predicated partly on the basis of requirements for an ideal cultch and partly on facility in manufacture. A curved disc form, partly modelled on the left valve of an oyster shell, proved most amenable to available manufacturing processes with the most suitable materials. A machine to make the discs was developed. A basic formula contained Type 1 Portland cement, calcium hydroxide, asbestos shorts, asbestos fines and gypsum. Zinc stearate to inhibit slime formation was included. It appears possible to substitute cellulose fibres for asbestos. Manipulation of the formula can dictate the time of disintegration. Various problems including lack of promotion and costs have inhibited development of a market for cultchettes.

All of these forms of cultch have their advantages and disadvantages, and choice depends on the type of culture to be practised. For example, limed roofing tiles would be economically prohibitive in British Columbia. Also, apart from economics, the cultch must be suitable for the type of seed ground available; there is no object in putting small pieces of cultch on soft muddy ground for they will become silted over, with high mortality.

Exposing Cultch

For collection of oyster seed there are three basic plans for exposing cultch to make it available to the larvae: (1) beach, (2) intertidal racks, and (3) suspension.

Beach

In this method loose shell contained in strings or bags is simply spread or laid on the bottom in the intertidal area in a location known by experience to provide good sets (Fig. 61). Shells are usually spread 3 or 4 (5–7 cm) deep. In most cases, this is an inefficient method but the efficacy varies with the type of bottom, intensity of the spatfall, and timing of the operation. If the shelling operation is to be on a large scale, spreading

FIG. 61. Oyster-shell cultch spread on the shore, Pendrell Sound.

must begin the moment a larval brood is reported to be in the water in order to provide sufficient time to expose enough cultch. If the larvae fail to survive, the shell can be lost for it may not be worth recovering owing to fouling and silting, unless the bottom is very hard and the shell is kept free of silt. Cement-dipped circles or strings or bags would be more efficient than shell in this type of operation.

Comparisons over a number of years have shown that suspended floating cultch is from 3 to 10 times as efficient as shore cultch. The main advantage of loose shell is low initial cost, for packaging is not necessary and it can therefore be handled in bulk. In suitable locations, it can be left for several years to pick up a spatfall.

As indicated previously, use of loose shell depends on the individual situation and the system has been used successfully in small scale operations in such areas as Ladysmith Harbour and Pendrell Sound. Either strings or shell bags are more satisfactory than loose shell for beach cultch since they may be recovered more easily. However, the base on which they are laid should be very firm or even on gravel there is a tendency to sink into it with loss of valuable setting area. This method is coming into extensive use in Pendrell Sound where the rocky shores form an excellent substrate. Adequate setting takes place in most years between the zero and 2 metre tidal level.

Intertidal Racks

Intertidal racks in the intertidal zone hold packaged cultch such as shell strings (Fig. 56c). In years of low larval abundance, rack cultch will collect a fair set while there will be virtually nil on bottom cultch. Preparations may be made well in advance and exposure can be delayed until the probability of a set has become high. In event of a set failure, the cultch can be recovered. Growth rate of spat is greater on rack cultch than on the bottom, and since there is little silting, survival rate is higher. Another advantage of racks is that, if they are reasonably well protected from wave action, the cultch may be left on the racks over winter.

There is however the cost of racks, in addition to packaging the cultch, so, in terms of investment, cost is somewhat higher than in the loose cultch system, though the return is much higher.

Racks are usually made of 4 cm by 8 cm lumber or cut poles, and are 1 or 2 m high. In most areas, unless the wood is creosoted, the racks last for only one season, owing to wood borer attacks. The system has been used successfully in Ladysmith Harbour and is a standard practice in Japan.

Raft or Longline Suspension

The suspension method is the most efficient and also the most expensive (Fig. 62a, b, c, d). Spatfall intensities, growth rate, and survival are greater than in the other two methods. In addition there are no tidal problems to consider when the cultching operation is in progress.

Oil drums, logs, creosoted or fibreglassed plywood pontoons, styrofoam blocks as well as plastic floats may be used to construct the suspension system of any size or shape. Lightweight concrete aggregate for float construction is also a possibility.

As in the rack method, exposure of packaged cultch can be delayed until there is reasonable expectency of a spatfall. Cultch is usually left suspended until October. One disadvantage of suspension is that spat are submerged continuously and thus not adapted to intertidal life where they are exposed for some time each day. To compensate for this, and to reduce difficulties of transfer from continuous submergence on the raft to intermittent submergence on the beach, seed should be moved during cool damp weather, if at all possible. Also, the time should be chosen to coincide with the beginning of a series of neap tides, when the lower and middle intertidal area is exposed for a minimum length of time. During transportation from the seed area to seed beds, cultch should be kept damp and cool by spraying occasionally with salt water. If these suggestions are followed, there should be little seed mortality, except that occasioned by mechanical damage, some of which is inevitable.

FIG. 62. Shell cultch operation, Pendrell Sound. A = scow load of shell strings; B = unloading shell strings which have already been hung on wooden poles; C = setting the string-loaded poles on the raft; D = shell strings hanging in the water.

Japanese Seed

While Japanese oyster seed is no longer imported into British Columbia as a result of breeding success in several areas, along with the advent of oyster hatcheries, an account of the subject is of historical and practical interest.

Japanese oyster seed destined for British Columbia originated entirely in the Miyagi area of Japan and is often known as Miyagi seed to distinguish it from seed grown in other producing areas such as Hiroshima, far to the south. Miyagi Prefecture is in northeastern Japan (Fig. 4) and Matsushima Bay, the main producing bay, is separated from the open Pacific only by a chain of islands. Summer water temperatures in Matsushima Bay reach 24°C with a salinity of 27‰. The collecting material used for export seed was entirely oyster shell. Since most of Japan's production of market oysters comes from raft culture, the characteristic shell is relatively small because of the short growing period and thin because of rapid growth. During winter, 70–100 of these shells are punched with a hole and strung on wires 2 m in length. Such shell strings are known as "ren". At the appropriate time in early summer, the "ren" are taken by sampan and placed on catching racks (Fig. 63) to collect the oyster spat. At the end of summer, these strings are removed from the catching racks to "holding" racks, which are placed at a tidal level which allows the strings to be uncovered for a part of each day. The strings are most often placed horizontally on these racks. The purpose of this "hardening" process was to condition the seed for the trans-Pacific shipment to the USA and Canada. It is thought the process hardens and thickens the shell edge so that the young oyster is able to retain the fluids inside the shell, thus preventing desiccation. The success of this hardening process led to the realization that it was important in preventing mortalities in suspended culture and it is now standard practice in Japan's own oyster industry.

Packing

The seed was packed for export during February, March, and, rarely, into April when it is quite cool in the Miyagi area. Low water temperature inhibits movement of oyster drills from the hibernation areas onto the shell strings. Further, growers receiving the seed are reaching the end of winter harvesting and ground is available for seed. Also, by this time, winter storms are over, so seed loss from this factor is eliminated. In addition, seed is available for the important spring growth period.

FIG. 63. Japanese seed-oyster racks.

The "ren" were removed from holding racks to the packing sites, of which there were about eight in the Matsushima areas. Each packing site had its own code number and this was stamped on all seed cases from that site (Fig. 64), so the source was identifiable. Shells were removed from the wires, placed in wicker baskets, and washed well with strong jets of water to dislodge mud and debris which might harbour predators. The shells were then examined for pests, such as oyster drills, and for quality (number and size of spat). Japanese seed was available in several forms, the descriptive characteristics of each form is given in Table 9.

To provide broken seed, shells were cut into two or three pieces with a machine like a paper cutter, or with a circular saw. The seed, regardless of type, was all packed into wooden boxes 90 × 45 × 30 cm and 1.3 hectolitres in volume.

After packing, the boxes were placed on racks in the intertidal zone for storage until a sufficient number were accumulated to form a cargo, which may take up to 6 weeks. The seed was lightered out to an anchored ship on which it was loaded as deck cargo and covered with rice matting (Fig. 65), shipments usually consisted of between 5 000 and 10 000 cases. The trans-Pacific voyage took 10–12 days, during which the seed is watered down at frequent intervals. Up to 40 days may have elapsed between time of packing and final planting of the seed on west coast oyster beds.

TABLE 9. Types of Japanese seed.

Type of seed	Average size (mm)	Number of spat per piece	Number of spat per case	Number of pieces
Unbroken	6–18	10 or more	12 000	700–900
Broken	6–18	5 or more	18 000	1 800–2 000
High count unbroken	6–18	–	40 000	5 000
High count broken	6–18	–	30 000	–

Inspection

A careful seed oyster inspection system was worked out through the years, the main purpose of which was to prevent, or at least keep to a minimum, the spread of oyster pests. Japanese oyster drills obtained a firm foothold in a number of British Columbia and USA oyster beds before introduction of the inspection system.

FIG. 64. Case of Japanese oyster seed.

FIG. 65. Deck cargo of Pacific oyster seed newly arrived from Japan with the rice-matting cover removed.

Major inspection was done at the packing sites in Japan, by a biologist from the Washington State Department of Fisheries, the regulations of which state require this type of inspection to prevent importation of pests. The State of California also sent a biologist to Japan to examine seed to be imported into that state. Seed destined for British Columbia was inspected by the Washington State authorities, for it was obtained through the Pacific Oyster Growers Association which negotiated purchase of large blocks of seed on behalf of its members. In addition, it was examined on arrival from Japan. Some large U.S. oyster growers purchased seed directly rather than through the Association. Cost of inspection was borne by the purchaser.

In addition to biologists who inspected the seed, the Oyster Growers Association usually had a representative in Japan, and his main task was to handle general liaison and look after shipping schedules and seed quality. Various Japanese governmental agencies also had inspectors on hand so that, theoretically at least, North American growers could be assured of high quality seed. It should be noted the onus for ensuring seed was of acceptable quality, particularly in respect to pests, was on the seed purchaser.

There are two main periods of oyster setting in Japan: one in early summer and one in late summer. The early set is destined mainly for domestic use and the second for export. North American growers preferred very small seed for they believed it survived better than larger seed.

Spatfalls in Japan occur with regularity, although there are annual variations in intensity and in degree of survival to time of packing. Each year, about 3 million "ren" (shell strings) are used to collect seed, and to make up one case (800 shells) of export seed between 20 and 35 "ren" were required, so a considerable amount of handling was involved. All of this involved manual labour and, with the rise in standard of living in Japan and labour costs, the cost of seed inevitably rose since resumption of the oyster seed business after World War II.

Pests

The danger of introducing pests with oyster seed has been noted. There are several species of potentially dangerous oyster drills, but the main problem occurs with the Japanese oyster drill, *Ceratostoma inornatum* (Fig. 66), which has unfortunately already

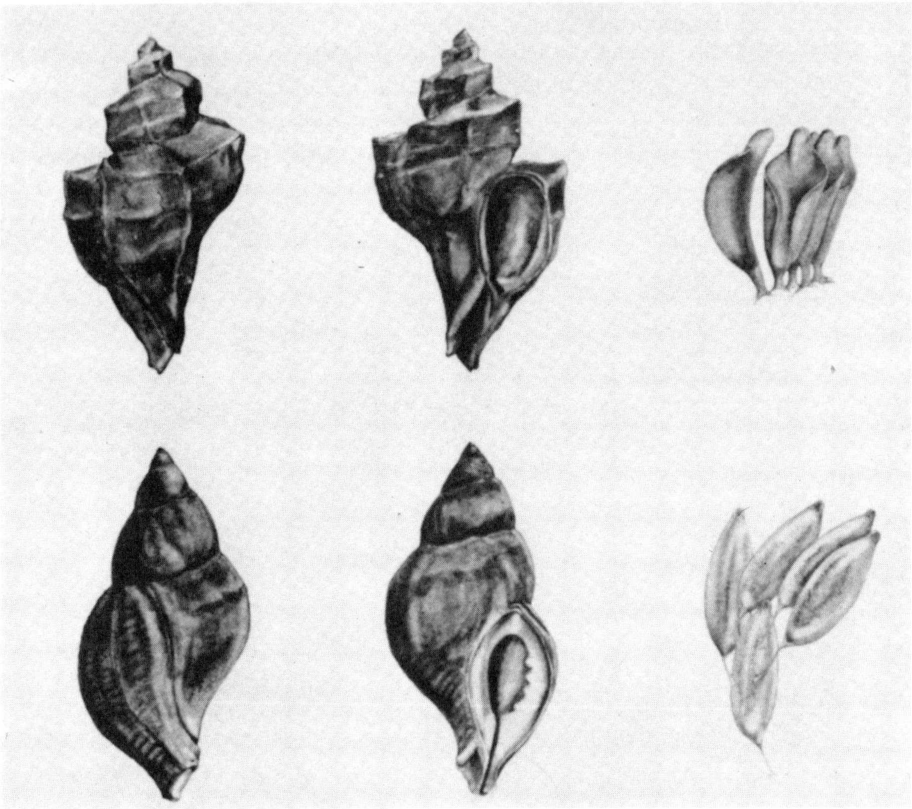

FIG. 66. Comparison of the shells (× 1.3) and egg capsules (× 3.6) of the Japanese oyster drill (top) with those of the native drill (bottom).

gained a strong foothold in some areas of both British Columbia and the State of Washington.

In addition to drills, another serious pest is the flatworm, *Pseudostylochus ostreophagus* (Fig. 67). This rather small and inconspicuous worm, wafer-thin, is able to drill a tiny, oval shaped hole in the shell of small oyster spat.

Mytilicola orientalis, a parasitic copepod (Fig. 68) which occurs in the lower intestine of oysters and mussels, also originated in Japan. In British Columbia it is more prevalent in mussels than in oysters, but appears, so far, to have caused no observable damage, except for reduced condition in mussels and oysters. Several years ago another species of *Mytilicola* in Europe nearly destroyed the large and valuable Dutch mussel industry through mortality and poor condition.

Along with the pests mentioned above, several other species of organisms have been introduced into British Columbia through the medium of oyster seed. Among these is the Japanese littleneck or Manila clam, sometimes called the ''baby'' clam, which is a useful addition to British Columbia fauna. Looked upon not quite so favourably is the Japanese weed *Sargassum muticum* (Fig. 69, 70) and the woodborer *Limmoria tripunctata*, one of the gribbles or pinworms.

Japanese Seed Case Equivalence

For many years the basic seed unit of oyster culture operations was the Japanese seed case. Since the advent of local seed supplies based on shell strings or bags there has been the problem of equivalance between those units and the Japanese seed case from the point of view of both value and productivity. The valve area of local shell (both sides) is

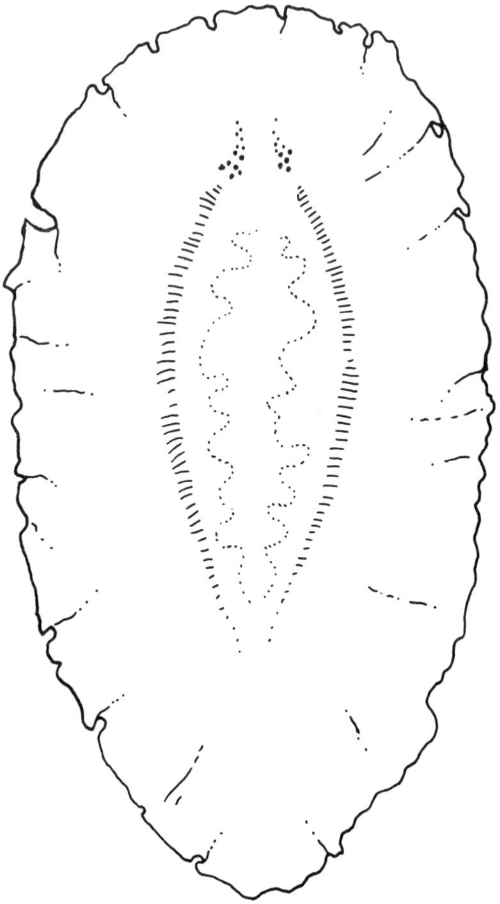

FIG. 67. The oyster eating flatworm (*Pseudostylochus ostreophagus*). ×10.

FIG. 68. The parasitic copepod (*Mytilicola orientalis*). ×22.

approximately 120 square centimetres as compared to the Japanese valve with 50 square centimetres. The seed case contained about 800 valves or 40 000 sq cm. The equivalent area of B.C. shells requires about 330 shells or 3½ shell strings or bags. From another point of view the Japanese seed case contained a guaranteed 12 000 spat or 15 spat per shell (1 spat per 3 sq cm). A British Columbia shell valve with this spat density would carry 40 spat or 4 000 spat per string or bag. In most years (1948–68) cultch in Pendrell Sound collected that amount or more.

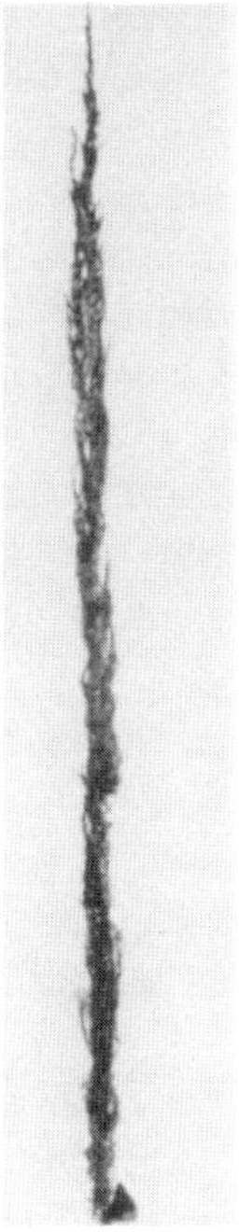

FIG. 69. The Japanese seaweed (*Sargassum muticum*) introduced into British Columbia waters with oyster seed. Whole plant with holdfast on pebble. ×0.1.

FIG. 70. Enlarged section of the Japanese seaweed (*S. muticum*). ×7.

In the past, as a compromise, 6 strings of 100 shells of B.C. seed were accepted as equivalent to one case of Japanese seed and now a 6 bag equivalence. This and other calculations in this bulletin is based on a 100 shell bag rather a 300 shell bag which is also now in use.

Oyster Feeding

Food

Though oysters have been much studied, relatively little is known about what actually constitutes useable food. Since the oyster is a filter feeder it must accept whatever food comes to it in the water in which it is living. It can select to a modest degree the food it ingests but, to judge from faecal material, this ability has limitations, for apparently many things which it takes into its stomach are not digested. What an oyster ingests is relatively well known and this includes bacteria, protozoa, a wide variety of diatoms, smaller plant-like forms both collectively called phytoplankton (Fig. 71), larvae of other invertebrate animals, and inanimate organic material lumped under the name of "detritus". This is the fine organic matter suspended in the sea and is composed mainly of minute fragments of plants and animals. In inshore waters, eelgrass and the large algae are thought to be the source of considerable quantities of this material. However, little is known of the value of various components of this material as oyster nutriment. There

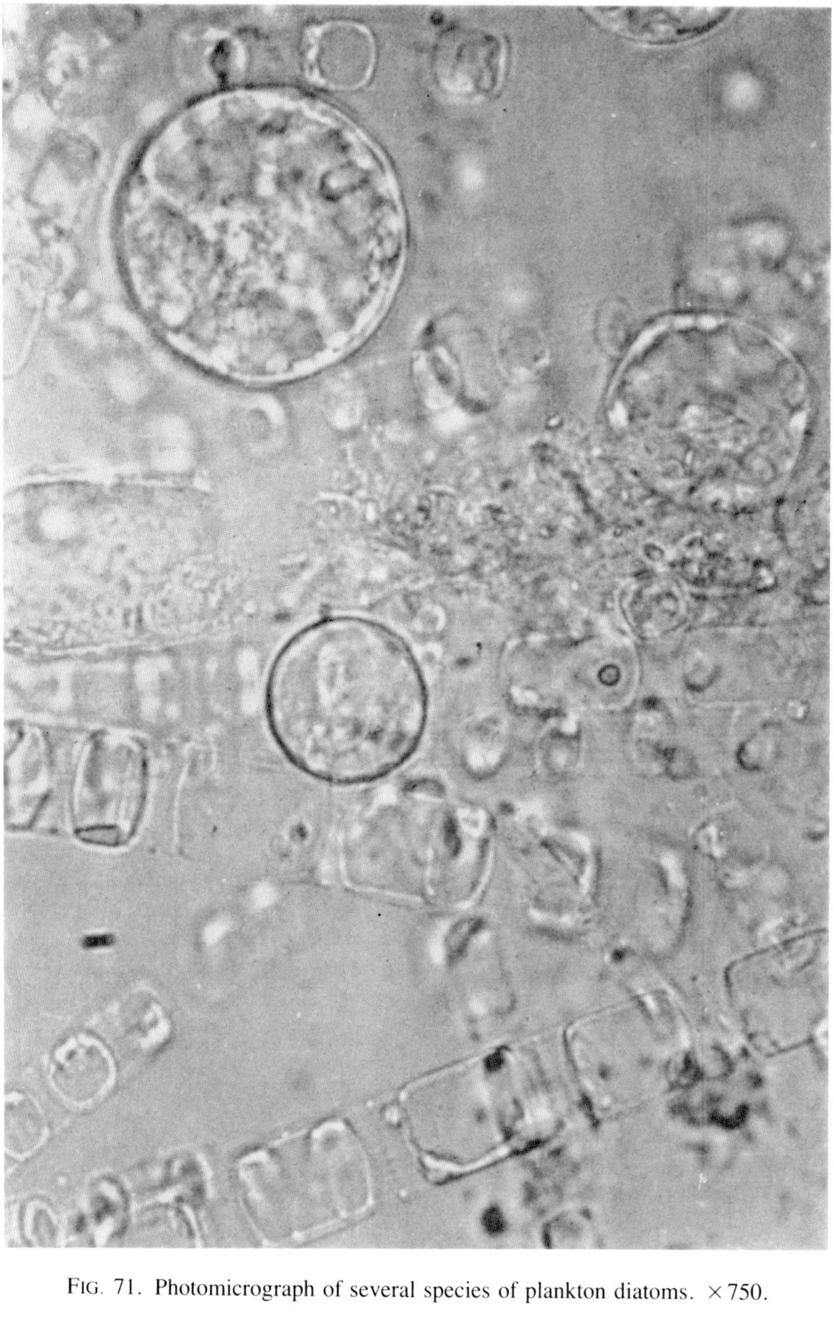

FIG. 71. Photomicrograph of several species of plankton diatoms. ×750.

is one school of thought which holds that diatoms and, indeed, only certain diatoms, are of paramount importance. Others believe detritus, possible by itself or with adhering bacteria, is all important.

Laboratory studies have shown oyster larvae are very particular and selective in what food they utilize and, of the numerous organisms used in experiments, only relatively few have been found satisfactory.

Fatness Cycle

Whatever the actual food, experienced oyster growers know the results of availability of good oyster food, when year after year, although with variation from year to year, oysters go through a regular fatness cycle. He knows this by experience, and from volume returns.

After spawning an oyster is extremely thin but by the end of October it has regained considerable weight and goes through winter in this condition. In late March or April, an oyster again begins to add weight and by May it is usually in its finest condition of the year, filling its shell with cream-coloured, even-textured meat. Most growers also associate improved condition with turbidity of the water, which comes about, usually in March, because of what is called the spring bloom of phytoplankton, caused by the right combination of light, temperature, and nutritients. Water over the oyster beds remains more or less turbid until November, when again it becomes crystal clear. Turbidity may be caused by rapid and intensive multiplication of microscopic marine organisms, by silt, or by suspended detritus, or a varying mixture of all three. Further, almost invariably, spring fattening is of greater magnitude than that in the fall.

The seasonal cycle of phytoplankton abundance as determined for 1 year in Ladysmith Harbour is shown in Fig. 72 and this approximates the picture in many parts of the world. Also is shown the seasonal change in condition factor (fatness) of oysters from a single bed. The spring increase in condition is shown clearly after a dramatic drop after spawning in early August. In general, the seasonal cycle of fattening of oysters in Ladysmith Harbour coincides with seasonal abundance of diatoms and with turbidity of the water. Whether or not this is a direct or indirect correlation is not known, but the relationship does exist and repeats itself year after year.

As indicated previously, however, level of fatness can vary greatly between years and it has been shown there may be a cycle when, for several years, the condition, during both spring and fall, remains well below average. Other instances occur when there is failure to fatten significantly in the fall. This places a grower in a difficult situation, because he does not wish to market oysters of inferior quality, and low meat return makes marketing unprofitable. If there are no alternative beds with better oysters, there is little to do but wait until the spring fattening, or purchase oysters from other growers to hold his market.

Another situation can occur, and often does, on many beds. This is when the summer is an exceptionally cool one and oysters fail to spawn; they then enter the fall season in a spawning condition, not ideal for market. They main remain "spawny" throughout the winter, and even in February and March typical spawn veins on the outside surface of an oyster may be seen. Usually by this time, however, they are quite firm, and if the body is cut the gametes do not run as freely as in summer. Oysters which fail to spawn may, alternatively, resorb gametes and reconvert it to glycogen. This may occur to a slight extent, when only the surface material is resorbed and a thin layer of glycogen covers the inner layer of spawn; or resorption may occur to a considerable extent, when most of the gonadal tissue is converted. The amount of resorption is possibly dependent on food available during the fall.

Factors Influencing Feeding

Feeding is also dependent to a considerable extent on water temperature; for more water and, consequently, more food, is taken in at higher temperatures than at lower

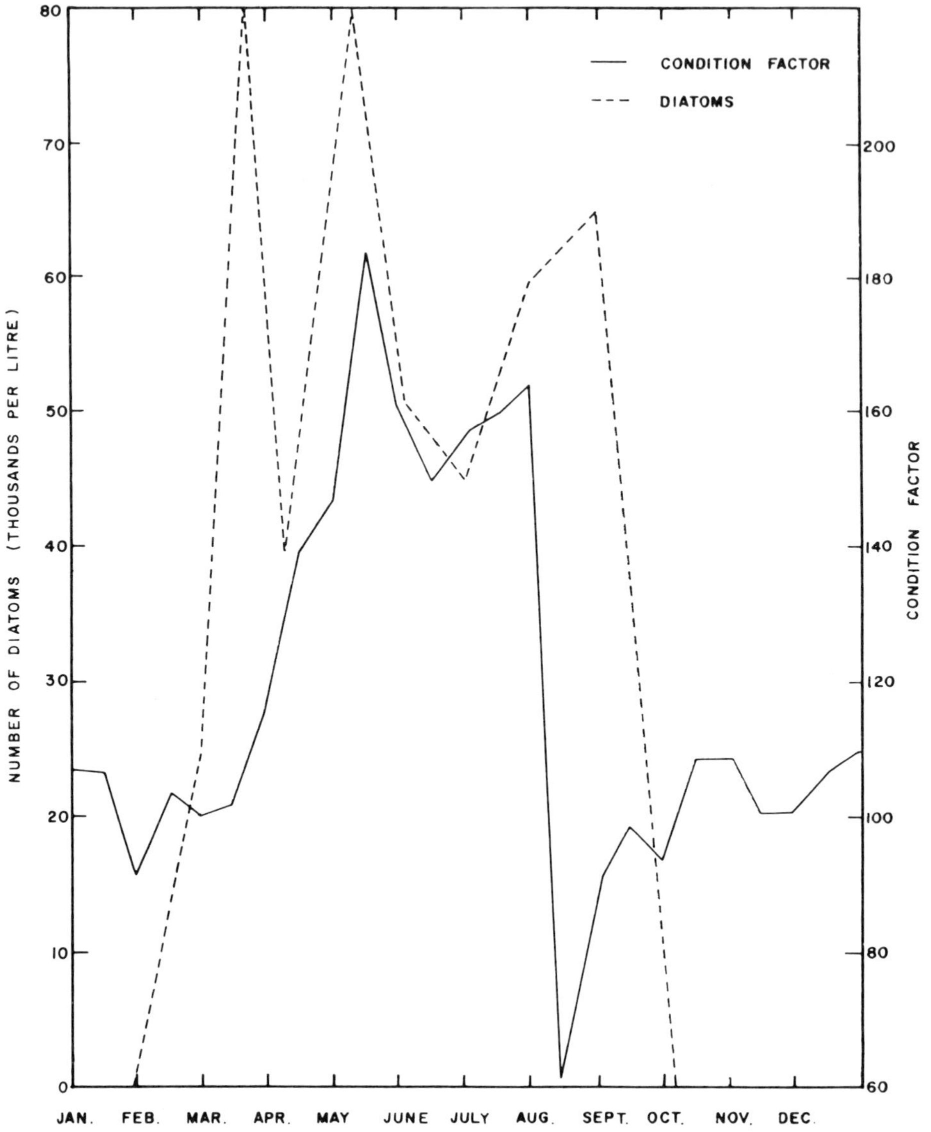

FIG. 72. Relationship between the standing crop (abundance) of diatoms and condition factor of Pacific oysters in Ladysmith Harbour.

ones. Transport of water through *Crassostrea* type oysters of market size is estimated at 30 L/h (although much less on the average) at optimum temperature, which for Pacific oysters appears to be about 20°C, and at optimum salinity of between 25 and 35‰. Condition of Pacific oysters held at salinities below 10‰ for more than 2 wk will tend to drop.

In addition, too much silt will cause oysters to reduce or stop filtration activities, as will an overabundance of microorganisms. Dense concentrations of marine organisms can discharge excretory products into the water in sufficient concentrations to inhibit or depress pumping action. Industrial wastes, such as pulp mill effluent, may also affect water transport in oysters, by its depressive effect on ciliary activity of the gills.

An overpopulation of oysters in enclosed bays is also thought to affect productivity by reducing the amount of food available to each oyster.

Growth

Growth in General

It is difficult to make definitive statements about growth of oysters, because it varies widely with tidal height and area in which they are grown as well as individual genetic differences. In addition, there are seasonal growth patterns which may differ from year to year.

There are two aspects of growth in oysters, one concerning shell and one the body. Shell growth appears to be largely dependent on temperature but at the same time the body must have sufficient energy, derived from food or food reserves, to convert calcium in sea water to shell and to provide the organic matrix. Body growth is largely the result of food supply but this also is in part, temperature dependent, thus, growth is essentially a summer phenomenon. It is possible that, even in summer, there may be a dearth of food which will lead to a cessation of body growth and, in turn, of shell growth. Shell growth, though, has been observed in oysters with emaciated bodies.

Growth in shell length and width is accomplished by activity of the outer surface of the outer fold of the mantle edge, while the inner surface of this same mantle fold secretes the covering periostracum. The inner shell surface (calcite-ostracum in edible oysters) is secreted by the outer surface of the mantle as a whole. Chalky calcium carbonate or calcite which constitutes the greatest part of the shell is believed to be secreted directly from sea water which contains an abundance of this material in solution. Harder material called aragonite forms the inner lustrous surface or nacre of pearl oyster shells or abalones. Conchyolin forms the organic matrix on which calcite and aragonite crystals are deposited. Thus, shell deposition or thickness is controlled by activity of the general mantle surface, and increase in shell area by marginal increment as a result of secretion by the outer mantle lobe.

Growth in Pacific oysters is extremely variable, both in rate and in amount. If an oyster is attached to a solid object, the left or lower valve will follow the contour of that object and this is called xenomorphism (Fig. 73). If oysters are grown in soft mud they grow long and narrow; this also happens when they are allowed to grow in clusters. When grown on very hard ground, such as gravel, the shape is round and deep, often with extensive fluting of the shell (Fig. 3). Rate of growth varies with the season; it varies from year to year during similar seasons; it varies with tidal exposure; and it may vary between two closely situated locations at the same tidal height. It is difficult, if not impossible, to present a growth picture that is representative of all of British Columbia, and each grower must learn growth characteristics of his own ground and its various parts.

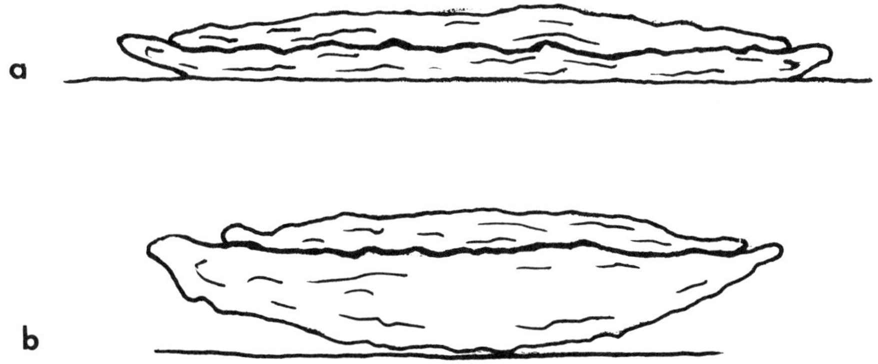

FIG. 73. Xenomorphism. a. Left valve of a *Crassostrea* when allowed to continue growth on a flat surface. b. normal shape with a cupped left valve.

There are various definitions of length, width, and thickness but the ones given here are generally understood by the oyster industry. It should be understood length as defined here is actually (biologically) height. Growth of oysters may be studied in terms of linear measurements (length, width, height) (Fig. 74) and/or volume. The latter measure is possibly the most valuable since it incorporates in one figure the three linear measurements. Further, it is a measure in which growers are most interested, for it is a basic measure of productivity.

There are also other means of measuring growth. If an oyster is weighed in water it is essentially the weight of the shell that is being measured, for the specific gravity of the oyster meat approximates very nearly that of water and, of course, the liquid contained within the shells makes no difference. A time series of such weights provides an indication of whether or not shell is being added, which may be in thickness, length, or width. Such data for an oyster growth study are shown in Fig. 75, and corroborate information from other studies that essential shell growth occurs between April and October. However, there is an indication that slight growth may occur during some winter months. This may be due to linear growth that is so slight as to be unmeasurable by conventional methods, or to increase in actual shell thickness by internal accretion. Growth studies of Pacific oysters in both British Columbia and Washington have indicated some winter growth.

The ratio, whole volume: shell volume has been suggested as a measure that might evaluate the potential yield of an oyster population. Whole volume is measured with the live oyster and includes valve volume as well as internal volume. Shell volume is measured with the body removed and represents only the volume of the two valves. The ratio indicates capacity of an oyster to produce meat, for it reflects the meat volume an oyster in peak condition could produce. In Table 10 are shown ratios derived from oysters from several localities. The high value of 2.12 for suspended culture oysters does indeed reflect the fact that these oysters do produce a relatively large quantity of meat for the size of the oyster. Oysters grown at Naden Harbour, Queen Charlotte Islands, on very firm ground had a round, deeply cupped shape and gave a ratio of 1.9. Ladysmith Harbour oysters grown on much softer ground did not have the nearly perfect shape of Naden Harbour oysters and gave a ratio of about 1.6.

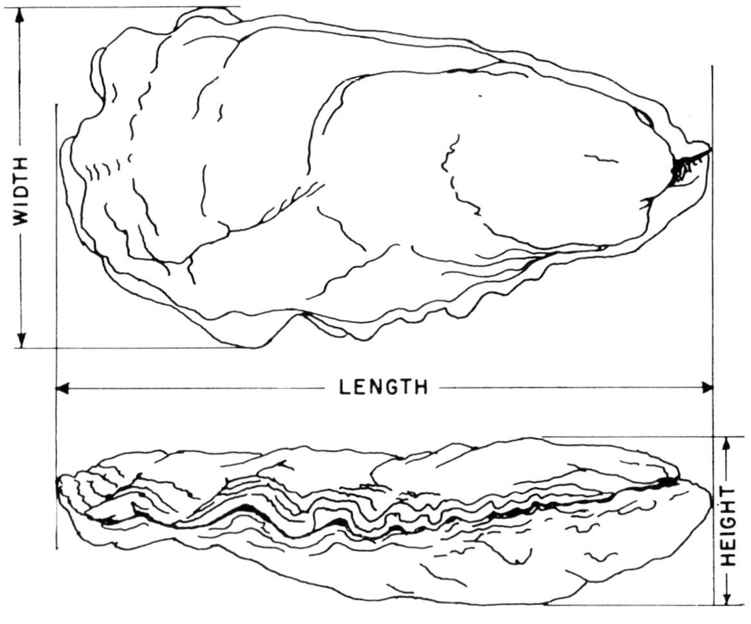

FIG. 74. Diagram to represent dimensional terms applied to oysters.

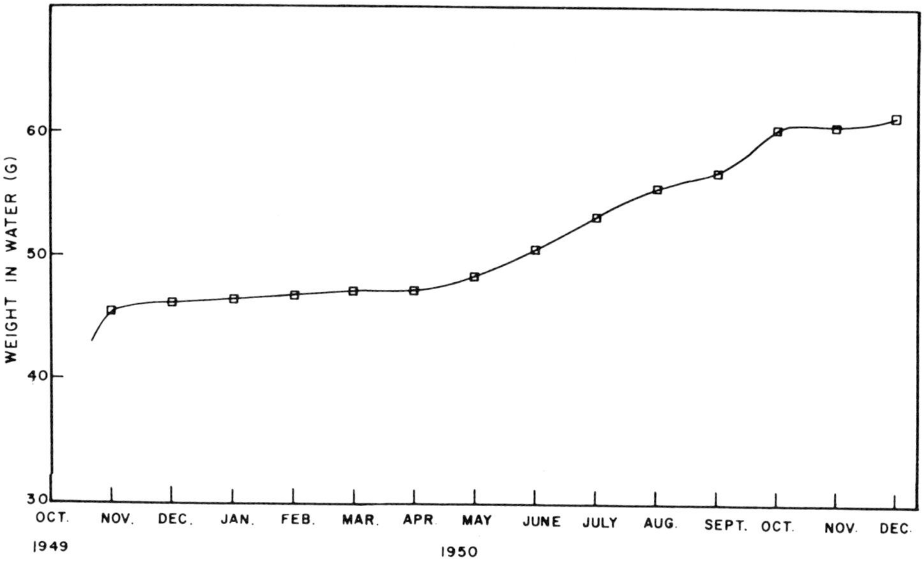

FIG. 75. Shell growth of the Pacific oyster (*C. gigas*) as determined by the weight in water, Ladysmith Harbour.

TABLE 10. Ratio of whole volume to shell volume of Pacific oysters at various locations in British Columbia.

Naden Harbour (Queen Charlotte Islands)	1.90
Seal Island	1.96
Henry Bay (Denman Island)	1.91
Chemainus	1.76
Thetis Island	1.86
Ladysmith (soft ground)	1.63
Ladysmith (hard ground)	1.80
Sherard Point (Crofton)	1.76
Eelgrass Flats (Crofton	1.55
Islands (Crofton)	1.68
Raft Culture (Ladysmith)	2.12

Growth Relationships

A useful source of information for oyster growers is the relationship between various measurements of growth. Of particular interest is that between size of an oyster and its meat content. Measurements of about 300 single well shaped oysters at the beginning of October produced the data shown in Table 11 and Fig. 76–79. Condition factor was not particularly high with little recovery after spawning. A few had not spawned and were still in a ripe condition. It will be noted there is little change in meat weight after the oyster reached a length of about 100 mm (4 inches) (Fig. 76). Absolute weights as shown here have little value in themselves since this changes throughout the year, but the relationship remains valid for there is a direct relationship between meat weight and internal shell volume as shown in Fig. 77. Thus many small oysters can yield as much meat as fewer large oysters. For instance in Table 11; Fig. 76, 100 oysters 100 mm (4 inches) in length yielded as much meat as 90 oysters 145 mm (6 inches) in length.

The economics of this situation is a balance between the extra cost of shucking more small oysters against the cost of holding oysters for the time required to grow an extra 45 mm. As mentioned elsewhere, theoretically, small oysters could fetch a better price than large oysters and it is almost certain market acceptability would be greater. It would be of value to growers to examine their oysters in this context.

TABLE 11. Relationship between length, whole weight, internal volume and meat weight in the Pacific oyster (*Crassostrea gigas*) Hammond Bay, British Columbia.

Length (mm)	Whole weight (g)	Internal volumes (cc)	Meat wt (wet) g
30–39	40	2	3
40–49	38	5	4
50–59	45	7	7
60–69	53	13	9
70–79	101	24	16
80–89	119	28	19
90–99	142	34	24
100–109	181	46	34
110–119	222	56	36
120–129	246	60	39
130–139	240	53	38
140–149	275	55	39

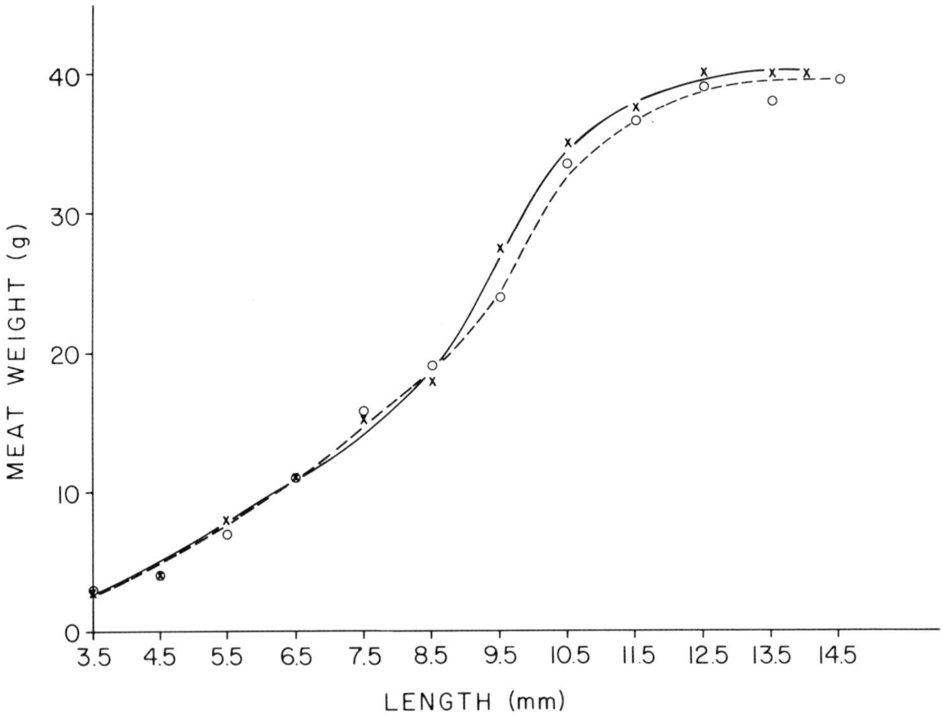

FIG. 76. Length–wet meat weight relationship of Pacific oysters. Hammond Bay. November, 1985.

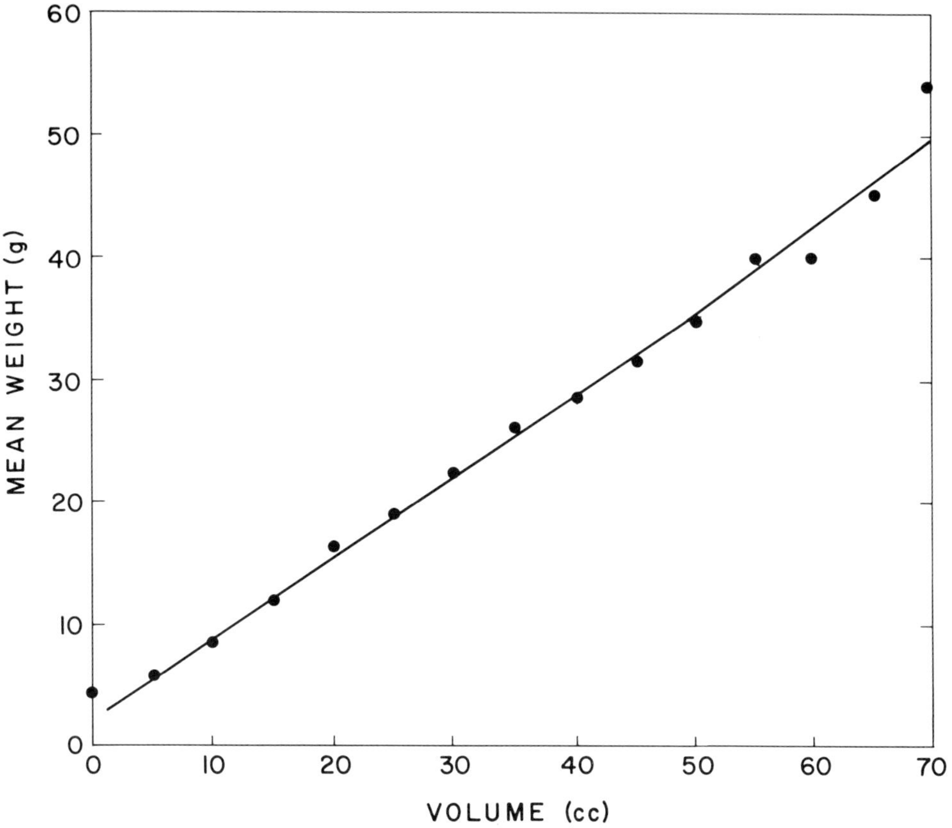

FIG. 77. Internal volume–wet meat weight relationship of Pacific oysters. Hammond Bay. November, 1985.

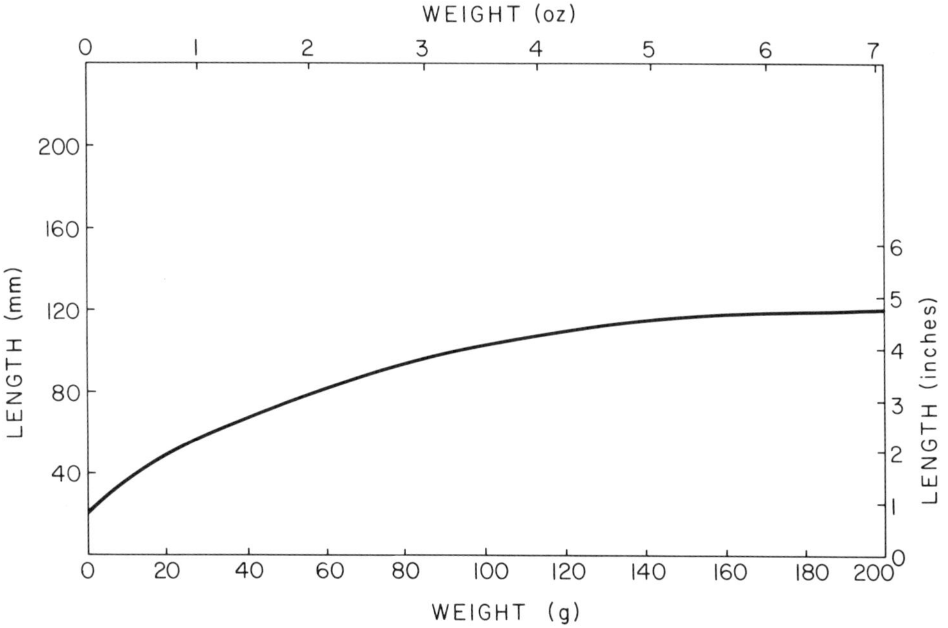

FIG. 78. Length–whole weight relationship of Pacific oysters. Hammond Bay. November, 1985.

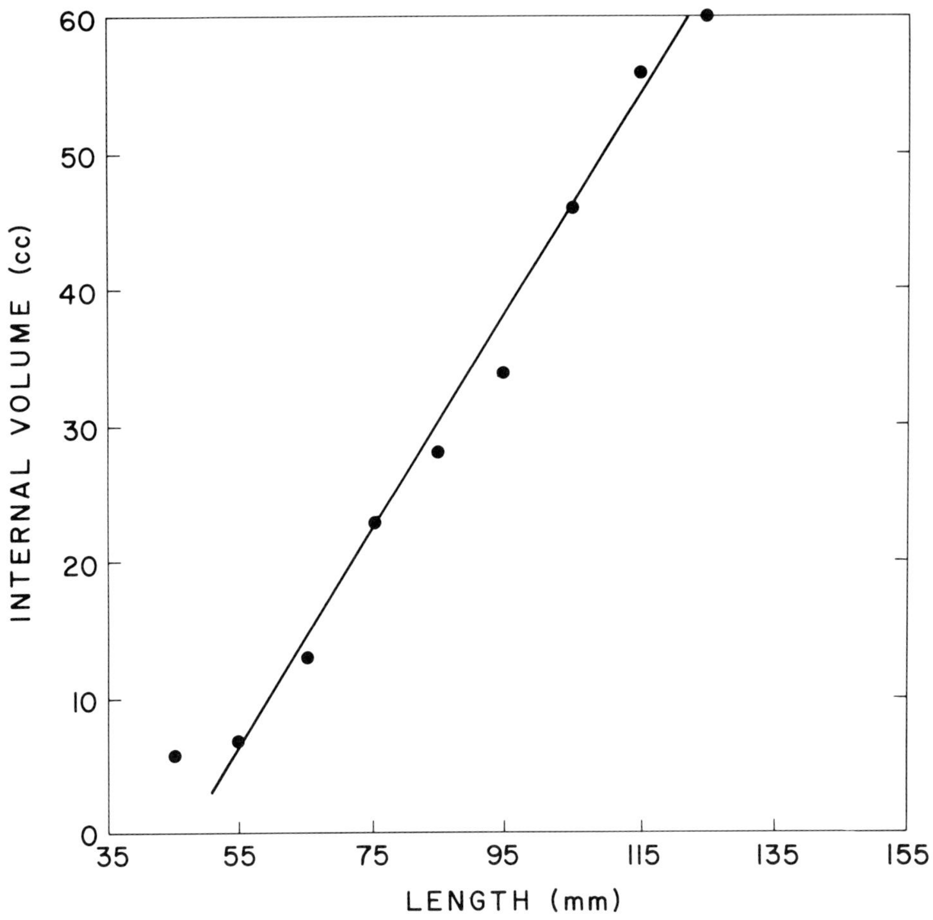

FIG. 79. Length–internal volume relationship of Pacific oysters. Hammond Bay. November, 1985.

Spat Growth

The growth of larvae has already been described.

Growth of early spat has been studied in Pendrell Sound during the period between setting and the following October. Measurements were made of linear dimensions of spat suspended from rafts at a point 60 cm below the surface. Results of this study are shown in Table 12 and Fig. 80. As shown in the table, there was no growth after the November 9 measurement, but by that time, at an age of about 3 months, the spat had attained an average diameter of 23.5 mm; and after another full summer of growth they were 90 mm in diameter.

Another study was conducted on an experimental oyster area (lot 164) in Ladysmith Harbour, when growth of oysters from three cases of Japanese seed was measured. Full data, including survival, are given in Table 13 and Fig. 81. When planted the spat were not greater than 6 mm in diameter and were placed at a tidal level of 1 metre on fairly firm mud–gravel ground. Initial weight and volume measurements were essentially of mother shell for the actual weight and volume of the spat at that stage were negligible. These data represent just over two growing seasons. Like most animals, oysters grow rapidly when young and more slowly as they become older. In oysters, reduction of growth rate occurs at 4 or 5 years of age.

TABLE 12. Growth of Pacific oyster spat, August 1955 to October 1956, Pendrell Sound, British Columbia.

Date	Mean diameter (mm)	Increase in diameter (mm)	% increase in diameter
1955			
Aug. 15	0.3	—	—
Aug. 18	1.1	0.8	1
Aug. 30	3.8	2.7	3
Sept. 8	7.9	4.1	4
Sept. 20	13.4	5.5	6
Oct. 6	18.6	5.1	6
Nov. 9	24.7	6.1	7
Dec. 14	23.5	0	0
1956			
Jan. 31	23.7	0.2	0
May 7	34.0	10.3	11
June 8	50.0	16.0	18
July 20	73.0	23.0	26
Oct. 11	90.0	17.0	18

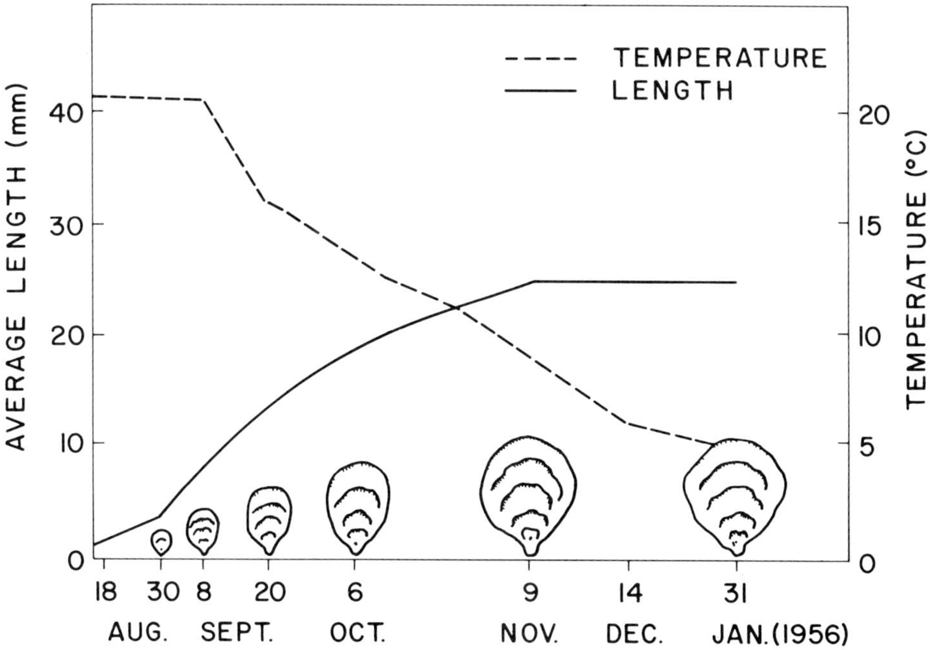

FIG. 80. Growth of Pacific oyster spat, Pendrell Sound, 1955–56.

TABLE 13. Pacific oyster growth in Ladysmith Harbour, Lot 164. Mean length, volume, and weight of oysters in three cases of Japanese oyster seed, April 1952 to June 1955.

Date	Length (mm)	Volume (litres) (per case)	Weight (kg) (per case)
1952	5	47.2	53
1953	53	410.1	224
1954	96	718.7	460
1955	109	845.8	710

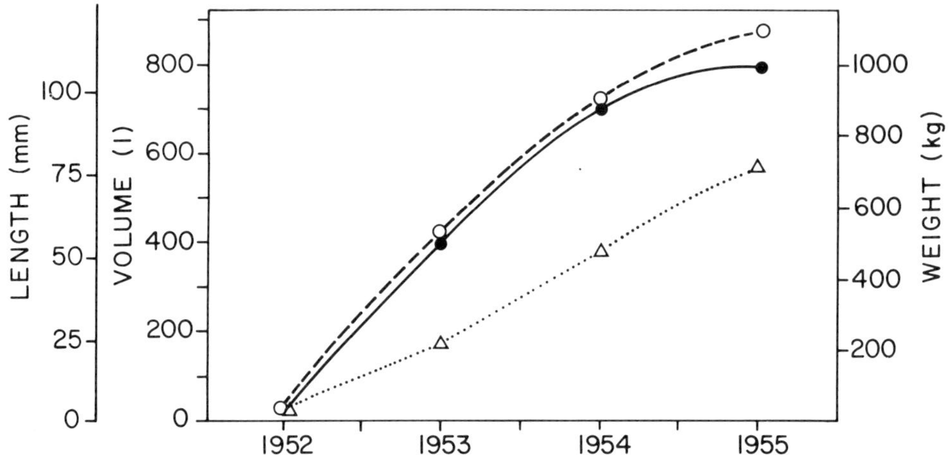

FIG. 81. Growth rate of Pacific oysters in Ladysmith Harbour. Mean shell length (+), case volume (⊙), and case weight [lb] (△) of a 3-case lot of Japanese oyster seed.

Growth Variability

To demonstrate variability of Pacific oyster growth with locality, data are given in Table 14 on oysters of similar size, grown under similar conditions, at similar tide levels at four locations each 1 mile apart along the shore of Ladysmith Harbour.

Table 14 and Fig. 82 show the growth rate, in length and volume, of Pacific oysters as measured from four year-classes of Japanese seed in contiguous beds in Ladysmith Harbour. Relatively even growth in length during the whole period may be noted, but of interest is the slow increase in volume until the last year when nearly 82% was added.

TABLE 14. Percentage increase in length and volume of Pacific oysters of equivalent size, Ladysmith Harbour, May 5–August 2, 1955.

Location	% increase in length	% increase in volume
Station A	12.5	44.5
Station B	17.7	75.4
Station C	15.7	46.6
Station D	8.5	27.4

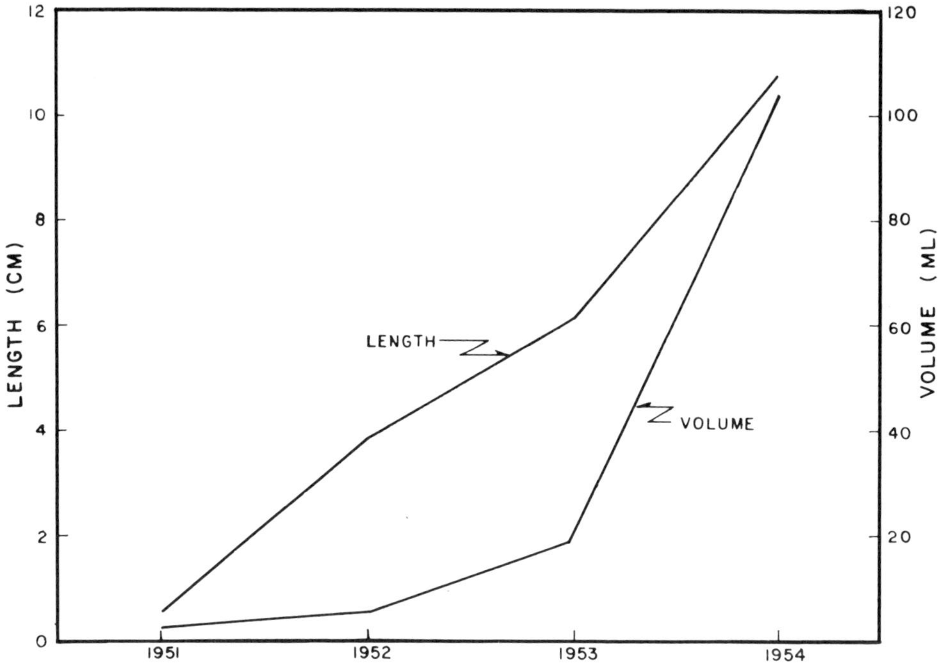

FIG. 82. Length and volume increase of Pacific oysters as measured from samples of four consecutive year-classes on the beds at Ladysmith Harbour.

TABLE 15. Growth rate of Pacific oysters from four year-classes, Ladysmith Harbour, 1951–54.

Year-class	Length (cm)	Volume (cc)
1951	0.50	1.0
1952	3.44	5.0
1953	6.18	19.2
1954	10.80	104.0

Seasonal Shell Growth

Seasonal growth of Pacific oysters was studied in Ladysmith Harbour. Oysters were marked and the length, width, thickness, and volume measured at monthly intervals. Results are shown in Table 16 and 17 and in Fig. 75 and 83.

There was no significant increase in length and width until April, in thickness and volume, until May. From then on, there were fairly regular increases in all measures until October. Shell growth began when the water temperature reached about 10°C and ceased at approximately the same point in the fall. Shell growth as measured by weight of the whole oyster in water, and as shown in Fig. 75, indicates a very slight increase in shell weight during winter.

These data represent averages and there was a wide variation in the amount of growth shown by individual oysters, for a few grew very little while others nearly doubled their size. Times at which various individuals began and stopped growing also showed wide variability.

TABLE 16. Mean dimensions of Pacific oysters at monthly intervals grown in Ladysmith Harbour, 1949–50.

	Length (cm)	Width (cm)	Thickness (cm)	Volume (cc)	Temperature (C)
November	6.90	4.74	2.89	29.2	9
December	6.82	4.66	2.96	29.3	6
January	6.90	4.64	2.92	29.8	4
February	6.78	4.65	2.70	29.5	6
March	6.86	4.62	2.87	30.3	7
April	7.03	4.72	2.90	30.8	9
May	7.28	5.00	2.97	32.4	12
June	7.51	5.22	3.07	35.4	18
July	7.91	5.51	3.23	38.5	20
August	8.19	5.61	3.33	44.4	19
September	8.65	5.78	3.47	48.3	15
October	8.75	5.90	3.48	51.3	10
November	8.77	5.94	3.50	51.6	8

TABLE 17. Mean monthly percentage increase in dimensions of Pacific oysters grown in Ladysmith Harbour, 1950.

	Average temperature (°C)	Length	Width	Thickness	Volume
April	9	9.0	7.8	—	—
May	12	13.2	21.9	12.3	7.8
June	18	12.2	17.2	17.6	14.7
July	20	21.2	22.7	28.1	15.1
August	19	14.8	7.8	17.5	28.8
September	16	24.3	13.3	24.5	19.0
October	10	5.3	9.3	—	14.6

Significance of the marked decrease in growth in length, width, and thickness during August (Fig. 83) is not known, unless it was associated with spawning, which, however, was not observed in the group being studied. It is noteworthy that shell growth continued during gonad (spawn) development and also during the so-called fall fattening period of September and October.

Correlation between onset and cessation of shell growth at about 10°C has already been suggested. It is known, however, that the Pacific oyster is able to pump water at temperatures considerably lower than 10°C. It may be that the temperature limits indicated are applicable only to shell growth and not to tissue growth, but more research is required to clear up these points. For instance, much of the rapid growth occurred during periods when the phytoplankton content of the water was low.

Growth was shown to begin rather slowly with onset of increasing water temperatures in April, and during the first 3 months less that 40% of the total annual growth in length, thickness, and volume occurred. However, nearly 60% of the increase in width occurred during this period. Length, thickness, and volume increased most rapidly during July, August, and September, thereby accumulating nearly 60% of the total growth in the 3 warmest months of the year. After this period, growth, in most instances, ceased quite abruptly, although the water temperature declined gradually.

Growth is also related to tidal height, and, as a general rule, the higher the tidal level, the slower the growth. However, differences are not as great with young oysters as they

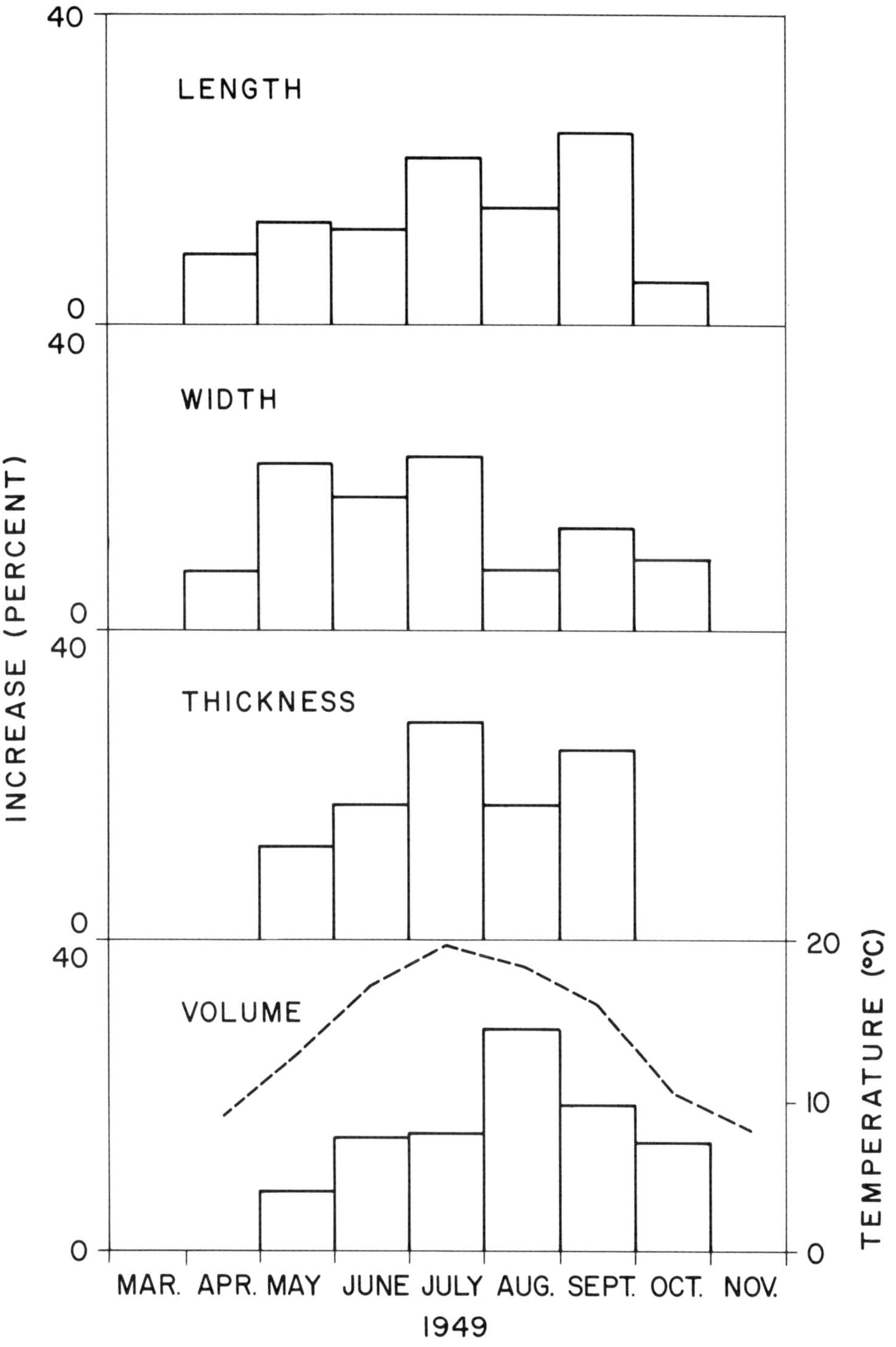

FIG. 83. Growth of Pacific oysters, Ladysmith Harbour, 1949.

are with older ones. This is fortunate because it is possible to plant spat at a relatively high tidal level where ground is generally firmer than at low levels, thus making for better survival, and yet still attain a fairly good growth rate.

TABLE 18. Mean daily hours of beach exposure at various tide levels for each month, Fulford Habour, 1964.

Tide level (m)	Jan.	Feb.	Mar.	Apr.	May	June	July	Aug.	Sept.	Oct.	Nov.	Dec.	Average
0.3	0.0	0.0	0.0	0.0	0.0	0.0	0.0	0.0	0.0	0.0	0.0	0.0	0.0
0.0	0.0	0.0	0.0	0.0	0.1	0.2	0.2	0.0	0.0	0.0	0.1	0.0	0.1
0.3	0.1	0.1	0.0	0.3	0.4	0.6	0.5	0.3	0.0	0.0	0.4	0.4	0.2
0.6	0.9	0.9	0.1	.07	1.3	1.3	1.2	0.6	0.3	0.3	0.8	1.0	0.8
1.0	1.9	2.4	0.7	.19	2.5	2.7	2.1	1.6	1.0	0.9	1.6	2.1	1.8
1.3	3.2	3.9	2.4	3.7	3.8	3.8	3.3	3.1	2.6	2.4	2.7	3.2	3.2
1.6	4.6	5.4	4.4	5.3	5.5	5.2	4.6	4.5	4.7	4.4	4.5	4.6	4.8
2.0	5.9	7.2	6.4	6.9	7.3	6.6	6.4	6.2	6.5	6.4	6.2	6.2	6.5
2.3	7.8	9.8	9.1	9.8	9.8	8.9	8.7	8.0	8.8	9.2	8.9	7.9	8.9
2.6	9.8	13.2	12.4	15.0	14.4	12.9	12.2	1.6	11.6	12.6	12.9	10.3	12.4
3.0	12.8	19.0	19.5	20.2	18.8	16.4	16.5	17.1	18.2	18.2	16.5	13.5	17.2
3.3	17.1	21.5	22.4	23.6	22.0	20.8	20.8	21.6	22.6	21.9	20.0	17.5	21.8
3.6	21.0	23.9	23.9	24.0	23.9	23.6	23.7	24.0	23.9	23.8	20.9	21.1	23.3
4.0	23.1	24.0	24.0	24.0	24.0	24.0	24.0	24.0	24.0	24.0	23.8	23.2	24.0
4.3	24.0	24.0	24.0	24.0	24.0	24.0	24.0	24.0	24.0	24.0	24.0	23.9	24.0
4.6	24.0	24.0	24.0	24.0	24.0	24.0	24.0	24.0	24.0	24.0	24.0	24.0	24.0

TABLE 19. Mean daily exposure at various tide levels for five British Columbia reference ports, as estimated from hourly height records.

Tide level (m)	Mean daily exposure (hours) at following stations in years shown				
	Victoria (1965)	Pt. Atkinson (1965)	Tofino (1965)	Bella Bella (1965)	Fulford Hbr. (1964)
−0.6	0.0	0.0	0.0	0.0	0.0
−0.3	0.0	0.0[a]	0.0	0.0	0.0
0.0	0.1	0.1	0.0[a]	0.0[a]	0.0[a]
0.3	0.3	0.4	0.2	0.1	0.2
0.6	1.0	0.8	0.8	0.3	0.8
1.0	2.4	1.7	2.1	0.9	1.9
1.3	4.3	2.8	3.9	1.9	3.3
1.6	6.5	4.0	6.1	3.2	5.1
2.0	9.6	5.3	8.7	4.8	7.0
2.3	13.6	6.7	11.5	6.7	9.5
2.6	18.8	8.4	14.4	8.8	12.6
3.0	22.6	10.6	17.4	11.0	17.0
3.3	23.8	13.3	20.2	13.0	20.9
3.6	24.0	16.6	22.6	15.2	23.2
4.0	24.0	19.6	23.6	17.6	23.8
4.3	24.0	22.1	23.9	19.8	23.9
4.6	24.0	23.5	24.0	21.8	24.0
5.0	24.0	23.9	24.0	23.1	24.0
5.3	24.0	24.0	24.0	23.7	24.0
5.6	24.0	24.0	24.0	23.9	24.0
6.0	24.0	24.0	24.0	24.0	24.0

[a]Less than 3 cm.

The potential effect of tidal height may be seen in Table 18 and 19 and Fig. 84, which show the amount of time ground is exposed at various tide levels. The date in Table 18 were derived in a different way from those in Table 19, hence the slight difference in values for Fulford Harbour.

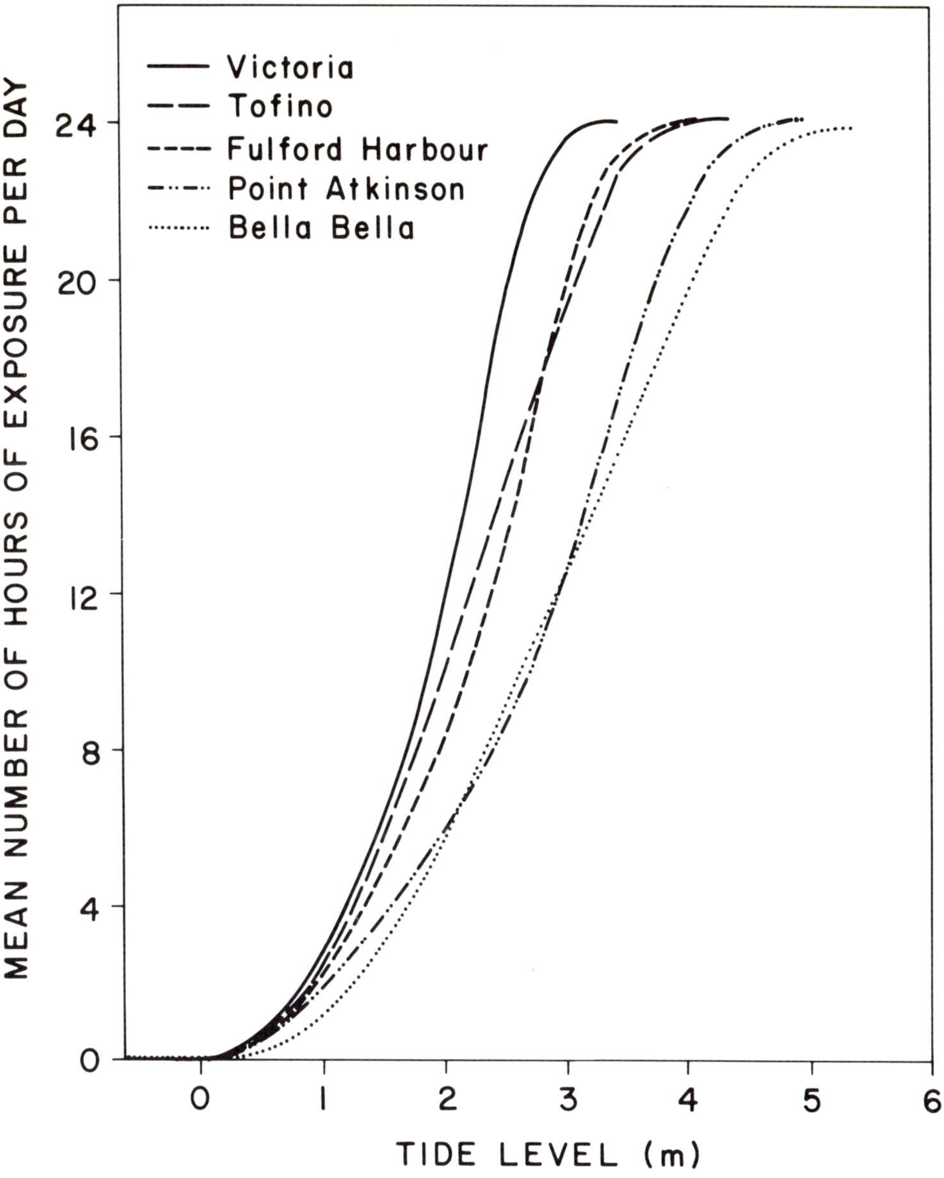

FIG. 84. Mean daily hours of exposure of beach at various tide levels for five British Columbia reference ports, as estimated from hourly height records provided by the Canadian Hydrographic Service.

Associated Organisms

In addition to the acquisition of a thorough knowledge of the oyster itself, growers should know something of the other animals and plants growing on an oyster bed along with, and sometimes on, oysters. These are an important part of the complex environment in which oysters live. Some of these organisms affect the oyster directly, others indirectly, but all exercise an influence on its well-being in one way or another. Some are predators such as seastars or drills, some are parasites such as the red copepod (*Mytilicola*) (Fig. 68), some, like clams are competitors, some use the oyster shell as a place of attachment, others as a place of refuge, and others as an incubator on which to deposit their eggs.

All culture sites do not have identical sets of animals and plants, for these vary with local hydrographic conditions. Many of these organisms are so small as to be invisible or inconspicuous, but what they lack in size they make up for in numbers. Among these are bacteria, one-celled animals of various kinds collectively called protozoa, attached forms of diatoms and larval or young forms, still microscopic in size, of many larger animals.

Of the diatoms, one in particular becomes so abundant in late winter and early spring that, where currents are not too strong, it covers intertidal rocks and parts of oyster beds and oysters with a brown mat of slime. This is the chain diatom, *Melosira*. To what extent oysters may feed on this is not known, but it harbours many species of Protozoa which are likely to be suitable oyster food.

Sponges

Several species of sponge occur on oyster beds and, quite often, on oysters themselves. These may be yellow, purple, or green, generally encrusted, with slightly raised water pores. They may compete to a limited extent with oysters for food but seldom occur in sufficient abundance to require corrective measures. In addition to the encrusting type of sponge there is another (*Cliona*) (Fig. 85) which bores into shells of molluscs and this is dealt with in the section on pests.

Coelenterates

This is a group of animals that includes jellyfish, hydroids, anemones, and corals. Only hydroids and anemones are associated with oysters. Hydroids resemble a small bush and occur mainly on suspended raft culture oysters. Of the anemones, a very large brown or white *Metridium* (Fig. 86), found mainly on piling, occurs frequently on suspended oysters but on beds only infrequently. The small *Diadumene*, (now *Haloplanella*) about 1 cm in diameter, green with vertical gold stripes is found on oyster shells, particularly in areas where Japanese seed has been planted, for it is an Asiatic species. On some oyster beds at the north end of Denman Island, there is a burrowing anemone called *Pachycerianthus*. None of the anemones interferes with the oysters well-being but they may be a nuisance in suspended culture.

Worms

There are several groups of worm that live on oyster beds and on oysters and these are flatworms, roundworms, and ribbon worms.

Flatworms are also called waferworms or oyster leeches. Those associated with oysters are generally quite small, not more than 2 cm in length and 1 cm wide, brown, and difficult to see on an oyster shell. They are wafer thin and appear to slide rather than creep. A Japanese species (*Pseudostylochus ostreophagus*) (Fig. 67) that has become established on the northwest coast oyster beds assumes pest proportions in some years.

Ribbonworms (nemerteans) are unsegmented, thin, highly extensible, and usually brightly coloured. They are found among clusters of oysters or mussels where there is a certain amount of mud. Those on oyster beds are purple, green, or chocolate brown with white transverse stripes. They feed mainly on other types or worms and are of no danger to oysters, although known to attack other molluscs.

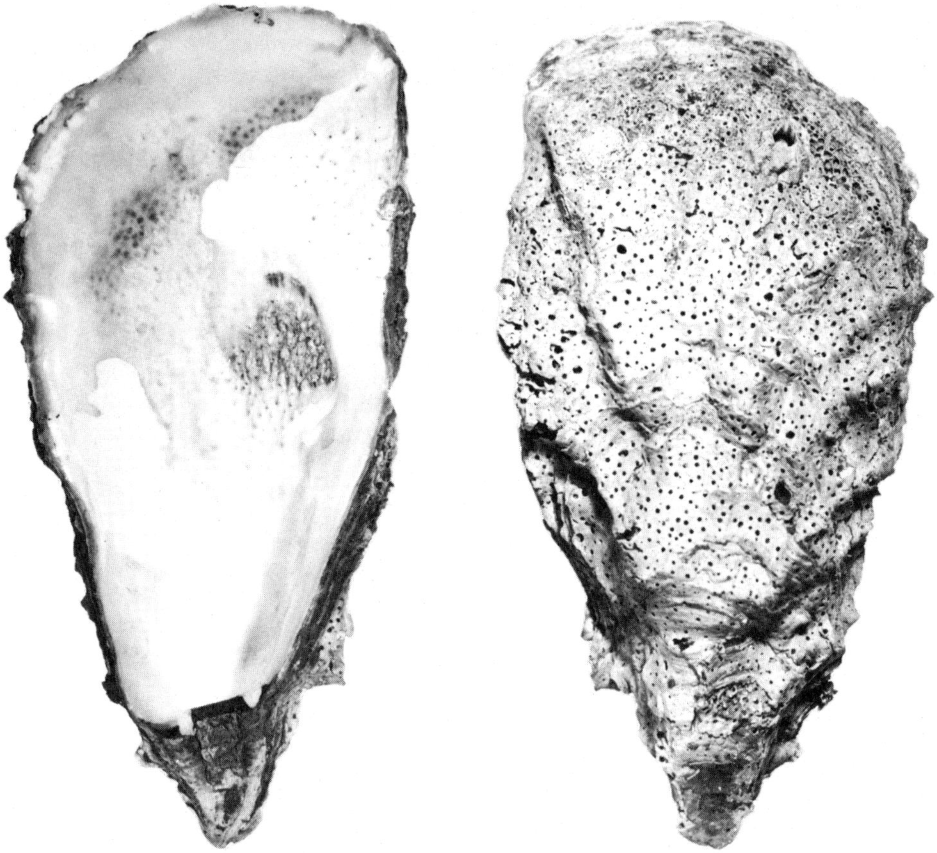

FIG. 85. Oyster shell pitted with burrows of the boring sponge (*Cliona*). ×0.75.

FIG. 86. The large sea anemone (*Metridium senile*), brown or white in colour.

The group known as seaworms (polychaetes) are segmented, and many species occur both in and on oysters. One group is free-living and is well known because of its usefulness as fish bait. A species, which attains a length of nearly 0.6 metre, normally lives burrowed in the ground but occasionally emerges from the burrows during the summer spawning season and may be seen swimming at the surface in the evenings.

Another group lives in calcareous tubes attached to oyster shells. Some called *Spirorbis* (Fig. 87b), have tiny white coiled shells; the larger sinuous shells up to 5 cm in length contain *Serpula* (Fig. 87a). Neither of these is dangerous to oysters, but another one, called *Polydora* (Fig. 88), lives near the shell edges and can cause harm and is described in the section on pests.

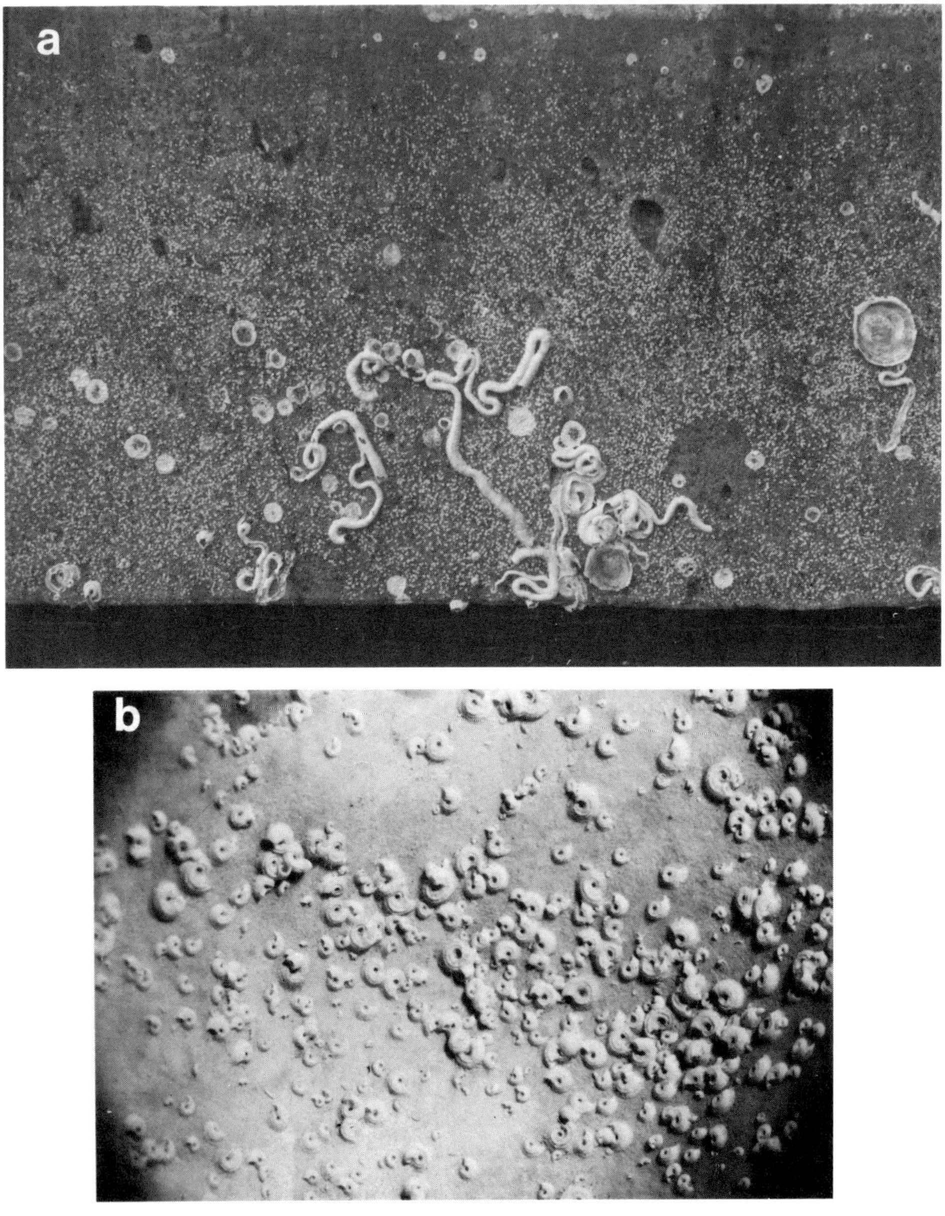

FIG. 87. a. Tube worms (*Serpula*), barnacle (*Balanus*) scars and numerous *Spirorbis* on a cement log. b. *Spirorbis*. ×4.

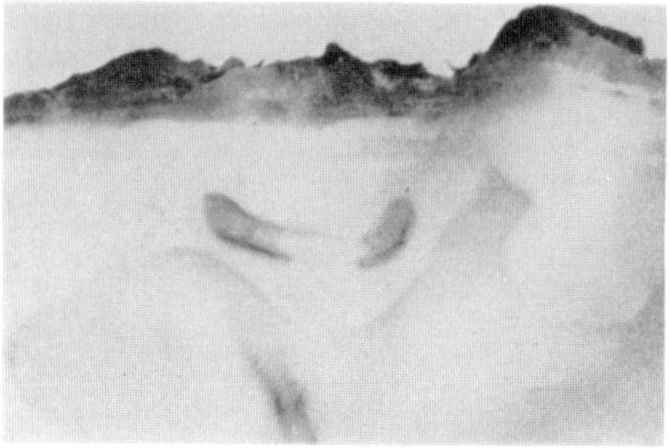

FIG. 88. Burrows of the boring worm (*Polydora*). ×2.

Several others live in membranous tubes buried in the bottom or on oyster shells with just a small portion protruding above the surface of the mud or out of the tube. One of these is *Euclymene zonalis*, which lives in muddy sand and has a tube less than 3 mm in diameter. In some years it multiplies in such great numbers that it causes the ground to become soft enough for oysters to sink into it. The only solution to this situation is to plough the ground to destroy the worms or allow it to lie fallow until the worms die off. So far, Boundary Bay is the only area where this worm has caused difficulty.

Crustacea

Many species of crustaceans, a group which includes crabs, shrimps, and sand fleas, occur with oysters. The three species of crabs which are able to do damage to oysters are described on page 107.

There are two species of *Hemigrapsus* (Fig. 89) which are very abundant in the intertidal zone. These shore crabs seldom exceed a shell diameter of 3 cm; one species, smaller than the other, is grey or green, while the larger one is generally red-brown. They are most often found under stones or clusters of oysters. Another rather hairy yellow crab (*Telmessus*) is not nearly as abundant as shore crabs, nor is the spider crab (*Pugettia*). These crabs are all scavengers and so no harm to oysters. In many ways they act as garbage disposal units on the oyster bed. Crabs of the genus *Cancer* are oyster predators and are dealt with in another section.

The barnacle is also a crustacean, although it does not look it except in the larval stages. The common shore barnacles (*Balanus glandula* and *Chthamalus dalli*) have very definite zones where they settle. If oysters are held below this zone there will be little barnacle settlement on them but if placed in the barnacle zone, heavy settlement, particularly from the larger *Balanus glandula*, may be expected. This species settles most extensively within a week or 10 days before or after April 1 in most areas of the Strait of Georgia, although there may be lesser settlements throughout most of the summer. Other than increasing the weight of the oyster, and causing a nuisance to shuckers, barnacles cause little harm to the table oyster (shucked) part of the industry, but may be a disadvantage to the half shell trade for the valves are marred by calcareous bases which are left after the barnacle is removed. However, the upper right valve is removed for presentation of the oyster to the diner and only edges of the left valve are seen. If there are barnacles on the left valve they are probably better left in place, adding a point of interest and a possible conversation piece.

Pill box bugs (*Exosphaeroma*), flattened dorsoventrally, are isopods found on the underside of oyster shells in those situations where there is a fair amount of fresh water rather high in the intertidal area. Amphipod sand fleas (flattened laterally) are found in similar locations but do not harm oysters. Two other isopods (*Limmoria lignorum* and *Limnoria tripunctata*) cause no harm to oysters but they are wood borers and can destroy

FIG. 89. Four species of crab commonly found on oyster beds. Edible crab or Dungeness crab (*Cancer magister*), ×0.25; red or rock crab (*Cancer productus*), ×0.25; raceful crab (*Cancer gracilis*), ×0.3; shore crab of which there are two species (*Hemigrapsus*), ×0.5.

untreated piling or other wooden structures quite rapidly. These animals are less than 6 mm in length and are also called gribbles or pinworms. Copper paint or creosote treatment of wood prevents their initial settlement.

Molluscs

Many species of molluscs occur on oyster beds and the more common ones are shown in Fig. 90. Full details on all clam species that may be found are given in Handbook No. 17 on intertidal bivalves, published by the Provincial Museum at Victoria. Clams and mussels are likely to compete with oysters for food to a limited extent but on most oyster beds they are not abundant. Mussels may become pests on certain beds particularly at the higher tide levels. In addition to clams, there are several species of snails, two of which are oyster predators. These drills are described in detail elsewhere. A common snail on most oyster beds is *Batillaria attramentaria* (Fig. 91), which is another species of Japanese origin. Fortunately, it is not harmful. A native snail (*Nucella lamellosa*) (Fig. 66), sometimes called the native drill because it does drill and eat barnacles, mussels, and clams, is normally harmless to oysters.

In addition to shelled snails, there are nudibranchs or sea slugs which are molluscs that have lost their shells in the evolutionary process. The sea lemon (*Anisdoris*) is a fairly large, bright yellow nudibranch about 5 cm in length. Another is the brown spotted

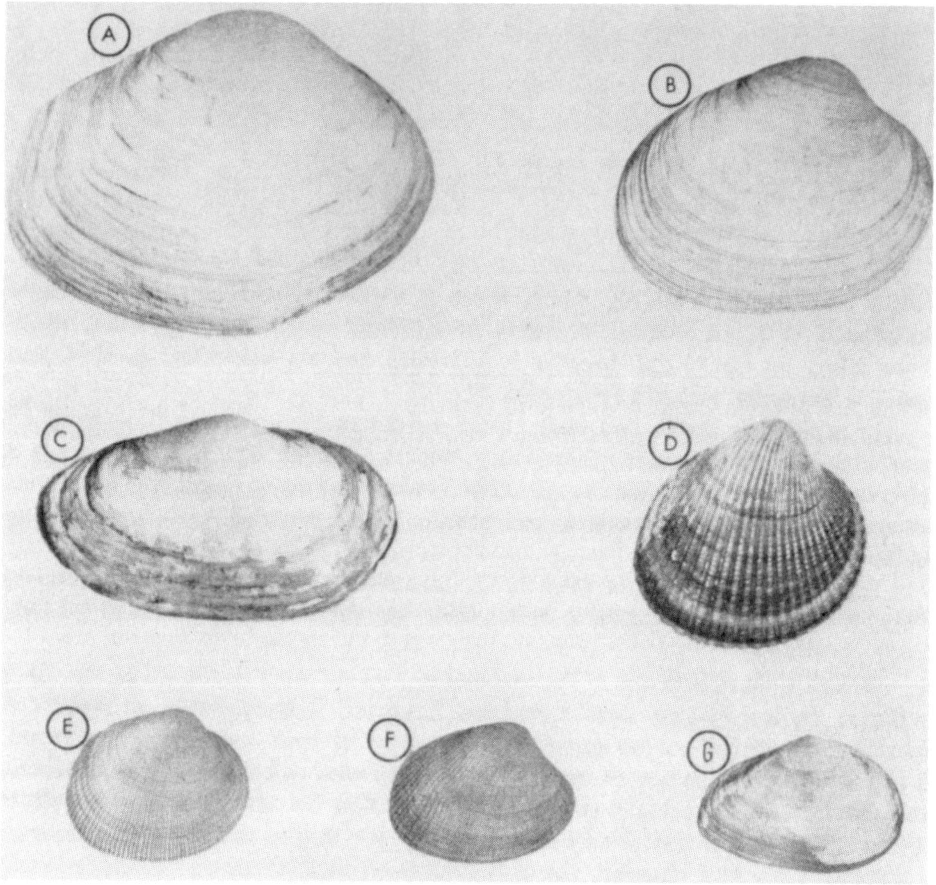

FIG. 90. Common clams found on oyster beds: (a) horse, (b) butter, (c) softshell, (d) cockle, (e) littleneck, (f) manila, (g) sand.

FIG. 91. *Batillaria atramentaria* (×1.3) **(A)** and one of several species of common chitons (×0.7) **(B)**.

Diaulula. Less abundant, and usually found stranded, are the transparent swimming *Melibe* and red *Dendronotus*.

Several species of chitons (Fig. 91), or coat-of-mail shells, with 8 overlapping shells, are often found clinging to oyster shells but these also are harmless for they feed by browsing.

British Columbia is one of the few oyster growing areas in the world that does not have the large slipper shell *Crepidula fornicata*. It is originally an American Atlantic coast species but has spread to many oystering areas, including Puget Sound. It inflicts no direct damage on oysters but attaches itself to oyster shells in considerable numbers, tending to smother the oyster. This *Crepidula*, or slipper shell, is shown in Fig. 92 so that it may be recognized. It is hoped that it will not become a problem in British Columbia. A harmless common slipper shell on British Columbia oyster beds is *Crepipatella lingulata*, also shown in Fig. 92.

The shipworm is another mollusc that affects oyster growers indirectly owing to its destructive attack on untreated wood in floats or scows. Two species occur in British Columbia and the native species (*Bankia setacea*) reaches a length of nearly one metre and is found throughout the coast. The other species is the introduced Atlantic shipworm (*Teredo navalis*) and at present is known to occur only in Pendrell Sound, where it has created difficulties for seed collectors there. Copper paint, creosote, or fiberglass are the only practical protective coatings against shipworms.

The addition of tributyltin (page 209) and other chemical antifouling paints is causing some concern where many vessels occur in the neighbourhoodof oyster operations. Antifouling paints operate on the principle of leaching out of the toxic elements from the paint coating.

Echinoderms

Echinoderms which occur on oysters include seastars, sea urchins, and sea cucumbers.

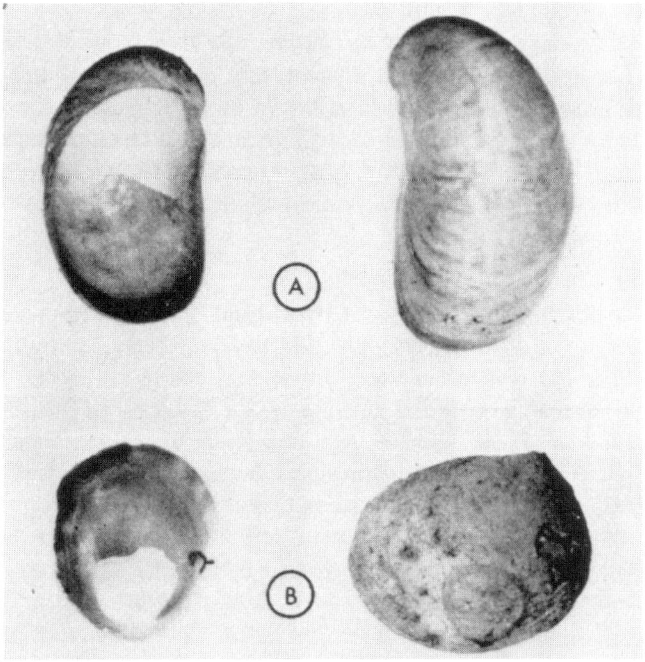

FIG. 92. Slipper shells associated with oyster beds. **(A)** *Crepidula fornicata* (× 1.5): does not yet occur in British Columbia. **(B)** *Crepipatella lingulata* (× 1.5): common on British Columbia oyster beds.

About five species of seastars occur commonly on British Columbia oyster beds. Of these only one, the leather star, *Dermasterias imbricata* (Fig. 93) is not an oyster predator. The others are dealt with in the section on predators.

On a few beds, the green urchin (*Strongylocentrotus drobachiensis*) is found, and, on beds with a good deal of nearly pure sand, the sand dollar (*Dendraster excentricus*) is seen. Neither of these is harmful to oysters as far as is known but the former feeds on mussels.

Cucumbers as well as urchins occur on raft culture oysters, but are not harmful.

Tunicates

These are sea squirts and will seldom be found on bed oysters but may be quite numerous in suspended culture or on seed strings, particularly in Pendrell Sound (Fig. 94).

Algae

On most oyster beds, in spring, there develops a flat, green-coloured seaweed called *Ulva*. Usually, it does not last long and seldom poses a serious problem. On other beds another brown flat-bladed seaweed, called *Punctaria*, also develops in spring. It can become so plentiful the oyster bed may be completely covered and the oysters hidden; it can interfere significantly if the oysters have to be picked up, either for harvesting or relaying. The simplest solution is, if possible, to use the bed either before or after occurrence of the weed.

FIG. 93. The four common sea stars found on or near oyster beds. Leather star (*Dermasterias imbricata*) ×0.2; sun star (*Pycnopodia helianthoides*) ×0.1; mottled star (*Evasterias troschellii*) ×0.1; ochre star (*Pisaster ochraceus*) ×0.2.

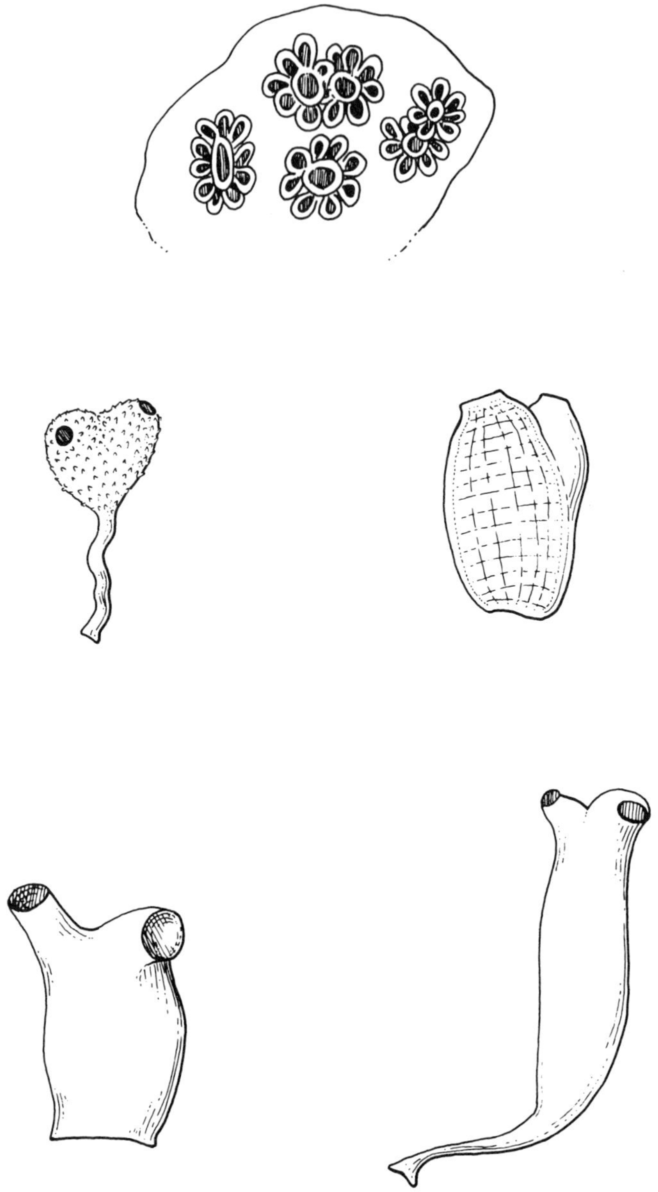

FIG. 94. Examples of tunicate (sea squirt) fouling species. Clockwise from the top: *Botryllus*, *Corella*, *Styella*, *Halocynthia*, and *Boltenia*. Approximately ×1.

Another weed, *Enteromorpha*, is green and is formed of long hollow strings which can form a very thick solid mat over an oyster bed in late spring and early summer. Normally, if there is warm weather during low-tide periods, the sun will dry it enough to cause it to float from oyster beds in quite large rafts. If the weather is wet, however, *Enteromorpha* may persist until it dies and rots. In this case, underlying oysters may suffer some mortality. Major mortalities from this cause have not been reported from British Columbia.

A prominent weed is the so-called Japanese weed (*Sargassum muticum*) (Fig. 69, 70) which was introduced from Japan with oyster seed, and which first gained a solid foothold in British Columbia waters between 1941 and 1945. It requires a rocky bottom just below zero tide level but some fronds may be found higher up. The main objection to this weed is the nuisance it occasions by becoming entwined in propellors of small boats. However, it may be of significance in the economy of shore waters because of the considerable amount of organic material it produces.

Eelgrass

A number of seaweeds (algae), as well as a single flowering plant, the eelgrass (*Zostera marina*), occur on oyster beds. Eelgrass is found mainly at lower levels of the beach and in British Columbia it is seldom a problem in growing oysters. Oysters may be grown successfully in eelgrass but a main difficulty arises in harvesting. However, a considerable portion of any eelgrass bed dies off during winter and if harvesting of oysters from an eelgrass bed can be delayed until early spring, there is usually not enough grass left to interfere significantly with harvesting. In the State of Washington, some beds have such a luxuriant growth of eelgrass that it has to be cut by specially designed drags.

Fouling

Fouling, or more properly, biofouling is the attachment of marine organisms, either plant or animal to the object of interest, whether it be oysters, oyster cultch, or a boat. It may be more of a nuisance than a serious problem so its importance may be over emphasized. There is a tendency for inexperienced oyster culturists to become unnecessarily alarmed over a level of fouling that is inconsequential. Also, it is often more economical to accept whatever deleterious effects fouling may cause than to undertake costly control measures. However, fouling can be a major problem in some areas. A rough rule of thumb indicates a difficulty when the volume of fouling approaches the volume of oysters or cultch. The type of fouling organisms may also be a factor as well as the market for which the oysters are grown i.e., the half shell market.

Fouling on oysters requires study because it may cause mortality, particularly in seed, reduce growth rate and in suspended culture cause floatation problems. It may seriously affect seed collection by reducing the surface area of cultch and some fouling organisms can feed on oyster larvae. The deleterious effects of fouling on cultch is one reason for spatfall predictions.

Fouling is generally less a problem in intertidal culture than in sub-tidal or suspended culture, for higher air temperature and exposure to sun and rain inhibit survival of soft bodied invertebrates such as tunicates or anemonies, the two main suspended culture fouling organisms. These organisms, among others, generally do not occur in low salinity waters as they are stenohaline, being restricted to a narrow, high salinity range. On the other hand, oysters, particularly the genus *Crassostrea*, are euryhaline, able to live and grow in a wide range of salinities. However, this genus often breeds best in areas with medium salinity such as 20–25‰. To take advantage of the euryhalinity of oysters and the stenohalinity of most fouling organisms, seed is collected and grown briefly in high salinity waters, then moved for final growth in water of lower salinity (<20‰) if these are available nearby, possible in most estuaries.

The main fouling organisms causing problems are sponges, anemonies, hydroids, bryozoa, tube dwelling polychaetes, mussels, tunicates, and algae. Acorn barnacles, chiefly the genus *Balanus*, and probably the most ubiquitous of fouling organisms, occur from the high intertidal to considerable depths and in great abundance, with settlement rates up to 25 per square cm. Fortunately most do not grow more than 1 cm in diameter. Some species attach with a solid plate, others only at the margins. Usually more than one species occurs in a given area so it may be necessary to identify differences so the breeding and distribution of each may be determined and taken into account in any scheme of culture and seed collection.

Sponges

Sponges may be encrusting or solitary and erect, the former more difficult to detach than the latter. In addition one group (*Cliona*) (Fig. 85) bores into calcareous shells creating a honeycomb of tunnels. Some may penetrate the nacareous layer causing the oyster to expend energy in producing more shell to cover the penetrations. Tunnelling may so weaken the shell that shucking becomes difficult. Some encrusting sponges may be controlled by exposure to sun in the early stages of development.

Anemonies

These are coelenterates, relatives of the jellyfishes and corals with a soft body and a tentacled oral surface (Fig. 86). The soft bodies with a broad holdfast are difficult to dislodge. They vary from about 1 cm to 15 cm diameter. The soft body makes it unable to withstand exposure to sun for any length of time. Anemonies do not harm oysters significantly and are more of a nuisance than a danger. They occur mainly in suspended culture particularly in the second year of suspension, enforcing the suggestion that this should be limited, if possible to a single summer. They cling tenaciously to oyster shells and are difficult to remove. They may be killed by exposure to hot sun with the time required variable according to the size of the anemone.

Hydroids

Hydroids, like anemonies, are coelenterates but with a peculiar life history (Fig. 95). The sexual phase is a small jellyfish (medusa) which produces a planktonic larva which, after settlement, develops into a branching bush-like hydroid which in turn produces the jellyfish asexually. Hydroids assume many shapes and sizes but the typical bush-like fouling species (*Obelia*) may attain a length of 30 cm. Hydroids may be killed by sun exposure but they may be also removed manually.

Bryozoa

Bryozoa are also termed polyzoa or moss animals and may be encrusting, or erect and branched. In both cases they are colonial, each individual in the colony being of almost microscopic size. Encrusting types are generally less than a millimetre in thickness but one colony may completely cover an adult oyster shell. They are not harmful to adult oysters but may overgrow spat less than about 1 cm in length. Erect branching forms are seldom more than 3 or 4 cm high and are not of concern.

Tube Worms

Tube worms are sedentarian polychaete annelids whose tubes may be calcareous, parchment-like, mud or sand (Fig. 96). They are not usually a major problem but at some sites and in some years there may be excessive outbreaks that may require drastic action. Air exposure of 24–48 h may be effective for most species. Another tube worm of importance is *Polydora* (Fig. 88), the mud blister worm. They burrow into the shell edge and are readily visible. The tube is partially filled with mud and sometimes the tube breaks through the nacre. The oyster is forced to lay down another layer of shell to cover it. *Polydora* is most often a problem in bottom culture in muddy areas. The first cure is to move to off bottom culture but a brine dip may be useful for both it and the boring sponge. A common tube worm in British Columbia is *Spirorbis* (Fig. 87b) encased in a tiny coiled calcarcous tube, 1–2 mm in diameter. It occurs both intertidally and subtidally, but are not particularly deleterious. In some instances they can nearly cover the surface of an oyster. Neither *Polydora* or *Spirorbis* are significant problems in British Columbia.

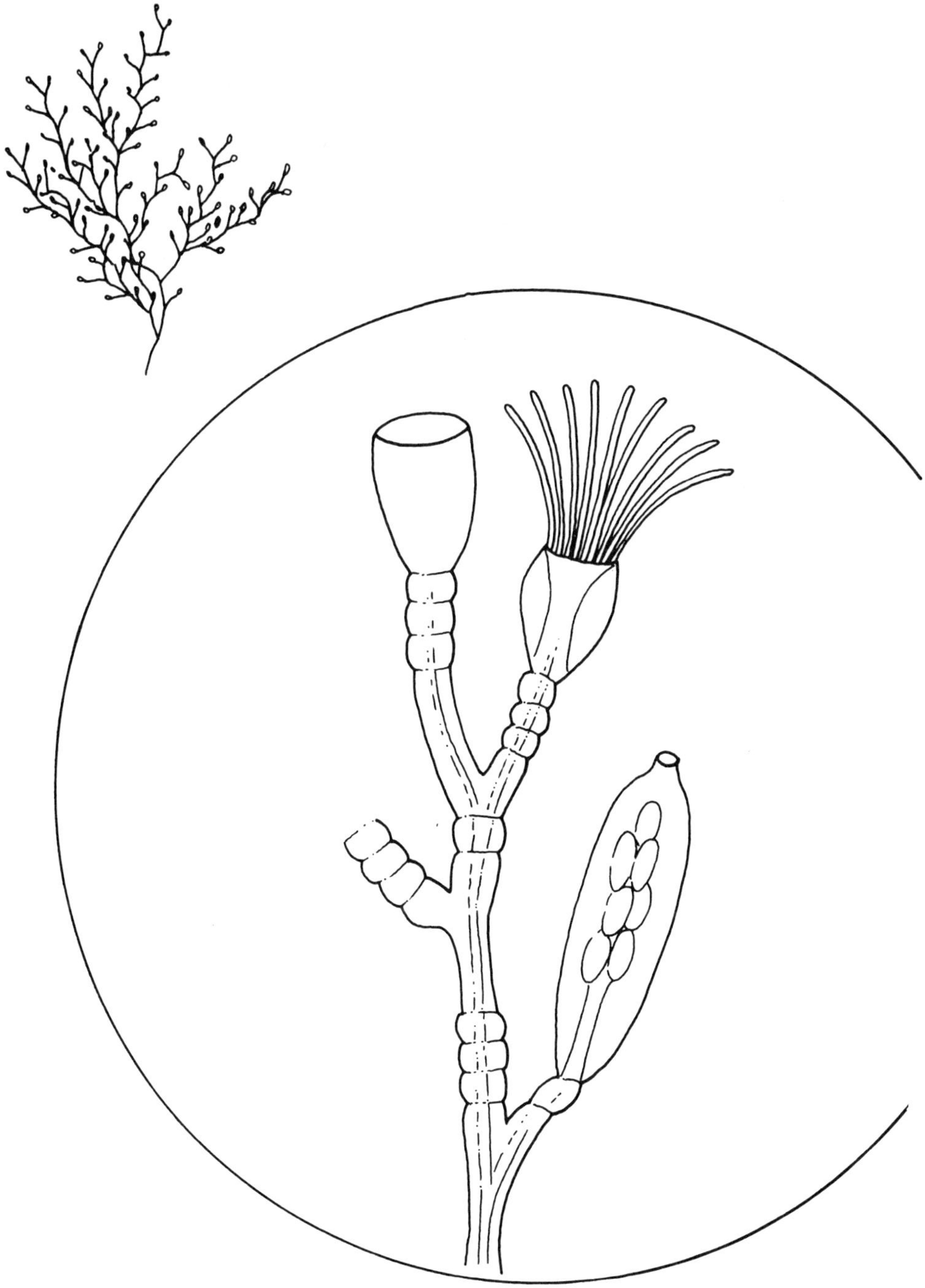

FIG. 95. The hydroid, *Obelia*. (×12).

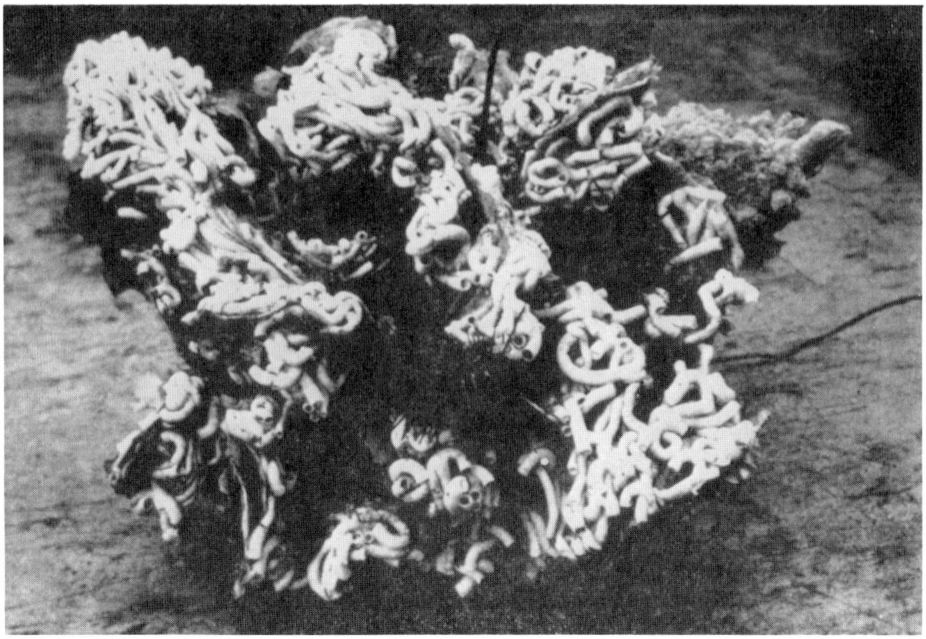

FIG. 96. Hecate Bay cluster showing worm tubes.

Mussels

Mussels are second only to barnacles in importance as fouling organisms, because the setting intensity can be very high. Rapid growth rate and high numbers can cause a high level of competition with oysters for food , as both are filter feeders. Smothering of spat may also occur. Knowledge of the breeding season and depth of distribution may assist in culturing around the problem by correct timing or position (depth) of culture activities. Probably the most economical method is by exposure to sunlight when the young mussels are less than 5 mm in length. This becomes a problem if the mussel breeding season is a long one. Alternative sites may be a possibility if an area with reduced mussel breeding may be found.

Mussels have a long breeding season in British Columbia and larvae have been found in every month of the year but in most areas studied there are spring and fall peaks. Mussels referred to here should not be confused with the large Californian mussels that occurs mainly on rocky exposed ocean shores.

Recently a chemical (xanthatin) has been discovered that will inhibit mussel settlement.

Tunicates

Tunicates or ascidians are either solitary or colonial (Fig. 94). Solitary types assume many shapes but all adhere to the substrate by a broad holdfast. The main body is contained within a test or envelope of varying consistency according to species, but most may be considered soft bodied and therefore susceptible to sun exposure. They may vary in size from 5 mm to 12 cm in length.

Colonial type ascidians consist of numerous small tunicate bodies encased in a fleshy encrustation up to 1 cm thick and capable of covering a whole adult oyster and because of the fleshy nature of the colony there is little effect from sun exposure. Some species of tunicates are cultured as food items in the western Pacific. The 2 cm long semitransparent *Corella* is a common tunicate on cultch and culture strings in British Columbia and is quite edible. Tunicates are not directly harmful to oysters although they may compete to a small extent for food. Like anemonies they are mainly a nuisance and this increases with the time in suspension, which, if possible should be limited to one summer.

Algae

Bacteria and diatoms can form a thin layer of slime, of no concern except on cultch, for if the slime film is heavy enough it may prevent spatting. If on growing oysters, it may be a partial source of food.

Probably more serious than the animal foulers are several species of algae already described (page 00). *Ulva* and *Entermorpha* are green and *Punctaria* and *Laminaria* are brown. *Ulva* is a flat thinned leafed green algae with a small holdfast and may be up to half a metre in length. In spring and early summer it may cover sections of tidal flats but in most cases does not pose a threat to oyster culture.

Fouling Control

The three main categories of fouling control are physical, chemical, and biological.

Physical methods include direct heat with a flame, or indirect by sun drying. Chemical methods use fresh water, hot water or chemical solution dips such as brine or copper sulphate. Biological control involves knowledge of biology and ecology of fouling organisms so the culture operation may be timed or placed to circumvent, or minimize fouling.

Physical Methods

These involve removal of suspended culture oyster, strings, trays or sticks from the water. This is a costly operation and undertaken only as a last resort when fouling become excessive.

Probably most drastic of these methods is the application of heat by means of some type of flame thrower. A delicate balance is required to ensure death of the fouling organisms and not the oysters.

Another method is use of water pressure with high pressure pumps and special nozzles similar to those for cleaning boat hulls.

Probably the most frequently used method is sun drying. It is most useful with soft bodies animals such as tunicates and anemonies, particularly young animals. It is successful only with very young barnacles. Usually one full days' exposure is sufficient, but the time depends partly on air temperature, humidity, size and type of foulers, and size of the oysters. There should be initial experimental trials with small quantities. The fouling organisms may not be removed immediately after death but may drop off several days afterwards. With small culture operations many organisms such as mussels and hydroids can be removed by hand.

Chemical Methods

These also involve removal of the oysters from the water with the additional task of placing them in tanks containing the chemicals. Fresh water may be classed as a chemical to marine organisms and only an hour or two of exposure is necessary to kill most fouling organisms, particularly soft bodied species. Hot water, owing to cost is almost out of the question but for exceptional circumstances.

There are a number of chemicals toxic to marine organisms but many are expensive and have many dangers, or limited effect. They must be used with caution. Probably the most useful is copper sulphate in a concentration of only one or two percent. Care must be taken in its use relative to other organisms such as fish. Exposure time of one hour or so for most organisms is adequate, but preliminary experiments with small quantities should be made.

Among the least expensive chemicals is brine where the bath should contain as much salt (NaCl) as will dissolve. In general chemical control is impractical.

Biological Methods

The biological method requires a knowledge of the biology and ecology of fouling organisms in the culture area with special emphasis on breeding season, vertical and horizontal distribution.

This many entail a specific fouling study for the experimental collection of fouling organisms. A properly designed experiment may demonstrate weak links in the progression of events that may be utilized to advantage in the culture process.

Information such as the breeding season of a barnacle species will indicate the cultch should be exposed after the barnacles set if it occurs during the oyster breeding season. In some circumstances in suspended culture, strings may be exposed purposely to collect the barnacle set, which in turn prevents settlement of more deleterious fouling organisms. Oyster seed should be large enough (3–5 cm) to prevent overgrowth by barnacles. Knowing the vertical distribution of a fouling organisms, it may be possible to place oysters in suspended or intertidal bottom culture or seed collection above or below the fouling zone, during the fouler's breeding season.

As previously mentioned, in estuarine areas where there is a significant salinity gradient, it may be possible to collect seed in high salinity areas where breeding takes place but where heavy fouling occurs; indeed in some contexts an oyster may be considered a fouling organism itself. Then the seed may be moved up estuary to a lower salinity area where fouling is absent or minimal but where oysters may still grow, for this ability to thrive over a wide range of salinities is a characteristic of the genus *Crassostrea*.

Fouling Study

Fouling should be part of the culture study process by keeping records of occurrence-time and extent of fouling as it occurs on seed collectors or growing oysters. In addition, specific studies may be carried out by exposing test materials of various types for different periods of time at various depths and at several sites. Preferably the test material should be oyster shell, or whatever material is used as cultch or both since these will be the substrates encountered by the fouling organisms. Wooden panels also provide information on marine borers attacks but suffer from disintegration if the attack is severe or if the panel is left exposed for a long period.

Exposure, whether intertidal, or at a fixed tidal level such as from a dock, or variable as from a raft where depth intervals remain the same, but alters with reference to mean low tide level, will depend on the type of culture.

Exposure designs are numerous and may be quite complicated as there is interaction between several variables such as, time, depth, site, season, materials, etc. A fairly simple design, consists of a set of six numbered shells or groups of shells. Shell No. 1 is removed and replaced at monthly intervals or whatever time period is desired. Shell No. 2 is removed at the end of the second month and so on down the line. Thus monthly and increasingly lengthy exposures up to six months are obtained. Initial exposures are timed to cover whatever period of the year is deemed critical.

Fouling organisms may be studied quantitatively by measuring weight or volume or by numbers in the case of barnacles. Without adequate literature, identification may be difficult but important groupings such as sponges and tunicates can be separated. Species may be identified as sponge ''A'' or sponge ''B'' and this will serve sufficiently well until specific identifications are made.

In this way important fouling organisms and their seasonal settlement may be determined. When the depth distribution is obtained, it may be possible to culture above or below this level. A knowledge of the fouling sequence and distribution opens avenues for some options in sites where it is a serious problem.

Predators and Pests

British Columbia is singularly fortunate in having relatively few oyster predators and pests in comparison with most other oyster areas of the world. Nevertheless, there are several, and growers must know how to recognize and to deal with them. All of them are invertebrates, and range from the one-celled Protozoa, one of which causes paralytic shellfish poisoning, to quite highly developed forms like seastars.

Protozoa

Protozoa are one-celled animals and one of these is discussed in connection with paralytic shellfish poisoning.

There are a number of nontoxic Protozoa which can discolour sea water when they are very abundant. Oysters can feed on these and when they do the gills may become slightly pink or red and when an oyster is opened and pressure applied, the red stomach contents run out of the mouth and the oyster is said to "bleed". Oysters in this condition are quite edible but not attractive, particularly when the liquid in the container becomes pink. In addition, "bleeding" oysters do not keep well even under refrigeration, so there is a risk of premature spoilage. Growers find it to their advantage to suspend operations during periods when red water occurs, for it usually only takes 2 or 3 days for the situation to clear.

Nevertheless, when coloured water or "bleeding" oysters occur, a check with authorities should be made regarding the possibility of paralytic shellfish poison.

Boring Sponge

In Atlantic coast oyster growing areas, the boring sponge (*Cliona celata*) (Fig. 85) is quite a serious pest, but in British Columbia it can hardly merit that description, although it does occur here.

Boring sponges attack molluscan shells, and can riddle a shell to create a honeycomb effect to such an extent the shell becomes very weak and can easily be broken; shucking is then difficult. Also, the burrows may penetrate the nacreous interior lining of the shell forcing the oyster to expend energy in producing shell to keep these perforations covered.

The yellow-coloured sponge is readily seen in the circular holes, about 1.5 mm in diameter, that perforate the shell surface.

As a rule, the boring sponge attacks only very old shells (10 years or older) in British Columbia, and this applies to both clams and oysters. In fact, it is difficult to find a living oyster that has been attacked by the sponge, as it is usually found in dead shells at very low tide levels. However, oyster growers should be aware of the organism and the damage it can do.

Flatworm

Flatworms (Turbellaria) also known as wafers or oyster leeches are small, thin (1 to 2 mm), more than 3 cm long, and 1.5 cm wide with a grey brown colour. While there are a number of native flatworms in British Columbia, the species of concern to oyster growers is *Pseudostylochus ostreophagus*, the latter name meaning "oyster eating". This is a Japanese species introduced with Pacific oyster seed (Fig. 67).

It breeds in summer, laying up to 8 000 eggs about 150 μm (0.15 mm) in diameter in brown patches 1 cm across. The eggs are incubated for a month and develop into swimming larvae. The whole life cycle appears to be not more than a year.

It is able to feed on oyster spat up to 1 cm by drilling an oval hole through the right valve. How it does this is not exactly known for there are no hard parts. A *Pseudostylochus* of average size can destroy up to 50 spat per month.

Control can be exercised by dipping the cultch in freshwater for an hour.

Pseudostylochus did not appear as a problem until 1971 when oysters and oyster seed from Pendrell Sound were shipped to France to replace the stocks of the Portuguese oysters (*Crassostrea angulata*) which had been decimated by disease. The high population level of flatworms persisted for several years in the Sound but since about 1975 it has ceased to be a serious problem. However, a watching brief is required. *Pseudostylochus* has not appeared in significant numbers in other growing areas in British Columbia.

Boring Sea Worm

Another fairly serious pest in some growing areas in other parts of the world does occur in British Columbia but is not a problem of consequence. This is the sea worm (polychaete) of which several species occur here, but *Polydora websteri* (Fig. 88) is the only one normally associated with oysters. These worms have free-swimming larvae which settle on the outside surface of oysters and make shallow burrows near the edge of the valves, where they are readily visible. They can make a mud burrow or can drill into the oyster shell itself and form ''U''-shaped burrows. Sometimes they may perforate the inner shell and the oyster then has to lay down a protective new shell.

Burrowing Shrimps

The mud shrimp (*Upogebia pugettensis*) and the ghost shrimp (*Callianassa californiensis*) (Fig. 97), are two species of crustaceans that can cause serious damage to oyster ground by making it too soft for oyster culture. *Upogebia* is generally grey-green with

FIG. 97. Burrowing shrimps: *Callianassa californiensis* — left; *Upogebia pugettensis* — right. ×1.

two clawed legs which are fairly small and equal in size. *Callianassa* is nearly transparent and the reproductive organs are coral-coloured. The two clawed legs are unequal in size.

Both species dig a "U"-shaped burrow with two openings, about 1 cm in diameter, at the surface of the bed and these are often indicated by mounds of sand. The burrows are lined with an almost impermeable lining of compacted mud so that when the tide drops the burrows remain filled with water; in many areas in some years the burrows are so numerous the ground is virtually a quagmire at low tide.

These animals posed a serious problem to native oyster culture, where dikes were required. Shrimp burrows with an opening on either side of the dike wall form perfect siphons and at low tide the water drains out of the dike. It was therefore necessary to lay down, in the dike, under the surface layer of gravel, a floor of plywood or plastic to prevent the shrimps from burrowing.

The very warm summer of 1958 caused a burrowing shrimp "population explosion" and considerable areas were rendered unsuitable for oyster culture. Populations subsequently diminished slowly. Control by plowing and rolling was found to be fairly effective.

Parasitic Copepod

In the list of unwelcome introductions with oysters from Japan is the parasitic copepod *Mytilicola orientalis*, sometimes called the red worm disease (Fig. 68). This is a crustacean, distantly related to crabs and shrimps. It is a small bright red organism with the maximum length of the female not much more than 10 mm, and of the male 6 mm. It occurs in the small intestine of various species of molluscs, usually near the anal opening, where the copepod holds its position by means of special hooks. There may be up to five or six *Mytilicola* in a single mollusc. Egg capsules may be found on females at any time during the year but they are most abundant during summer. Exact means of infection is not clear but it is certain that the mollusc is infected during the larval stage of the copepod.

Mytilicola has been found in British Columbia, mainly in the bay mussel (*Mytilus edulis*) but also in Pacific and native oysters. In Europe, it has caused significant damage through both mortality and reduced condition in edible mussels. So far there is no evidence that *Mytilicola* either causes mortality or has a serious effect on the condition of oysters, but the grower should be aware of the fact that it is a potential menace.

For Pacific oysters in British Columbia, the highest recorded infection rate was 20% and, for the bay mussel, 50%. There is a high correlation between areas where *Mytilicola* occurs and where Japanese oyster seed was planted.

Crabs

Three species of crabs can cause damage by opening and feeding on oysters and oyster seed by cracking shell edges with their powerful claws. Distinguishing features of these crabs are shown in Table 20 and Fig. 89.

TABLE 20. Identification characteristics of oyster predator crabs.

Cancer magister (Dungeness Crab)	*Cancer productus* (Rock Crab)	*Cancer gracilis* (Graceful Crab)
shell with fine granulations	shell without fine granulations	shell without fine granulations
large (to 17 cm)	medium (to 15 cm)	small (to 7 cm)
—	—	shell convex
red brown	dark red	gray or tan; red spots
last tooth often on shell edge largest	5 equal teeth between eyes	—

As a general rule crabs have not been a serious problem in British Columbia, although few beds escape some damage. They can be, and on occasion have been, destructive to young seed. The solution is to keep seed on relatively high ground if there is any possibility of danger from crabs. This applies specifically to the graceful crab which can be quite abundant. Another solution to the crab problem is a trapping program. Simple traps can be constructed quite easily and cheaply from laths or wire mesh (Fig. 98).

Oyster Drills

Oyster drills are among the most damaging of pests found on oyster beds. They are univalve snails which have, in the mouth, an extensible, toothed, rasping apparatus with which they drill in a clam or oyster shell a hole through which they can tear away the

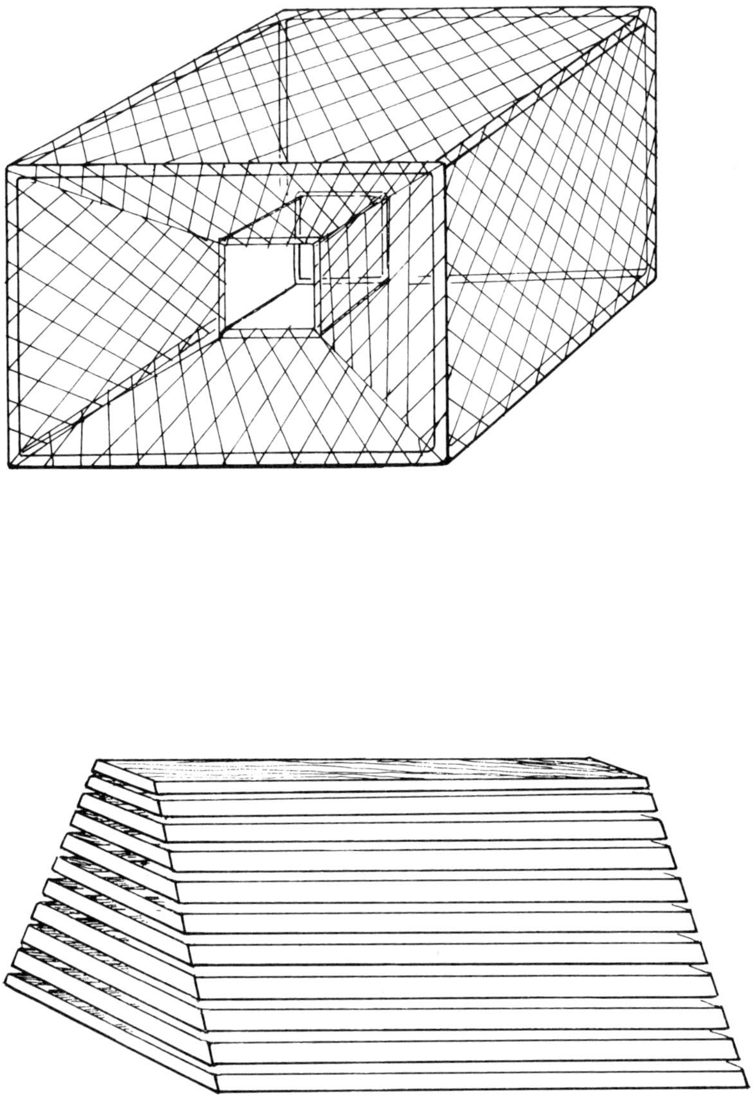

FIG. 98. Simple crab traps. *Top* — wire mesh. *Bottom* — wooden laths. (× 1/8).

FIG. 99. a. Hole drilled by the moon snail (*Polinices lewisii*) (× 1/2). b. Moon snail (× 1/2) and egg case (× 3/4).

meat. The most frequently observed drill hole is the countersunk circular hole often found on dead clam shells, and this is caused by the large, coiled moonsnail (Fig. 99). Oyster drills are much smaller and, as far as is known, only two species dangerous to oysters occur in British Columbia. Both of these were introduced: *Ceratostoma inornatum* formerly *Ocenebra japonica*, the Japanese drill, from Japan, and *Urosalpinx cinerea*, the Eastern drill, from the Atlantic coast. The latter species occurs only in Boundary Bay, and its abundance has dropped markedly since the cessation of imports of eastern oysters; it is no longer of consequence as a menace to oysters. There is also what is called the native oyster drill (*Nucella lamellosa*) which, however, attacks barnacles, clams, and mussels rather than oysters (Fig. 66).

Of considerable concern is the Japanese drill (Fig. 66, 100), which is known to occur in Boundary Bay, Ladysmith Harbour, Crofton, Thetis Island, and Comox. Specimens have been found at Sooke and at Henry Bay but significant populations do not appear to have developed in these areas.

Japanese drills breed by copulation, after which the female deposits about 1500 eggs in about 25 yellow capsules, each about 7 mm long, which look like grains of oats (Fig. 66). In Ladysmith Harbour capsules are deposited from April to November and nearly always on the underside of shell clusters or debris. Often, several females will choose the same spot, so quite large clusters of capsules may be found. Development in the capsules requires approximately 2 months, after which the young drills crawl out through an aperture at the top of the capsule. Only three of four drills are hatched from each capsule, for a great majority of the eggs are nurse eggs which serve as food for the embryos that are able to develop. After crawling out of the capsule, the young drill, less than 2 mm long is on its own and, very likely, it exists on plant material until large enough to drill shells. Fortunately there is no pelagic or free-swimming stage so it cannot be dispersed by water currents as is the case with oyster larvae. Little is known of its growth rate, but it probably reaches its maximum length of about 3 cm in 3 or 4 years.

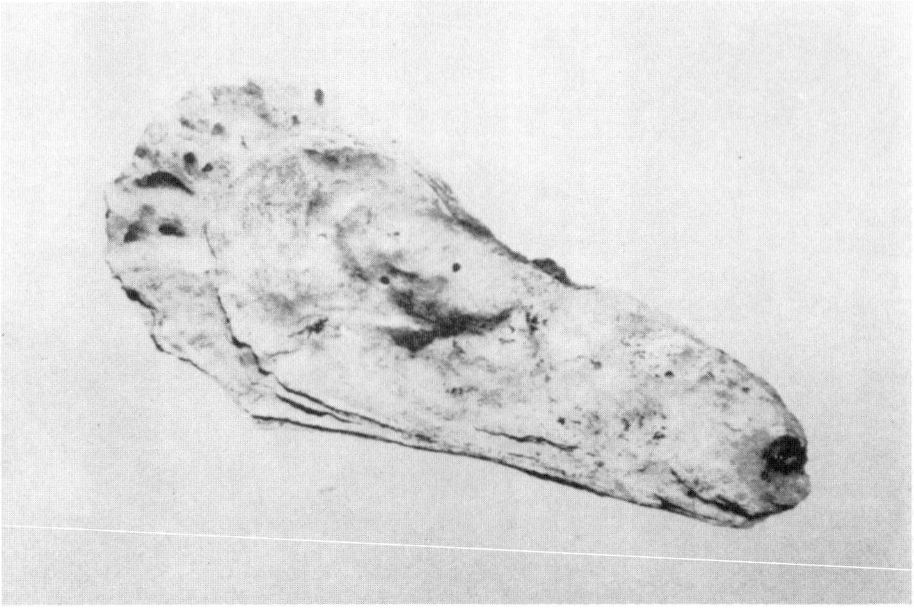

FIG. 100. Pacific oyster shell showing holes drilled by the Japanese oyster drill (*Ceratostoma inornatum*). ×0.8.

It requires about 2 wk for an adult *Ceratostoma* to drill through an oyster 5 cm in length while only a day or so is required for seed oysters up to 2.5 cm in diameter.

As described previously, considerable effort was expended to ensure that imported Japanese seed was drill free. Unfortunately a large part of the damage was done before a control system was instituted, but, even with this, there will always be a risk of accidentally introducing drills into presently drill-free areas. Also, there is the danger of drill-free beds being infested through inter bed movement of, either oysters or oystering equipment. Permits are required to move oysters or oystering equipment from one area to another, and certain grounds are designated as drill areas. For many years, organization and operation of the industry did not warrant such control, for there was virtually no interbed movement of oysters and with the industry under close biological scrutiny, any potentially dangerous situation with respect to drills was placed under safeguards. For instance, in Ladysmith Harbour, where drills were quite abundant by 1935, its distribution has remained relatively unchanged.

Once an area has become drill infected, it is difficult to clean it up entirely. It is possible, by properly working the ground and by manipulating planting, to hold the drills in sufficiently close check, damage is kept to a minimum. For instance, drills have difficulty in attacking oysters over 2 years of age because of their relative thick shells; so if drill-infected ground is to be used, only large oysters should be planted. Seed oysters should not be planted on drill-infested ground. Drills may also be held in check by a trapping program. The traps are simple, 2.5 mm mesh chicken wire baskets or envelopes approximately 30 × 45 cm baited with one or two clusters of year old oysters (Fig. 101). The traps are set in a regular pattern such as a zig zag or as a straight line, moved at regular distances around a central point. Traps should be examined at about weekly intervals and the bait replenished as necessary. Records of the catch of each trap are kept to obtain a measure of the efficacy of the trapping program.

When oysters from drill-infested ground are being harvested, every effort should be made to clean off the ground as completely as possible and, it at all feasible, the ground should be raked and all debris taken ashore. Drills can also be buried *in situ* by pulling heavy drags over the ground.

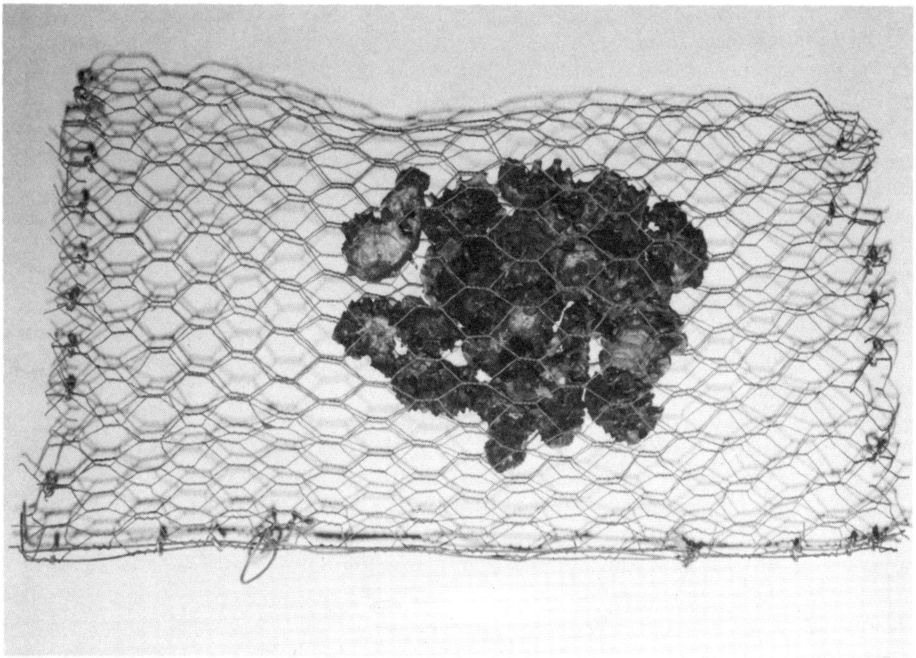

FIG. 101. Simple wire mesh drill trap. (× 1/2).

Although drills do not migrate extensively, two growers working on adjacent beds must cooperate in drill control activities, because the work of one grower alone will likely be ineffective.

Sea Stars

Probably the most important oyster predator in British Columbia, other than man himself, is the sea star. There are about a dozen common shore or shallow-water species in these waters but only about four of these can be classed as serious oyster predators. These are the ochre star (*Pisaster ochraceus*), the pink star (*Pisaster brevispinus*), the mottled star (*Evasterias troschelii*), and the sun star (*Pycnopodia helianthoides*), which are shown in Fig. 93.

They all open the oyster in much the same manner. In the initial step a sea star clasps an oyster with its suction-tipped tube feet with which it is able to apply pressure in excess of 4.5 kg against the oyster's adductor muscle, which is sometimes torn. However, it is thought that this pressure is not entirely necessary for the sea star has an evertible stomach which can penetrate the narrowest of apertures. Preliminary digestion takes place in the evertible stomach and when the oyster's adductor muscle is reached the valves gape, allowing the sea star to finish its meal. This is a point to be borne in mind for, as a result of this ability, a sea star on the outside of a mesh basket or container can feed on the oysters inside. It requires less than 24 h to devour an oyster.

Opinions are not unanimous as to whether or not sea star are attracted to molluscs. An attempt to grow oysters in a water depth of about 5 metres below zero tide level, in Departure Bay, failed owing to sea star predation. About 225 bushels of 7 cm oysters and 150 strings of 1961 Pendrell Sound seed were planted at this depth on an area about 75 m by 75 m on September 1, 1961. About once a week, sea stars were cleared from the planting area and from some 10 m around it by SCUBA divers. This was done quite regularly for 8 months and during this period, an average of 10 sea stars per week were removed. These were mainly *Pycnopodia* followed, in number, by *Evasterias*, and then

Pisaster. Between September 1, 1961, and January 23, 1962, about 200 sea stars weighing 181 kg were removed from the bed. The diving program had to be suspended during the second 8-month period and examination at the end of that time showed the adult oysters had suffered a mortality of 70% and seed practically 100%.

Whether incursion of a sea star into the oyster area was a result of direct stimulation or to accidentally running into it during random movements is not known, but the practical conclusion is that if there are oysters present, sea stars seem to find them.

Sea stars breed in spring and summer in much the same manner as oysters and have free-swimming larvae which allow them to become dispersed, so that after an oyster bed has been cleared of sea stars it may be reinfested by larvae, as well as by adults, from distant concentrations. Young sea stars, up to 4 or 5 cm in diameter, appear to be light-shy and are seldom seen except on the underside of rocks or among clusters of oysters. Breaking oyster clusters as soon as possible is an excellent way to control to sea stars by making living areas unavailable to them.

Growth of the ochre star (*Pisaster ochraceus*) has been studied in Ladysmith Harbour. Those set in August had reached a diameter (ray tip to ray tip) of 1 cm by October and were already attacking small oyster spat. By the following June, the length of a single arm was nearly 2.5 cm, and by the end of a year, 3 cm. In May of the following year, length of a single ray was 6.4 cm, making a total diameter of 12.7 cm, by which time they were capable of attacking oysters of any size (Fig. 102).

The ochre star has been found to travel up to 50 m in a 24-h period during a summer, but it cannot move nearly as rapidly as the sun star. There is a general tendency for all these sea stars to carry out small migrations for, as a general rule, they are found at lower tidal levels in summer than they are during winter, and this is probably a response to heat and desiccation.

In British Columbia, where most oyster culture is intertidal, oysters and sea stars can be reached at low tide so that control of this pest is relatively simple. They may be removed from the beds simply by picking them up and taking them ashore. They make

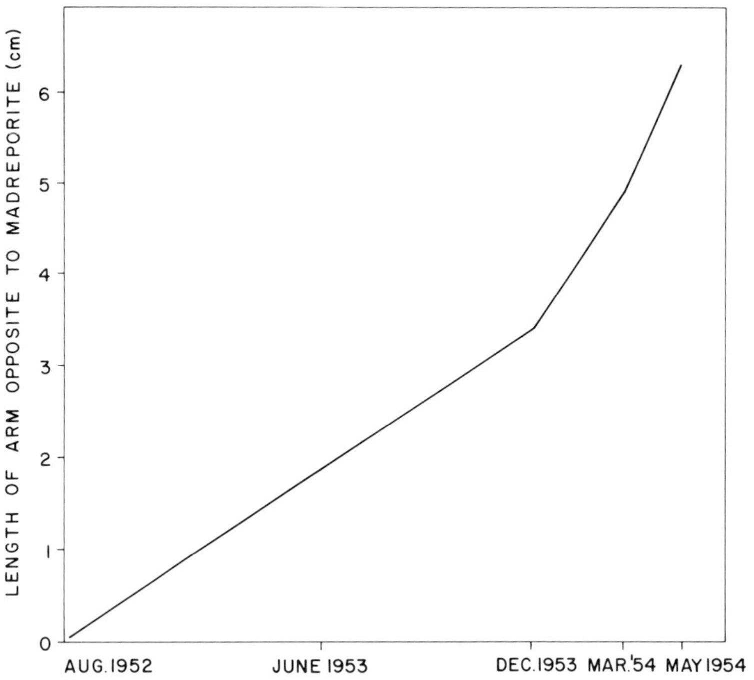

FIG. 102. Growth rate of the purple sea star (*Pisaster ochraceus*) in Ladysmith Harbour.

excellent garden fertilizer or they may be destroyed by applying a teaspoonful of quicklime or carbide to their backs. They are destroyed by the corrosive action of the chemical which does not affect any oysters with which it may come into contact. Mutilation is of little use in attempting to destroy them for a whole sea star can be regenerated from a single ray joined to a portion of the central disc. They are often seen with one or more short rays and this is generally an indication of regeneration of lost rays.

Birds

Ducks, particularly the Lesser Scaup, also called the Bluebill (*Aythya affinis*) and the Surf and White Winged Scoters (*Mellanita*) frequent oyster beds in large numbers during winter. A considerable portion of their diet at this time consists of molluscs such as clams and mussels but they are known and have been observed to feed on young Pacific oysters up to 2 cm in diameter. A large flock of these ducks is capable of doing considerable damage to a seed bed of single oysters.

Chemical Control

In the United States in recent years there has been some study of chemical pesticides in the control of oyster pests and predators. Many chemicals have been screened and some have been found to be efficacious in destroying various pests. Some of these chemicals are quite specific in ability to destroy only certain types of predators and this is important because of the need to protect oysters and other shellfish. Chemicals, mainly organic, have been used to kill oyster drills, crabs or shrimps, and seastars.

So far there has been some, though not widespread, use of such pesticides in oystering areas in the United States, research is still going on to discover how much pesticide can be taken up by shellfish, for oysters have been shown to accumulate, 70 000 times as much of the insecticide DDT in about a month as is found in surrounding waters. Also, more study is required on the effect of pesticides on the whole environment — how they affect plankton and how they affect all of the minute organisms associated with oyster ground. In other words, it is possible that the secondary effects of these poisons may be worse than the direct effects on the predators.

Often, but not always, it is possible to adjust cultural methods to reduce the effects of predators and pests but this requires a detailed knowledge of the biology and behaviour of these animals. There is little doubt that pesticides will, in the future, have an important role in the control of at least some shellfish predators, but with the present state of knowledge regarding total effects such as it is, it would be unwise to jeopardize what one already has. The oyster industry is waging a continual battle against pollution and the industry has be be wary lest it be branded as a polluter itself, by its use of pesticides in marine waters. Another problem in connection with pesticides is their occurrence in shore waters as a result of runoff from farm lands where insecticides or herbicides are used. How significant this is in the area of marine pollution is not completely known.

Disease

Disease in molluscs is probably the most difficult problem with which the culturist has to deal. Cells in molluscan body tissues are relatively small and difficult to study. There are seldom warning symptoms prior to the onset of a disease. Often it has run its course before becoming apparent. Warning signs may be appearance of "gapers" where the valves gape open but with the animal still alive or recently dead. This state doesn't persist for long, for scavengers such as crabs soon remove the meats and the shells are then termed "boxes" or "cluckers" from the noise when they are tapped.

Another sign is the appearance of pustules on the surface of the body or pus in blood sinuses. Pustules, however, may only be a sign of stress rather than a disease symptom. Emaciation of the body, when normally condition factor should be high, may be another indication of ill health. Often mortality has occurred before there is an awareness something is amiss. The difficulty is to determine whether mortality was due to disease or to other chemical or hydrographic factors which may have persisted only briefly.

Also it is rarely possible to actually cure disease in large populations of molluscs in the open marine environment. There are examples of decades of study of a single oyster disease without resolution of the cause although the industry has continued to operate.

In the unfortunate event of the occurrence of a disease the objective is to study and learn enough about its ecology so it may be possible to culture around it for minimal effect. Such a situation exists with the Denman disease in British Columbia which, after 25 years, the causative organism is not yet definitely known. It was soon found the disease was most prevalent at lower tidal levels in the intertidal zone and flourishes only during the months of April and May presumably being responsive to water temperatures. By planting the beds after May and harvesting before April, or by using the higher tidal levels, the main effects of the disease are avoided.

In the instance of *Perkinsus marinum*, a fungus disease of oysters in the south Altantic and Gulf of Mexico regions of North America, oysters are cultured to take advantage of the fact that mortalities and infections decline during colder periods of the year when water temperatures are below 25°C and are safe when grown in salinities less than 15‰.

A way in which disease control may develop is the Malpeque disease in *Crassostrea virginica* in Eastern Canada. The disease, which has only recently been tentatively identified, began in 1915–16 in Malpeque Bay and by 1939 had spread to all the oyster areas in Prince Edward Island. Initial mortalities were nearly 100% but by 1922 the few survivors in Malpeque Bay had reproduced well and formed the nucleus of a disease resistant stock. Oysters brought into Prince Edward Island always died. Attempts were made to keep the disease within the confines of Prince Edward Island and were successful for a number of years, but eventually it spread to the adjacent mainland. To enhance the speed of recovery of these stocks, disease resistant oysters from Prince Edward Island were introduced to provide an instant breeding stock and this procedure proved successful.

While there has been an awareness of some diseases such as Malpeque for a long time, most development in molluscan pathology has taken place in the last 20 years owing mainly to disease outbreaks that have occurred during that period. The reason for some of these outbreaks is not clear but most can be attributed to relaying indigenous species into a different environment or to the introduction of species from another country.

The introduction of *Crassostrea gigas* from Japan to the Pacific coast of North America without adequate precautions might be excused only because of ignorance of possible consquences as well as lack of an appropriate technology.

In this case a number of Japanese invertebrates were introduced with the oysters, several of which had deleterious consequences such as oyster drills, a nuisance seaweed, fouling organisms, a marine wood borer and possibly the so called microcell disease as well as the parasitic copepod *Mytilicola*. Positive results were the development of a modest oyster industry in B.C. and a fishery for Manila clams. In recent years, particularly with the advent of hatcheries and adequate means of quarantine, the safe, of nearly so, introduction of exotics is possible. A more difficult problem however is the effect of the new species on the indigenous biota. This may mean competing with, and possibly overriding and eliminating a native species; or on the other hand, its own elimination by susceptibility to native diseases.

The distinction between disease and parasitism is not always clear except that disease organisms are usually pathogenic while the effects of parasites are not necessarily lethal.

Fortunately, as far as is known, the Pacific oyster in B.C. is relatively free of disease organisms. This may in part be due to limited study for only one lethal organism is reported in comparison to six on the Atlantic coast of North America.

Amoebae of uncertain taxonomic status have been found in Denman Island oysters.

A bacteria caused condition known as "focal necrosis" is found in the Pacific oyster in Washington state and Japan but now recorded from British Columbia as "fatal inflammatory bacteraemia.

Shell Disease

Another example of disease is the one which attacked the Dutch oyster industry in the 1930's and 1940's. The inner surfaces of the shell of European flat oysters developed irregular green spots and warts. Oysters with a high level of infection failed to grow and mortalities ensued. Long study finally demonstrated that cause of the shell disease was a fungus distributed by water currents; the disease being prevalent in old decaying shells of other species of molluscs. By removal of the deposits of old mollusc shells and by bathing the infected oysters in a weak solution of a mercury base disinfectant, it was possible to bring the disease and mortalities under control.

The fungus has been been designated as *Ostracoblabe implexa*. It has been detected in flat oysters now being cultured in Eastern Canada although it is not causing serious damage at present. The European flat oyster has been introduced into B.C. but is not being cultured. *Bonamia ostreae*, a hemocytic protozoan parasite of that species (*Ostrea edulis*) occurs in Puget Sound, Washington.

Foot Disease

In the fall of 1956 an oyster disease occurred in Pender Harbour; 10% of the oyster population, of both Japanese and local origin, was affected. Most of the oysters concerned were about 4 years old, but only those on certain beds were involved, as were oysters transplanted from Drew Harbour on Quadra Island about 50 miles across the Strait of Georgia from Pender Harbour. The symptoms were practically identical to those described for the oyster disease known in France as ''maladie du pied'', or foot disease, which occurs in both the European flat oyster (*O. edulis*) and the Portuguese oyster (*C. angulata*) and is attributed to a fungus. The adductor muscle scars were partly covered with black proturberances up to half an inch high and nearly an inch in diameter. The thin outer covering of these proturberances was of a chitinous nature. On some shells the base of the abnormality was expanded considerably by a raised portion of nacre covering an extension of the puslike secretion contained within. The secretion contained a heavy bacterial concentration in addition to rather large circular bodies which constituted the main portion of the material. There has been no report of a similar occurrence since, although occasionally an oyster may be found with such a growth.

Summer Mortalities

In British Columbia there has been little evidence of disease in the Pacific oyster since it was first introduced. Very few mass mortalities have been reported, and these have all occurred in Boundary Bay during summers warmer than usual. Mortalities somewhat similar in nature have been reported from Washington State, and some of them have been serious enough to cause concern, but the cause has not yet been determined.

Denman Disease

In 1960 an outbreak occurred of a disease characterized by deep putsules on the surface of the body and mantle (Fig. 103, 104) and/or by pus-filled sinuses. Initially the disease was confined to an area between Henry Bay and a point about 3 miles southward on Denman Island in Baynes Sound (Fig. 2), with the main locus within Henry Bay itself. The mortality from this initial outbreak was estimated to be more than 30%. The disease develops in early spring, usually in April, when water temperatures are about 9°C, and are just beginning to rise from the winter low of about 7°C; it disappears by about mid-July when water temperatures have reached about 18°C. In this area oysters are nearly always in excellent condition and the dead or dying oysters also had a very high condition factor.

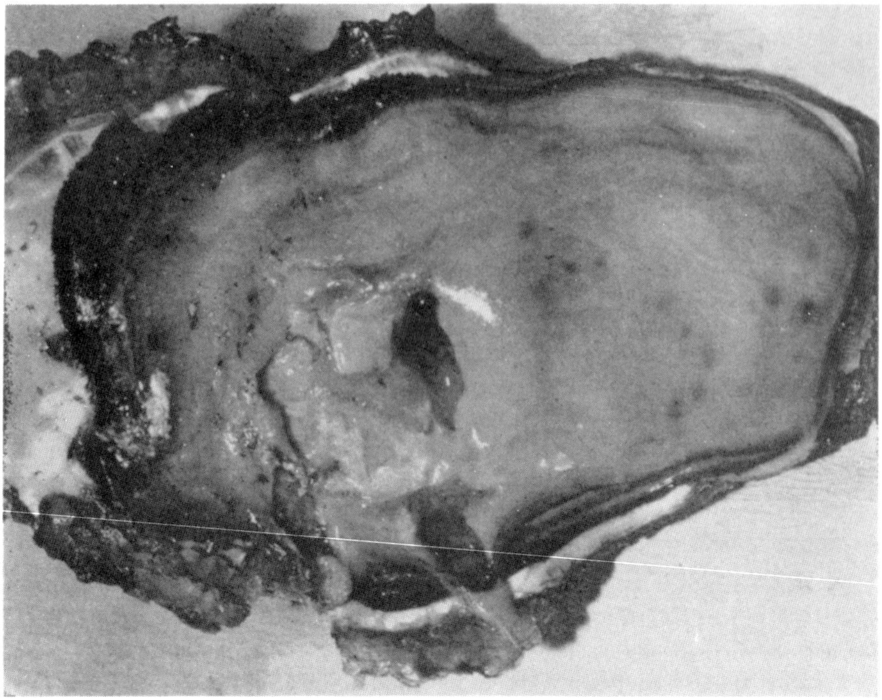

FIG. 103. Pacific oyster showing pustules typical of the Denman disease. ×1.

FIG. 104. Interior of a Pacific oyster shell showing the markings caused by pustules of the Denman oyster disease. ×0.7.

Other characteristics of the disease are:

1) Greatest mortality occurs in oysters at the lowest tide level.
2) Older age-groups (more than 2 years old) seem to be affected most.
3) The localized nature of the outbreak.
4) Recovery from the disease by about 10% of those infected.
5) Aquarium tests indicate the disease can be acquired by undiseased oysters from adjacent diseased ones.

During the period of the first outbreak in 1960, the disease was not observed on the opposite shore of Baynes Sound, only a few miles away, and it was not reported from elsewhere in the province. In 1961, much the same pattern as that shown in 1960 recurred at Henry Bay. However, it was also reported in Ladysmith Harbour where the average mortality over the whole of one bed was 13%, with 30% at the lowest tide level. The disease was also observed at Crofton where the average mortality reached 16%.

In 1962, the pattern was the same at Henry Bay but the level of infection (10%) and mortality was somewhat reduced. In Ladysmith Harbour, the bed which had been heavily infected in 1961 was only lightly attacked, but this could have been due to the fact that the grower sensibly moved the oysters from a lower to a higher tide level during winter.

In 1963, at Henry Bay, level of infection of 22% rose slightly over that of 1962 and in 1964 it was still higher with a maximum of 30% infection in mid-March. There were no reports of the disease from other areas in either 1963 or 1964, but significant levels now occur in Nanoose Bay.

At the time of the outbreak and subsequently, a number of studies were undertaken to determine the cause of the disease.

A bacteriological examination was conducted by microbiologists at the Nanaimo Biological Station with negative results.

A search for *Dermocystidium* (now *Labyrinthomyxa*) and *Hexamita*, organisms associated with oyster mortalities was conducted. Several amoeboid protozoans were isolated but not identified. In addition prepared material was provided scientists at the major North American fisheries laboratories working on oyster diseases. So far there are no definite conclusions regarding the cause of the Denman oyster diseases. A protozoan designated generally as a "microcell" is believed to be implicated. The occurrence of pustules may not be a primary manifestation of the disease. It has been learned, though, that oysters may recover from the disease and that, in the following year, they may or may not contract it again. While the cause is unknown, enough has been learned about the cycle of the disease to permit oyster growers to cope with it, if the disease reaches significant proportions. The fact that greatest infection occurs at or near zero tide levels indicates that any oysters at lower tide levels should be moved to the 2-m level (Strait of Georgia levels) prior to the period of infection in March and that low grounds should not be planted before June.

In addition, because there appears to be little mortality in oysters up to 2 years of age (although they may have the disease), it is possible to hold oysters under that age limit at the lowest tide levels.

Growers should keep a careful record of any mortalities out of the ordinary and, if exceptional, they should be reported.

In Table 21 is given the levels of infection of the Denman disease from single sampling during the peak period.

Transplantation

In 1977 a Federal-Provincial Transplant Committee became the sole agency controlling importation of exotic species because of concern regarding disease, predation, competition and genetic drift. A transfer permit from the Agriculture and Commercial Fisheries Branch is required before introducing or moving oysters from one area of the province to another.

TABLE 21. Denman disease infection rates of Pacific oysters at Henry Bay, 1960–1985.

Year	Infection rate (%)	Year	Infection rate (%)
1960	35	1971	*
1961	35	1972	39
1962	12	1973	24
1963	22	1974	*
1964	20	1975	*
1965	18	1976	24
1966	12	1977	*
1967	11	1978	*
1968	16	1979	*
1969	24	1980	20
1970	25	1981	27
		1985	25

*No data.

Pacific Oyster Culture

Oyster culture has a long history, dating back to the time of the Romans who prized oysters highly. In most areas where oyster culture has developed, it started out as a fishery of naturally occurring stock. With the native British Columbia oyster attempts at culture occurred only in the declining years of its fishery, too late to halt its virtual disappearance in areas such as Boundary Bay and Ladysmith harbour, two of the original producing areas.

Basically there are few distinct methods of oyster culture, but there may be many minor variations of these.

TABLE 22. Culture methods in British Columbia.

Bottom		Off Bottom	
Intertidal	Subtidal	Intertidal	Subtidal
			(suspended)
		Tray, string, stake, stick	Tray, string, stick

Types of Culture

Bottom Culture

Bottom culture, as the name implies is carried out on suitable substrate either intertidally or subtidally, and worldwide the most productive of culture methods in terms of volume. In Europe both sub-tidal and intertidal bottom culture is the usual method while in eastern North America virtually all bottom culture, the main system, is sub-tidal. At one time most culture in Japan was carried out on the bottom but in recent years there has been an almost complete turn to suspended culture, intertidal for seed and subtidal string culture for growth.

Tray Culture

Tray culture is little used and other than its current use in B.C., the only country using tray culture as the prime method is Norway where there is virtually no oyster bottom and the species (*Ostrea edulis*) does not lend itself well to string culture. In Australia and New Zealand it is used as an adjunct for finishing oysters too small for sale after harvesting from the main system of stick culture. Production of oysters from tray culture *per se* throughout the world is small.

Stake Culture

Stake culture was practiced by the Romans as well as the Japanese and a modified form using bricks or rocks was and is used in China, South East Asia and Australia.

Stick Culture

Stick culture, described later, is the principal method in Australia and New Zealand.

String Culture

Suspended string culture, now used extensively in Japan, Korea, and Cuba was also used by the Romans.

Bottom Culture

Bottom culture is the natural evolution from the way in which unexploited populations are maintained on natural beds. These are maintained by spatfalls on parent live oysters or dead shells. This often created oyster ''reefs'' with generations layered one above the other with those on the lower layers eventually dying by smothering or competition for space and for food.

The culture procedure replaces the reef with its mixture of age classes with a situation where, if possible, these are kept separated in different areas but there is usually some superimposition of young oysters by settlement on older ones. Most oyster growing areas in the world are naturally located in or near breeding sites but in some cases seed producing areas are partially separated from growing areas by hydrographic factors like salinity. Such a situation exists on Chesapeake Bay where seed collected in Virginia's low salinity James river, is used on distant higher salinity growing grounds.

In the generalized bottom culture system the bed area is divided into sections, each holding oysters of successive ages to market sized oysters which will be harvested.

The ideal oyster growing situation is one where breeding grounds are completely separated from growing areas and this is mainly so in British Columbia, with only four distinct breeding areas, with Ladysmith Harbour the only one closely associated with growing grounds.

This separation of growing areas from breeding sites simplifies the culture scheme and makes possible exact productivity measurements for the number of oysters harvested may be compared to the number of seed planted for there are no additions by local spatfalls. At one stage, on Ladysmith Harbour growing grounds, there was a mixture of five year-classes superimposed on the original planting.

Intertidal Bottom Culture

Site Selection

Site selection for oyster culture depends on many factors of varied importance, some physical, some biological, and some social and weight given to individual factors depends on the type of culture envisaged. Listed below are some of these, of which temperature, salinity and several others must be accepted for what the natural seasonal variations bring but most are factors of general suitability rather than specific criteria. For bottom culture the essential ones are bottom consistency (material), tidal height and wave action, for which some measure of selection is possible.

1. bottom consistency
2. tidal height
3. wave action
4. water quality
5. temperature
6. salinity
7. predation
8. pollution
9. ice
10. tidal flow
11. turbidity
12. eel grass
13. navigable waters
14. access
15. surveillance

Bottom Consistency

Consistency of the bottom must support seeded cultch or the adult oyster as well as hold it in position. The bottom may be of rock, sand, gravel, mud, or any combination of these, often with a mixture of broken clam shell.

Level sandstone type rock is suitable but must be relatively even to permit harvesting. It has the disadvantage that slight wave action moves the oysters, either rocking them to knock off shell growth or piling them into windrows. Boulders may be used to construct low fences as wave barriers on stony ground. An example of this may be found on the north east shore of Denman Island.

Pea gravel (1 cm in diameter) bottom forms an ideal surface. It occurs generally on level ground, gives adequate support with minimal silting and permits development of well shaped oysters with fluted shells and no harvesting difficulties. There may be some tendency toward movement by wave action but this may be reduced by fairly dense planting.

Sand is generally firm enough to support oysters, but particularly in backwaters there may be areas with a high water content that are too soft for adequate support and this may be aggravated by the burrows of mud shrimps. Riffled sand is firm enough but the riffling indicates a considerable amount of wave action. Stability is the important quality for a sand bottom.

There are probably more oysters grown on mud than any other substrate, mainly because of its ubiquity. The difficulty is assessment of limits of softness of mud that will support the oyster as well as limiting silting mortality. This is done by restricting mud substrates for planting seed. An oyster 4–5 cm or more in legth can survive and grow well on fairly muddy ground. For mature or near mature oysters, a rough rule of thumb requires a 60–70 kg person to sink not more than 10–15 cm when walking over the bottom. Oysters grown on mud tend to elongate and to have smooth shells.

There is a general tendency for intertidal ground to be firmer at the higher tidal levels than lower down partly because of the compacting action of waves.

Tidal Height

The Pacific oyster is a species which flourishes in the intertidal zone, that area between the low and high tide marks. Tidal heights are measured as the vertical distance between the level of the sea and the chart datum which is that of the lowest normal tides. A tidal level of 3 metres means that it is the vertical distance above the level of the lowest normal tide in that particular area. In British Columbia, in areas where breeding has oocurred, the position of greatest abundance is found to be between the 2- and 3-m tide levels (the Strait of Georgia). Although this is the area of greatest survival, maximum growth does not occur there but, instead, is normally greatest close to zero tide level. Most oysters are grown in the 1- to 2-m zone as measured in the Strait of Georgia, where the average tidal range is about 5 metres. Satisfactory growth may occur throughout the lower third of a tidal zone of this magnitude. Therefore, most of the ground selected should be between zero and the 2-m tidal levels. Some good ground may be found above this and can be used for storage and, possibly, for growing seed.

In Tables 18 and 19 and Fig. 84 are shown the average number of hours per day an oyster above a given tide level would be exposed during the year. These data are for areas covered by four reference ports in 1964, and for 1965 for Fulford Harbour where the average daily exposure during each month of the year is given. It may be noted the length of exposure for each 0.3 m/tide after the 2-m level increased markedly; thus, after an average exposure of 5 h/day, growth rate drops markedly. With the same type of information for other reference ports it is possible to make comparisons of tidal levels and exposure times, as for example betwen the Strait of Georgia and Prince Rupert. In the latter area the average tidal range of 8 m is nearly twice that in the Strait of Georgia (5 m) and with a somewhat different tidal pattern. For example, at the 2-m tide level at Point Atkinson there is an average exposure of 4 h while at the same level at Tofino there is an average exposure of 6.1 h.

Wave Action

A third factor to be considered in selecting oyster ground is the degree of protection from wave action. If there is an excessive amount of wave action, oysters are tumbled about and usually washed up along the beach into piles or windrows. Tumbling knocks off fragile shell edges, especially during the growing season, and even at high tide excessive wave action can cause turbidity by stirring up bottom sediments. This causes oysters either to stop feeding entirely or to expend considerable energy separating mud and sand from edible particles. It is difficult to state in specific terms the maximum amount of wave action that would permit oyster culture, for this has to be decided on the basis of experience.

It is not often possible to find optimum levels of the three factors together in the same ground and usually it is necessary to compromise on at least one of them. Since most potentially good oyster ground is already under lease, assessment of new ground should not be too difficult. Oysters may be grown, after a fashion, almost anywhere but when considered from the commercial point of view every condition less than optimum makes it more difficult to produce them economically and competitively.

Only experience can determine the amount of permissible wave action for it depends partly on the slope of the beach and type of wave, whether it is the short choppy wind induced or the larger ocean swell. Also frequency of occurrence of excessive wave action must be considered like the cross inlet winds (williewaughs) that occur occasionally in Baynes Sound and Saanich Inlet. Amount, type, and position of flotsam at the high tidemark will also provide an indication of the level of wave action.

Water Quality

Pacific oysters are extremely hardy, able to withstand wide variations of temperature and salinity. The species occurs over a wide range of latitude in Japan (30°N to 45°N) and has been successfully introduced into Australia and new Zealand. It has replaced the diseased Portuguese oyster (*Crassostrea angulata*) as one of the two major species grown in France and is grown as far south as Morocco. It has also been grown in Scotland and experimentally on the Atlantic coast of North America. On the Pacific coast, it has been grown commercially between Morro Bay and southern California (35°N) and Ketchikan, Alaska (55°N), and, experimentally, north of there, as well as in Chile.

In British Columbia, this oyster lives and grows in waters with a temperature range between 5°C and 24°C, and is able to withstand freezing air temperatures (−20°C) when uncovered by the tide. It grows in water with salinities as high as 33‰ and in river-mouth areas it may be in freshwater for up to eight hours a day. Water with the necessary temperature and salinity range is not difficult to locate; the main problem is to find fertile water in which oysters will grow rapidly and fatten well. Capacity of water to produce oyster food is variable and wide extremes occur within short distances. This type of water cannot be easily recognized and is generally located only by trial and error. In addition, the water must be free of industrial and sewage pollutants.

Water quality is determined by the major components of temperature, salinity and other chemical constituents such as nutrient salts and food as well as deleterious substances such as pollutants. Most of these are fixed factors for any one site and little can be done to alter their values. In general, water quality must be accepted as it exists for any specific site. If there is already a population of molluscs it may be accepted that water quality is reasonably suitable. At least temperature and salinity levels should meet the requirements of the species to be cultured.

There is a conception that knowledge of nutrient and food studies is necessary for a successful oyster culture. However desirable this information may be, it is most difficult to acquire, requiring a time consuming, long term effort with costly equipment and well trained personnel. As with other water quality data the best it can provide is a possible explanation for events. The oyster itself, by its own reaction to the environment provides the best measure of water quality suitable for its needs, for it integrates most of the factors as reflected in growth rate of shell and condition of the body. These are factors for which a record should be maintained by the oyster grower. If the site is oceanographically unique in some way such as being influenced by fresh water, temperature and salinity records might be kept.

Along most of the British Columbia coast availability of ground and the physical attributes such as bottom consistency and wave action take precedence over water quality.

Temperature

Although water temperature has a considerable effect on the growth rate of oysters and on the breeding cycle it has limited effect on site selection in a given general area in B.C., for often other factors have a greater bearing. The Strait of Georgia has somewhat higher temperatures than the west coast of Vancouver Island and the north coast and this is reflected in a higher breeding success there and in growth rates. But within the Strait of Georgia itself there is a general temperature uniformity as it affects oysters, except for very few special localized areas like Pendrell Sound, and this in itself is not a particularly good area for oyster culture in its broadest sense. Areas adjacent to tidal passages generally have somewhat lower temperatures owing to turbulence and upwelling and should be avoided if possible.

However, an oyster grower, while having to accept the temperature regime peculiar to his site should be aware of fluctuations and should maintain records for a year to two, preferrably with a temperature recorder which provides information on diurnal changes.

The Pacific oyster is said to be an eurythermal species able to withstand a wide range of temperatures and can withstand freezing if undisturbed.

Salinity

In British Columbia, salinity may vary considerably, and the euryhaline Pacific oyster can withstand a wide range of salinities, including very low ones for brief periods, but higher salinity areas are preferable. Salinity is one of the factors that influences flavour. As with temperature, growers should have a salinity record, particularly if the site is adjacent to a significant river flow.

Predation

Because of accessibility, bottom culture oysters are quite susceptible to bottom-living predators, but by the same token they are liable to a measure of control.

In British Columbia the main intertidal predators are gastropod drills, sea stars, and birds such as ducks, which can swallow small seed whole. British Columbia is fortunate in not having fish predators such as rays which require fencing, spiking or off bottom culture.

Oyster drills and sea stars are dealt with in the particular section on predators and pests.

Pollution

Pollution, whether sewage or industrial is usually apparent by the proximity of its source. The difficulty lies in determining the distance at which pollution for a specific site becomes ineffective and the potential for possible increases in the future. Common sense will most often indicate whether a growing site may be in present or future jeopardy.

Ice

In certain areas there is the problem of ice, which, fortunately, arises only during severe winters at infrequent intervals. Freezing on the British Columbia coast is often associated with periods of little or no wind, and if this succeeds a heavy rainfall a layer of freshwater develops on the water surface and ice can then form. As the tide falls the ice rests on the bottom and if it is soft the oysters are forced into the mud; if the bottom is firm, oysters may become frozen into the ice and when the tide rises they are lifted off the bottom. When the ice breaks up considerable numbers may be carried away and finally dropped onto other beds or into deep water, as observed at both Ladysmith Harbour and Boundary Bay.

Tidal Flow

Tidal flow is of concern mainly where there is a large tidal range at or near narrow passages where rapid (up to 10 knots in B.C.) flows occur. This will cause movement of oysters if in close proximity to current flows. Probably the main effect of an oyster bed close to a narrows is the low water temperatures regime caused by turbulence and upwelling which brings deep cold water to the surface thus inhibiting shell growth.

Rapid currents affect suspended string culture as it may cause strings to tangle.

Turbidity

Unless turbidity, usually caused by wave action on muddy ground or the silt load from nearby rivers, causes silt deposition great enough to cover oysters on the bottom, it is usually not of concern. Many of the great oyster producing areas of the world occur in areas of relatively high turbidity. British Columbia is almost unique in the low turbidity of most inshore waters. It is not known where dividing lines may be, between silting enough to cause excessive energy consumption caused by the need to separate silt from the food particles, or food particles adhering to silt particles and contributing to molluscan nutrition.

Eelgrass

In addition to the desirable characteristics of oyster bottom described above there are other situations that may be considered. At the lower tide levels, on ground that is otherwise suitable, eelgrass very often grows in profusion. Oysters will grow satisfactorily among eelgrass which, however, can interfere with working them. Eelgrass has a seasonal cycle, growing rapidly and extensively during summer and dying down to a considerable extent during winter. In most areas, if work on the oysters can be confined to winter months, there is normally little difficulty, otherwise it may be necessary to cut the grass as is done in parts of the State of Washington where a long cutting blade is dragged over the bed at high tide by a tugboat, the cut grass being carried away by water currents. So far, in British Columbia, little use has been made of eelgrass ground, probably because there has been sufficient ground of other kinds available.

Navigable Waters

Oyster growers, especially those using floating structures must comply with regulations of the *Navigable Wates Protection Act*. If there is a possibility that such structures might interfere with navigation or passage of vessels, clearance should be obtained from the Department of Transport.

It is essential that rafts or long lines be kept as far away as possible from navigation fairways because of the effect of swell from passage of vessels on structures and the fact that proximity attracts attention. Precautions such as placement of lights, reflectors, or signs will depend on the local situation.

Access

Access is a consideration in bottom oyster culture. Water access to leases is always available, but during winter, low tides are at night and stormy weather may create difficulties. This is alleviated with shore access which also permits movements of equipment such as tractors onto the beds. Matters are simplified if the leaseholder is the upland owner.

Surveillance

If possible an oyster lease should be under surveillance to ensure continued safety of floating structures and to prevent vandalism or theft, which are the owner's responsibility. These factors may be assessed according to local conditions. Sixteen experimental oyster culture rafts, positioned throughout the coast for two years in more less isolated areas, were unmolested with only one possible exception.

Culture

The broad principle of intertidal bottom culture is collection of seed in the summer of year one and placing it on protected seed or storage ground during the fall of that year for protection against winter weather. In March of year two, the spat are planted on seed ground where it remains until it reaches a size at which the clusters may be successfully separated. Depending on growth rate and type of ground on which it has been grown, this may be done in the fall of year two or spring of year three. At this time the seed may be transplanted to growing grounds, where it usually remains until it reaches harvest size, generally after three summers of growth.

Ground Selection

Once an area has been selected for intertidal bottom culture it is necessary to allocate areas for seeding and growing. With seeded cultch placed on the bottom, two main sources of mortality (excluding predation) are competition for space and silting. Little may be done about the former, but ground selection can eliminate or reduce silting mortality. This is proportional to softness of the ground so selection of as firm a bottom as possible is obvious. This is most often found at higher tidal levels so it may be necessary to balance possible reduction in growth rate associated with higher tidal levels with reduction in seed mortality. Ideal bottom is one consisting mainly of pea gravel with adequate protection from wave action.

The following factors should be considered in the selection of seed ground.

1. The amount of suitable ground at various tidal levels in proportion to the amount of seed to be planted.
2. Bottom consistency. Firm bottom reduces mortality and produces cupped oysters.

3. Seed at lower levels is more susceptible to attack by predators such as sea stars, crabs and drills.

4. The time available to work on seed at low tide levels is less than at high levels. It is possible to plant seed in one area and transfer to another for final growth.

Seeding procedure

In British Columbia after collection, naturally set seed is usually allowed to remain *in situ* at least until October, whether it be on the bottom, on racks, or on rafts. This permits the spat to attain near maximum growth for that year, for by the end of October significant growth will normally have ceased. In addition, air temperatures in October are low, but with little danger of freezing and fairly high humidity, so exposure to air does not seriously affect the spat when being transported, as it might when the weather is warm and dry. Seed is normally transported directly from suspension in the seed area to seed ground, or storage area so that it suffers a drastic change removal from a condition of continuous submergence to one of intermittent exposure to air.

Seed is most often moved in October before winter frosts. For seed caught in suspension it should take place during a neap tidal series when tidal ranges are small so the seed is not exposed for long periods which exist during the long runouts of a spring tidal period.

Unless the cultch, either shell strings or bags, is to be placed on a rocky bottom, the shells should be removed from the wires or bags. If left intact, bags or strings will sink, in time, even into firm bottom resulting in some mortality.

Seed is nearly always shipped as deck cargo, either on a scow or vessel, but it may be carried in the hold providing the air temperature is low enough. If at all possible, suspended seed should not be moved during periods of freezing temperatures since this will cause some mortality. The transportation period, although it can be as long as four or five days, should be kept as short as possible. Handling should not be unnecessarily rough, for this will cause shell breakage leading to mortality, particularly with seed about 15 mm or more in diameter.

Seed Ground

While most types of oyster ground may be used for seed, some are more suitable than others. Apart from predators, the two main sources of mortality in seed are competition for space on the cultch and silting, the latter being a function of the type of seed ground. Silting mortality is directly proportional to the amount of silt. The obvious requirement, is for a firm bottom, and this is usually associated with gravelly ground. Gravel and similar firm ground occur most often at higher tidal levels, where lapping of waves as the tide rises and falls prevents deposition of fine silt which nearly always occurs near low tide levels, unless there is strong current action. While growth rates are usually lower at higher tide levels, the differences in growth rate of seed at various levels up to the 2.1–2.4 m tide mark are not very great. Since high ground is of little value for growing, its use as seed ground is advocated, provided it has adequate protection from strong wave action. An area which has been used in this way is Comox Harbour (Fig. 105).

Indeed, in British Columbia, there are many intertidal areas at a level too high for normal oyster culture, which could be utilized for seed growing. In time, because of the small amount of oyster ground available in this Province, it is inevitable that there will emerge growers whose sole occupation would be growing seed up to a length of about 5 cm for sale to other growers with normal growing and fattening ground.

Planting Density

Presuming growth to maturity is to be carried out *in situ* in the planting area, seed is normally planted at a density equivalent to 500 bags or strings per hectare, assuming each holds approximately 100 Pacific oyster shells. This level means 50 000 pieces/ha (20 000 per acre) or 4 per square metre.

FIG. 105. Heavily planted seed bed for Pacific oysters, Comox Harbour.

A minimum commercial set of 15 spat per shell may have a survival of 5 spat on average oyster ground owing to natural mortality, silting, competition for space on the shell, cluster separation, and excluding predation. This number provides a harvest of approximately 22.5 t at 4 400/t equivalent to 9 000 L/ha depending on the oyster size and condition factor in a culture time of 3 to 4 years.

However, this is a relatively light planting density which may be doubled or more without undue crowding. A planting yielding 50 to 60 oysters per square metre assists in preventing movement of oysters by wave action and permits efficient harvesting. Moderately dense planting does not seem to affect growth or fattening excessively and makes optimum use of the production area. Plantings of 2 000 bags or strings per hectare are known.

However, if the cultch is heavily set at about 50 spat per shell or more, the production will be much greater and up to 25 oysters per cluster can survive after the first summer of growth. In this case cluster separation is necessary. A less dense planting would suffice although heavy plantings would allow cluster breaking by harrow rather than by hand.

If the spat count is significantly greater than 25, consideration might be given to cutting the shell into smaller pieces, the size of which will depend on the type of ground receiving the seed. Small pieces of cultch theoretically reduce competition for space on the shell but may increase silting mortality. The firmer the ground (less silt) the smaller the pieces that can be planted and this may reduce the cluster breaking effort. The shells may be cut by a gang of stone mason saws set in series. For very firm ground the shells may be reduced to even smaller pieces by crushing, especially if the spat count is very high in the region of 100 or more. However, experience with broken Japanese oyster seed casts some doubt on the gain to be obtained by this practice (see page 136).

It should be appreciated that because of all of the variables mentioned, these planting densities are only best estimates and an individual grower must assess his own situation and keep careful records for evaluation of results.

Alternatives in utilization of seed grounds are:

1. Plant at a low density such that when clusters are separated on the spot the seed will represent the number of oysters that may be matured in the area without transplanting.
2. Plant at a high density and transplant all the oysters to other growing ground.
3. Plant at a high density and transplant only some of the oysters, leaving the remainder to mature on the spot.

As with all beginning oyster culture efforts the seeding operation is initially one of experimenting with variables of bottom type, tidal height and density.

Spreading

Seed may be spread at low tide after being dropped overboard at the appropriate spot at high tide. To reduce labour the high tide drop should be carefully arranged. An alternative and superior method is to remove the shells from strings or bags and place the shells in a pile on a top float or scow, from where they are spread by shovel or other mean as the scow is moved slowly over the seed bed. Mechanical spreaders have been developed, but since this is usually a once-a-year operation, the expense is not warranted unless large quantities of seed are being planted. A relatively inexpensive means of spreading seed is to wash it overboard by means of water pressure from a hose which can also be used for unloading any type or size of oyster. The shovel method of spreading seed, while seeming primitive, is effective and if done carefully gives a spread of uniform density.

The following is a summary of factors which influence planting and growing of seed. Each factor must be assessed in the light of local conditions and method of culture in use.

1) The amount of suitable ground at various tide levels in proportion to the amount of seed being planted.
2) Firmness of the bottom; this affects the shape and survival of seed. Firm ground increases survival and induces the growth of a round, deeply cupped oyster.
3) Unbroken seed is more readily moved by tidal and wave action than is broken seed, but broken seed is more readily silted.
4) Seed at lower levels is more susceptible to attack by predators such as crabs, sea stars, and oyster drills, for these species normally occur there rather than on higher ground.
5) The time available to work on seed at low tide is less than at a higher tide.
6) Desirability of transplanting.
7) The natural habitat (where survival of natural set is highest) in most British Columbia areas is between the 2- to 3-m tide levels and it is here that survival is maximal.

Cluster Separation

Cluster separation should be practised for there are many advantages in the development of better shape, growth, and fattening, along with greater ease and efficiency in opening.

Cluster separation is the process of breaking up clusters of oysters (Fig. 106) developed from a single cultch unit into as many individual oysters as possible. This is normally done in the late summer of the year following planting. It would be preferable to do this in late winter before spring growth adds new and fragile edges. Handling breaks these off and they must be replaced. It is said that the Atlantic oyster compensates for the loss of shell edges by relatively speedier growth, but this has not been demonstrated for the Pacific oyster. Usually, however, the period of breaking is regulated by the time available to the grower and this occurs most often during summer after harvesting.

Some mortality during the breaking process is inevitable, and it may reach as much as 25%, depending on the degree of separation required, as well as on the care taken. The extent of mortality will also depend on the type of bottom on which the seed has been grown, for soft-bottom seed is usually easier to break than hard-bottom seed. Size

FIG. 106. Cluster of Pacific oyster seed spatted in Pendrell Sound in July 1965 and photographed in December 1966.

of the oyster is also a factor, and there is an ideal size where seed have grown away from the mother shell but have not yet grown too much together. This point can be determined only by experience. Consistency of the mother shell is also an important factor, and one reason why growers prefer suspended culture shell cultch is that they believe the thin soft mother shell makes easier breaking as compared with the thick hard shell of local seed. This is true only to a certain extent for, at the right time, seed will peel off a hard shell quite easily. Skill and care on the part of breakers is necessary.

Briefly, excessive mortality may be due to:

1) Breaking too finely
2) Carelessness
3) Seed too small or too large relative to the hardness of the shell.

Corrective measures for the first two factors are obvious. With the the third it may be necessary to delay breaking for a time to allow further shell growth of small seed. If the seed is too large there is little that can be done except to accept the mortality or to allow the seed to grow to maturity without breaking.

The simplest way to break is to pick up the cluster and pull the oysters apart by hand, possibly aided by prying with a breaking iron which is a pointed steel bar about 30 cm by 3 cm by 5 mm. Other types of tools are used according to individual preference. After pulling or prying the loose oysters apart, a sharp blow on the mother shell is usually sufficient to separate those remaining. Breaking to obtain single oysters only is the ideal, but impossible to attain without excessive mortality. Clusters of two or three oysters (doubles or triples) are satisfactory.

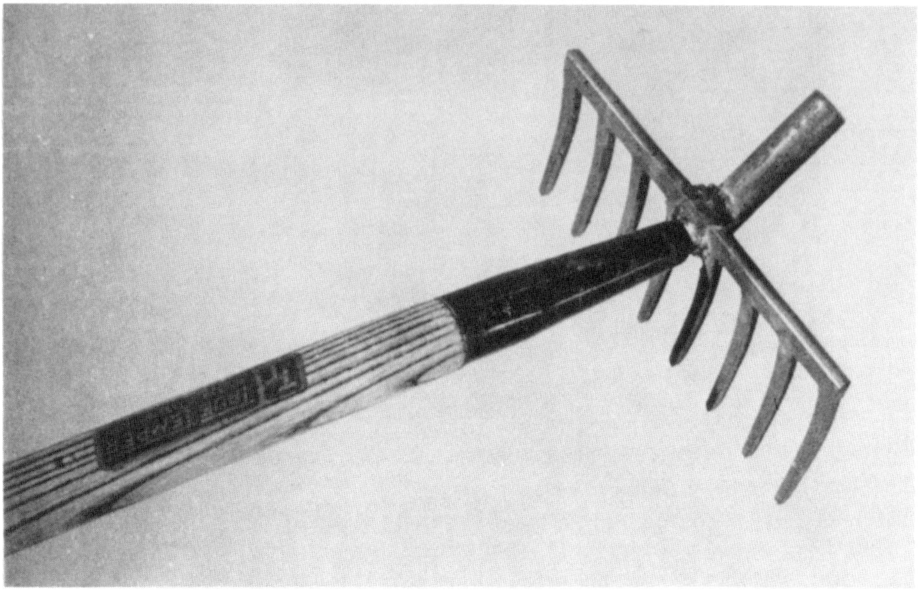

FIG. 107. Cluster-breaking rake.

A small rake, about 15–20 cm wide, to the back of which has been welded a hammer made of a short piece of round iron about 2 cm in diameter and 3 to 6 cm in length, is also used for breaking (Fig. 107). The rake is used to pull the cluster out of the mud if necessary and to turn the cluster over to permit a blow on the bottom of the mother shell by the hammer. This method is not as efficient as hand breaking but it is fairly rapid and less laborious, as it is not necessary to pick up each cluster.

A third method of separating clusters is to tow a spring-toothed English harrow over the seed bed at high tide (Fig. 108). This lifts the clusters out of muddy ground and performs a reasonable job of breaking. Its main advantages are speed, low cost, and independence of low tides, and it apparently does not cause excessive mortality.

The cost of breaking is variable, depending on the method used and extent to which separation is carried out. Minimum time of breaking by hand can be placed at about 2 h/6 bags of seed.

Separation of seed on cemented veneer rings may be done in two ways. One is to cut the circles into small pieces before planting, but this is feasible only if good unsilted seed ground is available. Otherwise it is imperative that the circles be removed from the wires because they will sink, in spite of light weight, into every type of bottom except solid rock. The single circles are then spread on the bottom quite densely. Breaking is done in the usual way, as described for shell. Local shell cultch seed may be crushed or broken in some manner prior to planting, and in this way it comes to resemble Japanese broken seed. Several types of crushers have been used: in one such machine loosely meshing gears are placed at the bottom of a hopper into which are fed loose shells. Another method is to have the cultch shells carried on a rubber belt under a rotating gear wheel which presses down on the shells and crushes them as they pass under it. Shells can also be cut into pieces by circular saws. While some mortality is caused by such methods, theoretically it is less than breaking mortality and increases total survival by reducing competition for space on the shell.

The inevitable mortality due to breaking, silting, and competition for space is compensated for by purchasing excess seed equivalent to the amount of expected mortality from these causes. For example, if all spat in a bag or string were to survive a production

FIG. 108. Spring-toothed harrow used for breaking clusters of oysters.

of abut 50 L would result. However, a production of about 16 L per bag is the normal yield, so for a production of 450 L, 24 bags of seed are required instead of six. While not all mortality can be prevented, a good proportion of it can be, if all the causes are examined critically with this end in view.

Improving Ground

It has been emphasized that it is seldom possible to find oyster ground with all the necessary requirements; one or more of them is usually lacking in some degree.

Mention has been made of the difficulties in growing oysters on soft ground. Hardening soft ground can and is being done. Most of it is done on a long-term basis, most often by utilization of shell from the shucking house. The actual method of treatment varies with the softness of the bottom, but the general principle is a primary layer or layers of whole shell followed by a covering coat of crushed shell. If available at a reasonable cost, gravel may also be used.

It is also possible to layer the ground with veneer cuttings or thin plastic sheeting before applying a top coat of crushed shell or gravel. Soft sandy ground may be hardened or treated only in special cases, such as ground softened by mud shrimps. Another method of hardening ground is by means of drainage ditches but this is not often used. In certain circumstances it may be necessary to destroy eelgrass. This may be accomplished by covering it with plastic or roofing material until the plants die.

Problems of wave action may be overcome by the use of breakwaters. Stone is often used if it is immediately available; anchored boom sticks help to arrest the main force of breaking waves. While many of the above suggestions are feasible on a practical basis, the economics must be carefully considered before embarking on major efforts along these lines, and small-scale pilot experiments should precede extensive programs. At present, it may not be economical, but it is possible that in the foreseeable future tide levels on oyster ground will be adjusted by lowering the level of high ground and by filling in low ground. By lowering high ground, especially, considerable acreage can be brought into use.

Attempts to improve the productivity of oyster ground have been made in Australia by sinking containers of fertilizing material in various parts of the bottom. Linseed oil meal with superphosphate has been used although any nitrogenous manure will serve. Some oyster bottoms are normally rich in nitrogen and do not require more. It is said, however, that all bottoms are improved by the addition of phosphate. The purpose of adding these nutrients to the water is to induce plankton production. Assessments of the efficacy of such fertilizer additions are difficult to make.

Subtidal Bottom Culture

Some species of oysters typically occur subtidally like *Crassostrea virginica*, the Atlantic oyster which ranges from the Canadian Maritimes to the Gulf of Mexico. The culture process is much more difficult than that required in intertidal culture. Usually dredges or tongs are required for sampling and harvesting. Fouling is more intense subtidally than intertidally and predators more difficult to control. In Long Island Sound in eastern USA, special vessels drag gangs of string mops over the bottom to harvest sea stars which are entangled in the mops which are lifted to the surface and dipped in a tank of heated water. Another method of sea star control is to distribute powdered quicklime over the oyster beds. The chemical destroys the sea stars without harming oysters. Predation by sea stars has long been recognized as a serious impediment to subtidal bottom culture in British Columbia (page 111) and probably the reason there has been no development of this system. However there is the possibility of the existence of specific areas where this method may be prosecuted.

It should be mentioned United States oyster production of 280 000 t (1983), the largest of any country, emanates chiefly from subtidal bottom culture on the Atlantic and Gulf Coasts.

Productivity

Productivity of the Pacific oyster may be defined as meat production in relation to amount (number) of seed planted. It is the result of interaction of such factors as mortality, growth rate, and condition or fatness. These factors are dependent on type and quality of seed, and the particular oyster bottom used, as well as on seasonal and annual variations. If productivity is known or can be estimated, potential production may be calculated by multiplying productivity by the number of units of seed per hectare and by the number used. As an example, it may be found that a certain 4 ha bed has a productivity of 20 L per bag of seed, on the average, from plantings of 450 bags/ha. The estimated potential production will be 4 × 450 × 20 L or 36 000 L at the end of the 3- or 4-year growth period.

It would be of value for oyster growers to understand something of the factors that control productivity. Among these is oyster mortality.

Mortality

In considering problems of mortality and survival, care must be taken in determination of the baseline on which calculations are made. Using egg production as a criterion, the final survival is infinitesimal, for in nature only a relatively small proportion may be successfully fertilized. It is known that under ideal conditions larval mortality is high so the number of spat is a small proportion of the original many millions of unfertilized eggs. But it is at this point the grower begins to assess the mortality situation. In Pendrell Sound it is possible to obtain settlement of more than a 1 000 spat per shell of average size. It is manifestly impossible for this many spat to survive even with special treatment of the cultch. Thus the level of mortality finally occurring should not be assessed on the basis of 1 000 spat per shell. If the shell is grown to a point where most will be marketable, not more than about 25 oysters can survive. It seems that when considering shell cultch this should be the approximate baseline on which to calculate survival. If the cultch receives special attention such as cutting the mother shell into small pieces, the number surviving may be greater for theoretically there will not be high competition for space. Alternatively this is so if the cluster is separated when the spat are no greater than 25 or 30 mm. Thus maximum potential survival should be the base line for calculating mortality rather than total original number of spat. In a vexar bag of 100 shells with a spatfall in excess of 25 spat per shell the grower can expect a survival of not more than 2 500–3 000 oysters and this might be the realistic basis of calculation.

There are four main causes of mortality in Pacific oysters.

1) Natural mortality
   (a) Age
   (b) Predators
   (c) Disease
   (d) Exceptional circumstances
2) Competition for space
3) Silting
4) Cluster separation

Natural Mortality

This may be due to factors such as old age which, in normal cultural procedures, does not constitute a problem since Pacific oysters have been known to attain an age of 20 years or more and the usual age for harvesting is, at most, 5 years but more commonly about 3 years.

Predators such as sea stars, oyster drills, and crabs, may contribute materially to mortality rates, but some measure of control is possible as discussed on page 00.

Disease can be a serious mortality factor but, so far, British Columbia waters have been relatively disease-free. Disease is not easily controlled, especially when it first strikes, but often culture methods can be modified to materially reduce the effects.

Other types of mortality which may be considered in the "natural" category are loss by ice and movement by wave action from freak winds; in two instances in British Columbia, tidal waves caused destruction of oysters in Clayoquot Sound on the west coast of Vancouver Island, the first wave on May 22, 1960, from an earthquake in Chile and second from one in Alaska on March 28, 1964.

Competition for Space

Competition for space (crowding) on cultch is a most important source of mortality in the present system of oyster culture. This means that only a certain number of oysters of a given size can be attached to a given area of cultch; the larger the oysters the fewer per unit of cultch. Experiments to determine the effects of crowding started with spat of approximately equal size but with different numbers per unit of cultch. These spat were grown by the suspended culture technique, so that types of mortality, other than that due to competition for space, had no opportunity to operate, or had minimal effects. With 14 spat per shell, 10 oysters survived to a length of 11 cm, with 34 spat per shell 14 oysters survived, and with 87 spat per shell only 20 oysters survived. Thus, to double survival rate a six-fold increase in the original number of spat was required.

If one of these units had been cut into several pieces before the start of the test, the sum of the oysters on the pieces would have been greater than the number surviving on the larger unit of equal area. The reason is that the length of the perimeter relative to the surface area as well as the actual surface area itself affects survival. This is because the additional perimeter permits oysters to spread further away from each other and from the ones in the centre, and is one reason for separating clusters at the end of the first year, although by this time competition for space has already has considerable effect.

The time when small pieces of cultch are most disadvantageous is at initial planting. However, reducing the size of cultch is a method of reducing competition for space only, and other types of mortality may have greater effect on smaller pieces of cultch than on larger pieces.

Still another way in which the grower may reduce his mortality due to competition for space is to purchase larger spat, where mortality due to this factor has already taken place to a considerable extent. But the economics of the situation has an important bearing, for any advantage depends on whether seed is purchased on the basis of number of spat, or volume of cultch, or both, and on the price range of different types of seed.

Silting Mortality

The considerable difference between numbers of seed planted and numbers of oysters harvested led to experiments to determine the cause or causes of this discrepancy. Japanese seed with known numbers of spat were grown under conditions as close as possible to commercial conditions. Counts of surviving oysters were made annually and on comparable lots of seed; mortality during the first year averaged 91% whereas during the second and third years after planting there was only a 2% increase in mortality as shown in Table 23. This particular seed lot had a very high count (up to 1000 000 spat per case).

One way in which to increase the number of spat in a unit of seed is to increase the spat count per piece, and another is to decrease the size of the pieces so there will be more of them in the unit. But the smaller the piece of cultch, the greater may be the mortality due to silting, so the advantage of using small pieces to reduce competition for space is lost unless the seed is given special treatment by placing it on very firm ground with little silt.

TABLE 23. Survival and mean length of Pacific oyster Japanese seed in Ladysmith Harbour, 1952–55.

Date of count	Survival (%)	Mean length (cm)
April 1952	100	0.6
April 1953	9	6.6
June 1954	7	9.4
June 1955	7	10.3

Even in this experiment, where cultch pieces were very small, there was probably some mortality from competition for space, but it would have been minimal. There was no evident predation or sign of disease, so silting was indicated as the main source of this very significant mortality during the first year after planting. It is likely that the major mortality occurred within the first few months.

An example of the effect of silting mortality is demonstrated in an experiment (Table 24). Sixteen random groups of 10 spatted shell cultch were allotted randomly to 8 trays (1 metre square) held off the bottom and 8 plots (1 m × 1 m) on the bottom. These were arranged in 4 groups of 4 at 4 tidal levels between the 0.3 m tidal level and the 2 m tidal level, for a 7 month period during the summer growing season. The bottom was muddy with a gradient in softness between the two tidal levels.

TABLE 24. Comparative survival of Pacific oyster seed planted in trays and on the bottom.

Tidal levels (metres)	Percentage Survival	
	Tray	Bottom
1.5	66.2	61.1
1.0	56.2	35.7
0.5	70.7	29.4
0.3	64.4	22.2

The survival gradient from the upper level to the lower is quite evident. The factors may be described as follows:

Bottom mortality = natural plus competition for space plus silting = 62.9%
Tray mortality = natural and competition for space 35.6%
Silting = 27.3%

Substituting the experimental values and cancelling out the common factors of natural mortality and competition, an estimation of 27.3% is shown for silting mortality. There was no evidence to suggest appreciable natural mortality. So the tray mortality of 35.6% may be accepted as due to competition for space. In Table 25 is shown the spat counts per 100 pieces of shell cultch on each of the trays where there was no silting mortality and indicating the competition for space is least with lowest spat counts per unit of area.

TABLE 25. Spat counts and percent survival of spat per 100 shells held on trays.

Spat count (100 shells)	Survival	% Survival	% Mortality
2 386	1 236	51.4	48.6
2 269	1 186	52.3	47.7
3 133	1 224	57.4	42.6
1 956	1 325	67.7	32.3
1 953	1 194	61.4	38.6
1 489	1 087	73.8	26.2
1 330	952	71.5	28.5
1 238	994	80.2	19.8

Average survival on the bottom was 37% as compared with 64% on the trays. The tray mortality of 36% could be attributed mainly to competition for space, for there was no apparent sign of any other type of mortality. The bottom mortality of 63% could be attributed to a combination of competition for space and silting. It is likely, however, that a considerable part of the silting mortality occurred prior to the time when sufficient growth had taken place to initiate competition. On the softest (lower) bottom, mortality reached 78%, while on the hardest (higher), it was only 48%. This then is an indication that seed ground should be hard ground which is generally found at higher tide levels, where wave wash keeps silting to a minimum.

Cluster Separation

Mortality from this cause is discussed on page 127. That it may reach a level of 25% is of importance, and every effort is required to reduce it. Smaller pieces of cultch and careful breaking are means of reducing the loss. New types of cultch that will reduce breaking loss are required.

Seed Productivity

Bottom culture productivity studies were conducted with Japanese seed since that was in standard use at the time.

Studies were conducted on:

1) Comparison of individual case units of broken and unbroken seed from different sites in Japan.
2) Similar comparisons based on larger units.

Individual Case Productivity

From 50 cases of seed, 20 were chosen at random. These were arranged in two rows of 10 with broken and unbroken seed cases alternating in each row. To eliminate horizontal and vertical differences in the ground the rows and columns were placed close together and the seed planted on relatively small plots of 4 m by 4 m on March 23, 1949 on the Provincial Shellfish Laboratory's experimental ground at Lot 164 in Ladysmith Harbour. The gravel mud ground was firm and at the 0.6 m tidal level.

In May 1950 the clusters were separated by hand. Breaking mortality measured between 15 and 20%. Each case required about 2 h breaking time, but this is probably longer than normal commercial practice since more than usual care was taken. Not all pieces of broken seed needed breaking, but since there were more pieces, the time for separating clusters in both types of seed was approximately the same. At the time of breaking, the oysters were transported down the shore in the same relative positions to a level of 0.3 m lower and the plots increased in size to 6 m by 6 m. This gave a planting density of more than 250 cases per hectare, considerable more than the normal commercial rate.

At the time of planting a random case of unbroken seed contained 850 pieces of shell cultch with a mean spat count of 11 spat per shell giving a total of 9 350 spat. A case of broken seed contained 3 000 pieces of shell with a mean of 9.3 spat per piece or a total of 28 000 spat.

At the time of cluster separation, May 1950, the average weight of oysters from a case of broken seed weighed 190 kg and a case of unbroken 150 kg.

In May 1954 the 20 cases were harvested and oysters counted, weighed and the volume measured with results shown in Table 26.

TABLE 26. Productivity in number, weight and volume of individual cases of broken and unbroken Pacific oyster seed from Japan and grown in Ladysmith Harbour, 1949–54.

					A. Number of Oysters						
	4 381	3 131	4 676	6 295	4 254	3 272	2 168	4 320	3 003	3 991	39 491
	(M-4-B)	(Wun)	(M-4)	(Wun)	(W-4-B)	(Wun)	(M-4-B)	(Wun)	(M-4-B)	(Wun)	
	3 477	5 542	4 311	4 481	4 730	5 457	2 609	2 887	3 883	3 564	40 941
	(Wun)	(M-4-B)	(Wun)	(M-4-B)	(Wun)	(M-4-B)	(Wun)	(M-4-B)	(Wun)	(M-4-B)	
TOTAL:	7 858	8 673	8 987	10 776	8 984	8 729	4 777	7 207	6 886	7 555	80 432
Treatment Totals — broken — 40 413											
— unbroken — 40 019											
					B. Weight of Oysters in Tonnes						
	0.58	0.47	0.59	0.52	0.44	0.44	0.32	0.54	0.30	0.46	5.16
	0.84	0.94	0.86	0.91	1.00	1.08	0.66	0.86	0.47	0.57	8.07
TOTAL:	1.42	1.41	1.45	1.83	1.52	1.52	0.98	1.40	0.77	1.03	13.23
Treatment Totals — broken — 6.68 t											
— unbroken — 6.70 t											
					C. Number of litres (dry)						
	887	539	958	1 526	745	539	426	781	781	745	8 129
	1 313	1 420	1 207	1 313	1 455	1 668	994	1 065	745	887	12 070
TOTAL:	2 220	2 019	2 165	2 839	2 200	2 207	1 420	1 846	1 526	1 632	20 199
Treatment Totals — broken — 10 153											
— unbroken — 10 045											

Statistical analysis indicate there was no significant difference in yield from broken and unbroken seed in this experiment, either in volume or number or in weight of oysters. This is apparent from the treatment (broken, unbroken) totals in each case. In number of oysters, the differences in yield between broken and unbroken seed was only 394 oysters; in weight 19 kg; and in volume, three bushels (108 L). There was no significant difference in the yield of rows in respect to numbers, but the lower row (lower tide level) yielded significantly higher than did the upper row in respect to weight and volume. Thus a difference in tidal level of 0.3 m may produce a significant difference in yield (weight) as a result of differential growth.

That this experiment demonstrated no difference in yield between broken and unbroken seed may be explained by the fact that, although broken seed has a higher guaranteed minimum number of spat, the mortality in this type of seed is greater due to burying and silting of small pieces on average oyster ground. This is confirmed by the high mortality of spat on small pieces from cement-coated egg trays. Also, the differential count indicated by the guaranteed minimum may not always exist.

A most striking feature of this experiment was the extreme variability in yield in the same type of seed from the same Japanese packing site and this was shown clearly at the end of the first year where, in both broken and unbroken seed, the highest yield in volume was more than double the lowest.

Since there is so much variability in oyster ground in respect to rate of growth, measurements directly influenced by that factor are unsuitable for drawing general conclusions on yield. However, the yield in number of oysters may be expected to be little affected by rate of growth and this may be used as a reliable criterion of productivity. The range in productivity in respect to number of oysters was from 2 168 to 6 295 with an average of approximately 4 000. This is for a 5-year growing period; however, with indications that mortality was insignificant in the latter 2 or 3 years. The yield in weight varied from 301 to 1 083 kg, from a mean of approximately 1 500 bags or strings per hectare. At this intensity the planting amounted to at least 250 cases/ha. As explained previously, this intensity was necessary to reduce variation in soil and fertility conditions and may account in part for the relatively slow growth.

Productivity from Larger Units

To provide a further test on productivity and to compare broken and unbroken seed in another way, eight cases of broken seed planted in a group, on March 22, 1949, were compared with eight cases of unbroken seed planted together in an adjacent plot. The plots, each 20 m by 10 m, were located on medium hard ground between the one and 1.3 m levels.

In February, 1951, clusters from these plots were separated and moved to larger plots 25 m by 25 m, between the 0.2 and 0.5 m tidal levels. In May, 1954, when harvested the eight cases of unbroken seed produced approximately 290 bushels weighing 6 050 kg, with an estimated 31 610 oysters. The eight cases of broken seed yielded approximately 318 bushels weighing 6 634 kg, with an estimated 34 661 oysters. In this case, the broken seed yielded about 10% more than unbroken seed in weight, volume and number. In numbers, the yield per case was little different from the experiment with individual cases. The average production in weight was higher by about 113 kg per case than the single case experiment, but this may have been due to better growing conditions.

Since counting was nearly impossible in this case, the number of oysters was estimated from samples and therefore may not have a high degree of accuracy. Production in numbers was higher by about 150 oysters per case than the single case experiment.

Unseparated Clusters

To compare broken and unbroken Japanese seed, whose clusters were not separated, two cases of seed, one broken, and one unbroken were planted on 4 × 4 m plots at the 1-m tide level on March 22, 1949.

On July 1, 1950, the two cases were moved to the 0.3 to 0.6 m tide level on adjacent plots 12 m by 6 m. At this time, the broken seed weighed 95 kg and the unbroken seed weighed 90 kg.

On January 19, 1953, the oysters were harvested and the unbroken seed produced 4 800 oysters in 1 300 L, weighing 780 kg and the broken seed yielded 4 260 oysters in 1 854 L weighing 725 kg.

Broken seed developed smaller clusters than unbroken seed, as well as a considerable number of single and double oysters. There is little doubt that if little or no breaking is to be done, broken seed is superior to unbroken seed, although, generalizations of this nature should be made with care because of variation in initial counts and in this particular instance the comparisons involved only two cases.

Intensity of Planting

A preliminary test to determine the extent to which intensity of planting influences yield was carried out using Japanese seed of 1949.

On June 17, 1950, a case of broken seed whose clusters had been separated were divided into 4 equal portions. Two portions were planted on 4 square metre plots and two of the portions on 20 square metre plots. In May, 1954, the yield from the two small plots was 567 kg and the yield from the two larger plots was 770 kg. The smaller plots had a planting intensity of 750 cases per hectare, while the less dense planting was 150 cases per hectare.

As would be expected, there is a real difference between the two yields and the total yield points up once again the large degree of variation in productivity from various cases of seed.

Results of these experiments indicate there is little difference in productivity between broken and unbroken seed on the type of ground on which the tests were carried out. There is little difference in the cost of breaking the two types of seed with an adequate break. If no breaking is to be done (it should always *be* done) somewhat better results are derived from broken seed since the clusters are generally smaller and there is a proportion of single and double oysters.

It appears that one of the deciding factors in the choice of unbroken or broken (cut) seed is the degree of silting and burying that occurs on the seed ground.

Condition

Size of meat, either by weight or by volume, between two oysters of equal shell dimensions may be very different. The seasonal cycle of change from glycogen (animal starch) to spawn has already been described, and the plumpness or extent to which an oyster fills its shell cavity may be due to the presence of either glycogen or spawn, or a mixture of both. Strictly speaking, a "fat" oyster is one which is plump with glycogen. However, the oyster industry describes a plump oyster, whatever the cause, as a "fat" oyster. The word "condition" would be much more apt to describe the plumpness of an oyster, if modified by the reproductive state.

In late summer, usually August, after an oyster has spawned, it is quite thin and watery, with a glassy appearance. This is because in prespawning oysters a large part of the body is filled with spawn (eggs or sperms), and when this is discharged there is little material left.

Almost immediately after spawning, if food is available (and this is not always so) the oysters begin to build up their depleted bodies by converting food to glycogen, which, in sufficient quantity, gives oysters in winter a plump, creamy appearance. By the end of November, there is little food left in the water and, as low temperatures decrease feeding rates, the degree of fatness of an oyster during the winter is mainly decided by then. Because oysters enter a near-hibernating state in winter, little, if any, of the "fatness" that was recovered by fall is used up. If for some reason the fall feeding season has been poor, then oysters may go through the winter relatively thin.

Another situation may exist when, as occasionally happens in British Columbia waters, oysters fail to spawn. In this case spawn is carried throughout winter and, although they may be plump, they are not "fat" in the true sense of the word, for the plumpness is caused mainly by spawn and not by glycogen. However, spawn may also be converted to glycogen in varying amounts. This, too, is generally completed by the end of November, so conversion of spawn to glycogen appears to be associated with success of fall feeding.

The degree of fatness attained by November changes little during the winter until about March. At that time, increasing temperature and light create conditions in which oyster food can flourish, and there is normally a rapid increase in fatness during spring. This is when Pacific oysters are nearly always in prime condition and fill their shell cavities with bodies composed largely of creamy, succulent, glycogen-filled meats.

In May, when water temperatures rise still higher, oysters begin to develop spawn and the glycogen-filled tissues are replaced by either sperm or eggs, according to sex. The fatter the oyster, the greater is the amount of spawn developed. Spawning condition is maintained throughout summer until spawning occurs and the cycle begins again.

While oyster growers know of this seasonal fatness cycle and observe it throughout the years in a subjective way, a quantitative measure is necessary for recording and comparative purposes.

There are several ways of quantitatively measuring the condition or fatness of oysters, although, in most instances, plumpness, or the degree to which oyster meats fill the shell cavities, is measured, rather than true fatness; for, as mentioned before, most methods do not distinguish between glycogen and spawn.

The industry, particularly in the United States, uses a gallon to bushel ratio in percentage terms. That is, if 75 gallons of oyster meat are produced from 100 bushels of oysters, then the return is "75%", and increase in this percentage value indicates an increase in condition. The number of oysters per gallon compared with the number of oysters per bushel (oysters less trash) may also be used for it stems from the following relationship.

$$\frac{\text{Total litres meat}}{\text{Total litres of shell oysters}} \times \frac{\text{Number litres of shell oysters}}{\text{Number of meats per litre}} = \text{fractional return} \times 100 = \text{percentage return}$$

In this way meat production is related to the size of an oyster for the measure of volume (bushel) is in reality doing this (number of oysters × size of oysters). It should be remembered that the unit of volume, whether it be bushels, cubic feet, litres or of some other measure, should represent net volume, i.e., excluding trash and including only shucking oysters.

Growers often mistakenly refer to number of oysters per litre as a measure of condition. This is of no value for this purpose unless reference is made to the size of the oysters, for the litre could be composed of 100 large oysters in a poor condition or 100 small oysters in good condition.

It has been the tendency in British Columbia to use liters per unit of weight (tonnes) rather than volume (bushels) as a measure of condition. In most cases, this is a fair approximation, although the weight-volume relationship in Pacific oysters is not strictly regular. However, quite a regular relationship exists between the total volume of whole oysters and the volume of the shell cavity (Fig. 109). This is used for a more precise measure of condition in oysters, known as the "condition factor".

Shell-cavity volume is the difference between volume of a whole, closed, intact oyster and volume of the two valves after the oyster has been shucked. Volumes are determined by displacement, or by weighing in air and water, the difference between these two values is volume in the metric system. After weighing, the meats are dried until all moisture is lost. Condition factor is obtained from the ratio.

$$\frac{\text{Weight of dry meat}}{\text{Volume of shell cavity}} \times 1\,000.$$

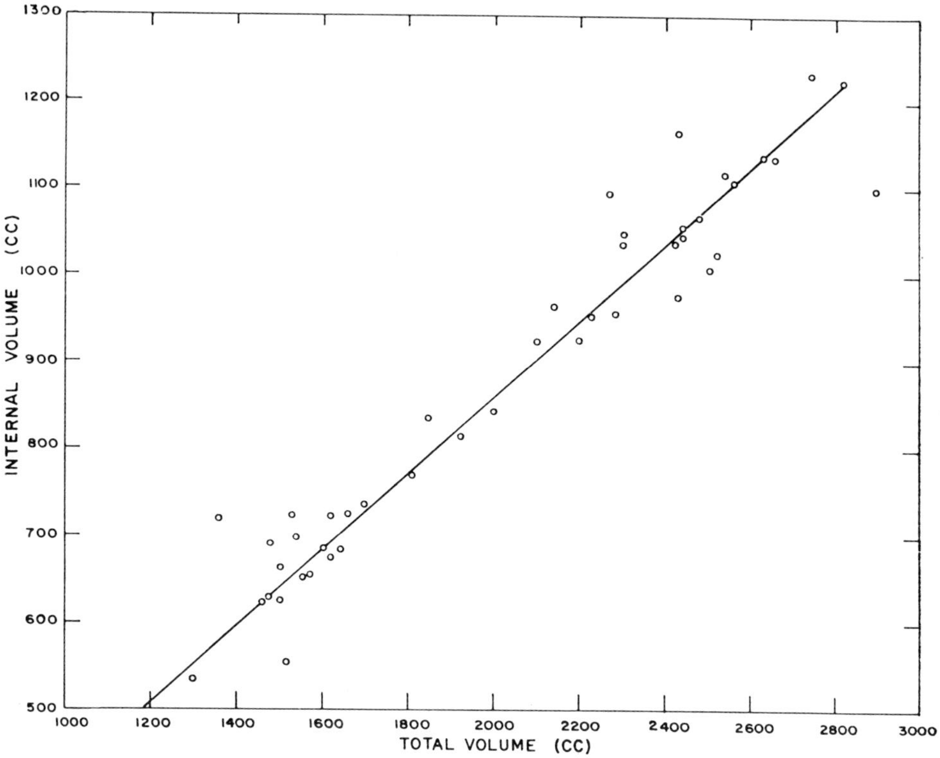

FIG. 109. Relationship between the total volume and internal volume (shell cavity) of Pacific oysters (*C. gigas*).

This gives a value that ranges from about 40 (very thin oysters) to about 150 (very plump oysters). Multiplication by 1 000 merely converts the ratio from a fraction to values readily comparable. The assumption is that a solid, meaty oyster fills the shell cavity, while a thin one does not. Meat production, whether it be in litres or kilograms, is related to shell volume, whether it be in bushels or cubic feet.

This condition factor is a somewhat artificial ratio. It is possible to use volume of the oyster meat and this has been done. However, the water content is variable and an oyster may be plump with water which is usually lost soon after the meat is removed from the shell. The rapid change of volume when much water is present does not provide comparable results so, to evade the problem of water, the dry weight of meat after water has been removed was adopted. Consequently, the greater the dry weight of meat relative to the size of the cavity that contained it, the better the condition of the oyster and the larger the numerical ratio.

A further specific method for measuring true fatness is determination of the glycogen content of oysters. This is a different type of measure for it does not determine plumpness. As a rule it is used only for basic scientific studies. Still another measure is the percentage of solids present, which is, in reality, a measure of the moisture content, but does not refer to size.

Studies at Ladysmith Harbour over a period of years have demonstrated the seasonal cycle of condition as measured by condition factor (Fig. 110). As described above, there is little change from November to March, a sharp rise to May, a levelling off until spawning when a sharp dramatic drop occurs, followed by a slow rise to November when the winter level is again reached. Stages vary in level, as well as in timing, from year to year. In some years, the winter level may be quite high, in others, relatively low.

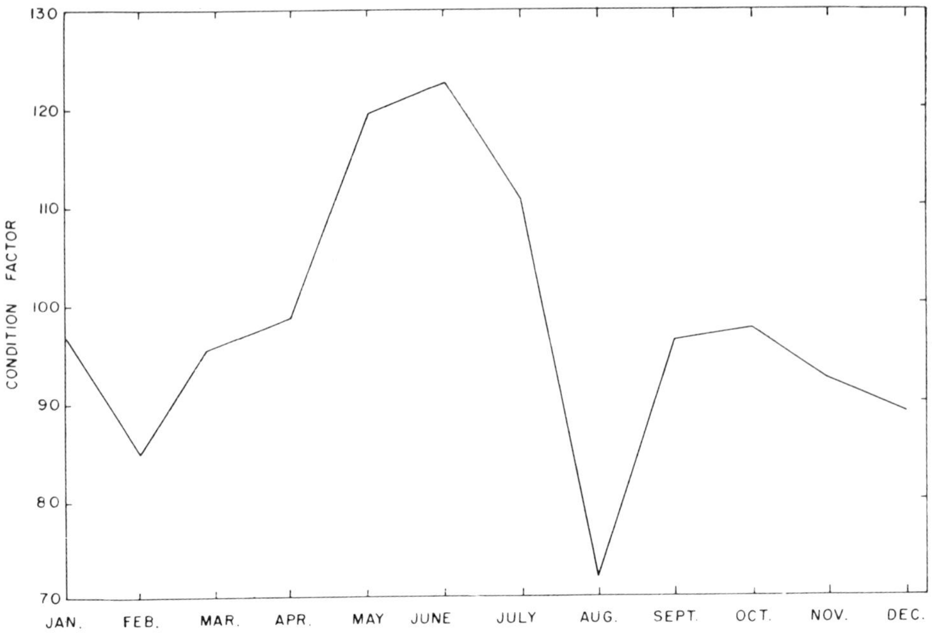

FIG. 110. Seasonal variation of the condition factor of Pacific oysters in Ladysmith Harbour, Mean of 148 samples, 1952–56.

Recovery to a reasonable level may occur rapidly after spawning or it may take several months. Just as land farmers experience good and bad growing seasons, so do oyster farmers.

TABLE 27. Average monthly condition factor of Pacific oysters grown in Ladysmith Harbour, 1952–56.

	1952	1953	1954	1955	1956	Average
January	105.7	68.5	94.6	92.0	99.1	91.9
February	99.8	—	62.7	90.9	—	84.4
March	91.4	—	77.2	93.1	90.7	88.1
April	128.2	—	73.3	—	75.1	92.2
May	160.3	98.9	109.8	118.9	91.7	115.9
June	149.1	103.4	99.2	134.4	108.4	118.9
July	156.7	92.9	114.1	140.0	73.6	115.4
August	90.4	55.8	83.4	91.1	107.2	85.5
September	96.5	—	114.1	92.0	—	100.8
October	109.5	76.3	128.0	98.2	81.9	98.7
November	101.6	78.5	96.2	70.4	91.4	87.5
December	190.0	62.2	92.8	96.0	76.7	87.3
Average	116.5	79.5	95.4	101.5	89.5	

Table 27 records data condensed from condition factor samples taken at various stations in the Ladysmith area and shows the variation from year to year, and from month to month. This picture of seasonal change is well known to oyster growers and indicates the condition factor is a fairly adequate quantitative measure of these changes.

A quantitative measure of condition permits growers to analyse accurately variations in productivity of the whole or parts of their ground, and gives an indication of whether or not they are utilizing all parts of the farm to best advantage. It gives an accurate check on whether or not variations in production are due to seasonal and annual changes in condition, or to other factors. If there are drastic changes in condition, they may be pinpointed accurately as to time, and possibly correlated with causes. It is impossible to rely on memory or on impressions in such instances. This leads to the question of what information may most usefully be recorded by the grower.

Some consistent volumetric measure of shell stock passing through the plant should be maintained: bushels, baskets, or tubs. Weight is satisfactory but, usually, volume is more easily determined. In addition, origin (i.e., part of bed) of shell stock should be recorded. Growers are already required to report production in litres or in number of dozens for half shell oysters to the Provincial Agriculture and Commercial Fisheries Branch so only the number of oysters per litre need be determined. This may be obtained from check counts of the number in a volumetric measure, or from counts of actual litre measures. If this were done on a daily basis, and it should take only a few minutes; growers would be sure to find the information most valuable. Further, at the end of the season it would be possible to calculate the actual number of oysters harvested. By comparing this with the number of seed planted, an estimate of survival and of the efficiency of the operation may be obtained.

Meat Yield

Since, in the calculation of condition factor, the whole volume of the oyster is known, it has been possible, by statistical procedures, to establish the average quantitative relationship among size (i.e., volume), condition factor, and meat weight.

Statistics are not required to show the oyster grower that there is, indeed, this relationship between oyster size and meat yield, and between meat yield and condition, for he knows it and operates his business on these principles. But the definite quantitative relationship may be of interest and for the Ladysmith area data (Table 28) this may be expressed by the equation:

$$E = 0.346X_1 + 0.112X_2 - 26$$

where E = weight; X_1 = volume; X_2 = condition factor.

TABLE 28. Condition factor data of Pacific oysters, Ladysmith area, 1952–56.

Condition factor	Total whole volume (cm)³	Weight drained oysters (g)	Average weight oysters (g)	Average volume (cm)³	Number of oysters
50–59	29 758.8	4 812.0	17.1	106.2	280
60–69	40 798.2	7 693.7	22.2	110.7	346
70–79	50 514.0	9 114.6	19.1	106.3	475
80–89	73 936.3	14 068.7	22.3	117.3	630
90–99	103 892.4	20 792.4	24.5	112.5	848
100–109	79 421.0	17 907.8	28.0	120.4	639
110–119	23 859.5	5 110.0	23.0	109.4	218
120–129	36 363.0	9 262.8	27.8	109.2	333
130–139	28 885.7	7 934.2	34.8	126.6	228
140–149	19 014.0	5 279.2	31.0	111.8	170
150–159	11 511.0	3 008.0	27.3	104.6	110
160–169	4 869.0	1 438.0	41.1	139.1	35
170–179	—	—	—	—	—
180–189	4 081.0	1 041.5	29.7	116.6	35
190–199	5 450.0	1 709.7	35.0	111.2	49
200–209	3 024.0	102.7	4.1	120.9	25
	214 samples				4 221 oysters

This means that with an increase of 1 cm^3 in volume (size) there is an average increase of 0.346 g of meat and with each unit increase in condition factor there is an average increase of 0.112 g in weight of meat.

Thus, an oyster of a given size would yield an average of 2.8 g more meat with an increase of 25 C.F. (condition factor) units in condition. In terms of the usual number of oysters in a litre this increase would be between 336 and 454 g. It can be seen that in terms of hundreds or thousands of litres this adds up to a considerable weight of oyster meat. The oyster grower, however, must continue to produce oysters and cannot avoid seasonal fluctuations in condition factor. But, it is necessary for him to examine his beds carefully to ensure that he knows and takes advantage of areas that produce oysters with the highest condition factor, not only on the average throughout the year, but at specific times.

Another way in which the importance of condition factor is shown is indicated in Fig. 111, which illustrates the relationship between condition factor and the production per bag or strings or case of seed in litres. This is based on a return of 4 000 oysters per 6 bags with 4 kg of meat per 4 L. An increase in condition factor of 25 points increases production per 6 bags or strings by nearly 14 L. The vertical position of the line on the graph is determined by the size (volume) of the oyster which, in this case, was 120 cm^3, roughly equivalent to an oyster 120 cm in length. This measurement is approximate, for the relationship between length and volume is, of course, not constant, for both width and thickness affect volume. It might be expected that larger oysters would show a relatively greater increase in meat weight with increase in condition factor, and the data do tend to show this when individual regressions are calculated for each size group. The maximum increase was calculated to be about 18 L (4.0 gallons) for each 25-point increase in condition factor for oysters in the 150–159 cm^3 volume group.

It may be realized, the condition of oysters is of great importance in terms of production, particularly with the Pacific oyster whose value is in terms of meat yield as compared with some other oyster industries or markets where the number of individual oysters is the criterion of production.

In addition, the appearance of an oyster, which is related to its condition, is an important factor for it is critical to consumer acceptance. Oysters in prime condition have a good appearance with a smooth, cream-coloured, rounded body and with thick, crinkly, creamy mantles. Thin, transparent, flabby mantles, or a body blotched or dark in colour or with the dark digestive gland showing, indicates oysters in poor condition. Although the grower soon learns to recognize large differences in condition, small differences are difficult to estimate on quantitative basis. It is most important to have adequately written records for memories are short and subjective judgements not always reliable. Records of some

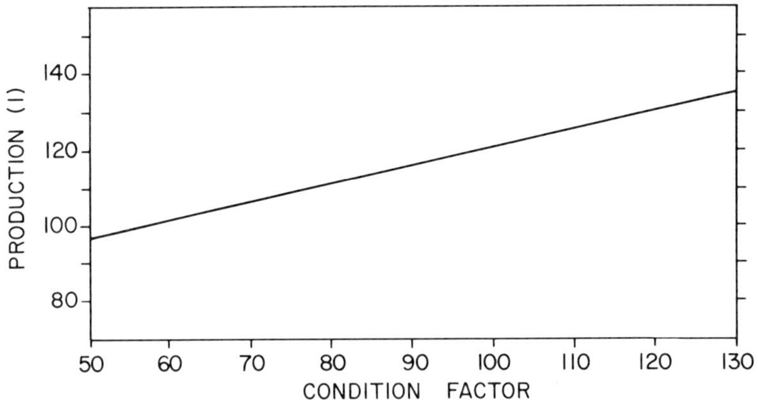

FIG. 111. Relationship between condition factor and oyster production in gallons per case of seed, for oysters approximately 12 cm long (120-cc volume), Ladysmith Harbour.

consistent quantitative measurement of condition should be maintained along the lines suggested above.

If the effect of interaction of mortality, growth rate, and condition as described are properly understood and adequate records maintained, the effect of each factor may be estimated, and the grower will be in an excellent position to make adjustments to his methods of culture to get the best out of his oyster ground and from the seed he plants on it.

It cannot be emphasized enough that oyster production is the result of what happens on the oyster beds and not in the shucking houses. It is on the oyster beds that effort and study will pay greatest dividends.

Harvesting

Hand Picking

There are many ways to harvest oysters and the method chosen depends on the type of ground, size of the operation, and availability of labour.

In British Columbia the most widely used method is picking by hand. Oysters may be picked into baskets of various sizes, directly onto a scow that has been spotted near the centre of harvesting operations, or into a wheelbarrow (Fig. 112). Hand picking allows a certain amount of selection if the oysters are not all of market size. Specially manufactured galvanized-wire baskets are available which are light to carry, the wide mesh permitting mud to be washed off the oysters easily. Rubber-covered-wire egg baskets are also suitable and available from poultry dealers. Loaded baskets are carried to a scow, raft, or boat which has been anchored near the harvesting area at high tide. Sometimes the baskets used are too large to be lifted by hand; in this case, they are buoyed and, on the next high tide, a boat or scow with a derrick lifts the baskets on board. If the ground is firm a fork-lift truck may be used. Small farm tractors pulling carts have also been used.

Many growers use rakes to pile the oysters into windrows. Forks are then used to transfer them from the windrows to baskets, sacks, wheelbarrows, or directly onto floating equipment spotted nearby. When a new area is to be harvested it is necessary to previously clear space of oysters for placing the floating equipment if the oysters are to be transported. The space is usually marked at low tide by means of stakes.

These are the picking methods that are used for low-tide operations.

FIG. 112. Harvesting wild oysters. This method is also widely used for harvesting cultivated oysters.

Tonging

The high tide methods are mainly mechanical but one, seldom used in British Columbia, is tonging. Oyster tongs (Fig. 113) are made up of a pair of flat poles hinged about three quarters of the distance along from the handles. At the working end of each pole is a half-basket arranged so that when the handles are drawn together the two half-baskets join to make an enclosure. The bottom edges of the half-baskets are studded with teeth. To operate, the tongs are lowered to the bottom and spread. The handles are then drawn together with an alternate lifting and drawing motion. In this way oysters between the jaws are lifted out of the mud and either rolled ahead of the rakes or rolled into the half-basket. When the handles are together again the tongs are drawn to the surface and emptied. Tongs might be used on occasions when tidal conditions make it difficult or impossible to harvest by picking.

Drag Dredging

Standard high-tide harvesting equipment is the drag dredge as shown in Fig. 114. The size, shape, and construction of the dredge will vary according to preference, size of operation, and type of bottom. The vessel carrying the dredge equipment may be self-propelled or towed, but it must have adequate power to operate the winches and booms.

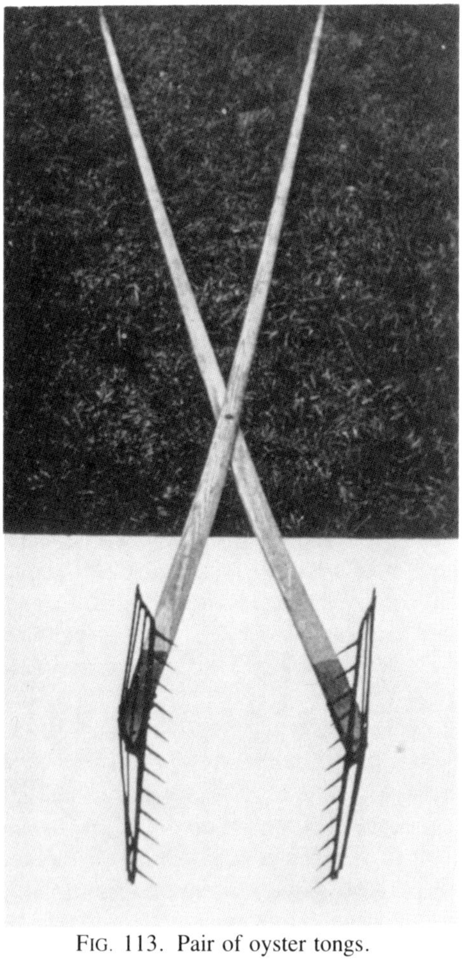

FIG. 113. Pair of oyster tongs.

FIG. 114. Drag dredge for harvesting oysters.

Hydraulic power is widely used. Drag dredges, as the name implies, are dragged over the bottom and the mouths are so fitted that the oysters are scooped into the dredge bag, which may be of rope or chain mesh. At the end of the tow, the dredge is hoisted from the water, swung over the deck, and the bottom of the bag is opened by tripping or lifting a locking bar. Either a single dredge, or two, one on each side, may be operated from a single vessel.

The drag dredge is not a highly efficient harvester for tests have shown that at best, it is able to collect little more than 25% of the oysters in its path. Further, there is a tendency for it to break oysters rather badly under some circumstances. On soft ground, it is often advisable to use a harrow or raking device to lift the oysters out of the mud before using the dredge. For most family-type operations the dredge is not an economical method of harvesting.

Hydraulic Dredging

For large harvesting operations, the Bailey dredge was developed. This is based on a hydraulic principle by which water pressure is used to create a flowing mixture of sand and oysters under a hood which rests on and makes a seal with the bottom. The moving oysteres are deposited on a conveyor belt which carries them to the surface, leaving sand and small particles behind. The Bailey dredge is anchored, and is towed about its anchor in ever widening circles by a tugboat. There is very little effect on the bottom, and the oysters are generally quite clean, but there is a tendency at times for the meats to be badly sanded. This is quite an efficient machine, capable of handing a large volume of oysters but only economical on this account.

A more recent development for oyster operation is the modification — by the Atlantic Biological Station — of the Chesapeake Bay escalator-type clam harvester (Fig. 115). The principle employed is also hydraulic and, in this case, water jets lift the oysters from the bottom directly onto a conveyor belt which carries them to the surface. The escalator is usually about 8 m (25 ft) in length and can operate in depths up to 2 m (6 ft). This is a very efficient gear and does no damage to the bottom. Three of these machines have been built in British Columbia. The cost would probably be uneconomical for the family-sized operation but several adjacent growers could profitably use a machine on a cooperative basis.

Transportation

There are no hard and fast rules governing the type of vessel used to transport oysters either from the bed to the shucking house or from one part of a bed to another. In British Columbia these vessels range from small skiffs to 60 t capacity scows. The size of vessel normally depends on extent of the operation. An average small grower uses a sink float, top float, or small scow. The latter is the handiest, for it can be easily poled, towed, or powered with an outboard motor; it has good carrying capacity and draws little water, which is an important advantage. Log-top floats are fairly expensive to build and are also costly in term of the effort needed to move them. Moreover, they do not last as long as

FIG. 115. Escalator-type oyster harvester.

lightly built plywood scows. Aluminum gillnet herring skiffs are now used in many oyster operations. Trucks or tractors are used on firm ground. Where wet storage is possible, sink floats (log float with the floor attached to the bottom of the logs so it is submerged) are most useful. However, fiberglass pontoons instead of logs may, in the long run, be less expensive and more versatile.

Gear

Normal hand operations require D-handled grain shovels (usually aluminum), garden rakes, and vegetable scoop forks. These will vary according to harvesting needs and personal preference.

Many methods of harvesting oysters have been tried and it is likely that as long as oysters are harvested new gear will be developed. Each grower must adapt to the needs of his own operation.

Raft (Suspended) Culture

Because of the precipitous nature of most of the coastline of British Columbia, there is relatively little intertidal oyster ground suitable for bottom culture. In the Strait of Georgia practically all of it is already under lease although there is a certain amount of ground on the north coast and west coast of Vancouver Island. Further expansion of the Pacific oyster industry, other than through more extensive and efficient use of ground, is possible only by alternative culture techniques.

One technique is the method of growing oysters on rafts, called either "raft", "floating" or "suspended" culture. The term "raft" will be used to denote suspended culture. This technique is employed in oyster-growing countries such as Japan, where in recent times there has been almost a full conversion to this method from bottom culture. Seed or young oysters are strung on rope or wire and suspended from rafts or long lines of various types where they are grown to maturity.

In British Columbia suspended culture lends itself admirably not only to professional oyster growers, but also to trollers or gillnet fishermen who would be provided with an occupation and an additional source of revenue during the off-season. The fisherman has the basic knowledge and gear to conduct such an operation which is compatible with fishing schedules, as preparations are made and strings placed on the raft in March before the fishing season starts. If the anchor system is sound, the rafts need no further care until the oysters are harvested in November or later.

The advantages of suspended culture are many. Compared to intertidal oyster ground, in British Columbia the area suitable for this type of culture is large, for there are hundreds of well-protected bays and inlets along the coast. Productivity per unit area may be 10 times that on the bottom, and in Japan, with the same species of oyster, it is said to exceed 750 t/ha. The time required to produce oysters of market size is reduced by at least 1 year or more and the quality is considered to be superior to those grown on the bottom. An important factor is that most potential suspended culture areas are well removed from domestic or industrial pollution. Also there is no significant predator problem.

Procedures

The basic culture method consists of stringing pieces of shell, on which have been spatted a number of young oysters, on lengths of wire or rope. Shells with young oysters attached should be 20–30 cm apart to allow unhindered growth on all sides (Fig. 116). The strings are suspended from some form of floating structure, such as a raft, or longline in water deep enough to prevent the strings touching bottom. Areas chosen should be reasonably well protected from wave action so the raft will not be subject to excessive

FIG. 116. (*above*) Section of raft culture strings. ×0.2. *Left* — before placing on raft in February 1967. *Right* — growth to October 1967.

motion. Under ideal conditions seed from the previous summer, placed on the raft in March, may produce the relatively large market-sized oysters by the following November, although this is seldom possible. Spatted sticks or PVC tubes may also be used in suspended culture (page 174).

Rafts

Although rafts (Fig. 117) are the basic means of supporting culture strings, there are others. The Japanese use what is termed a ''longline'' method, where strings are suspended from heavier lines supported by barrels and held apart by anchors. This method can apparently withstand somewhat greater wave action than standard rafts. It is also used successfully in Norway for supporting oyster trays, where lines are suspended from shore to shore in small coves, making submerged anchors unnecessary. This has been used successfully in one instance in British Columbia.

Another suitable method is termed ''boom stick'' where single logs are tethered end-to-end with anchors, pilings, or shore fastenings at each end (Fig. 118). The shell strings are simply draped over the logs, which should be relatively stable floaters that are unlikely to overturn. However, personal ingenuity may be allowed full play. Other floatation methods may consist of the basic log raft with cross-beams, oil drums, inverted lightweight concrete boxes that are supported by a cushion of air, fiberglassed or creosoted plywood pontoons, styrofoam logs, or metal drainpipes filled with extruded styrofoam.

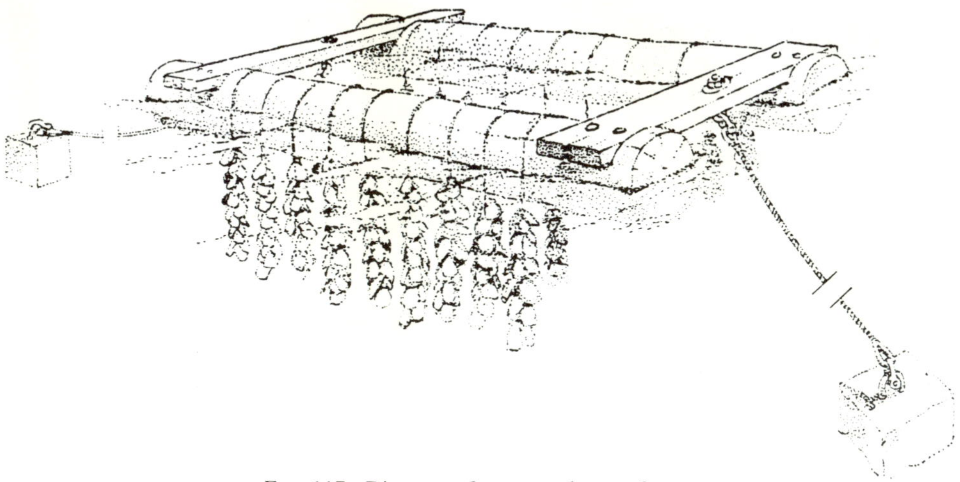

FIG. 117. Diagram of oyster culture raft.

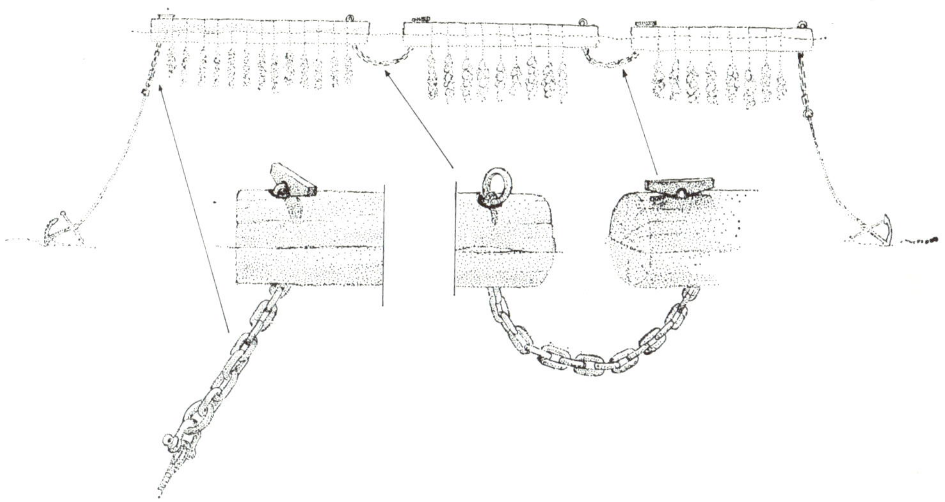

FIG. 118. Diagram of "boom stick" method.

Experiments with ferro cement logs (12.0 × 0.6 × 0.6 m) now with 15 years of use, have shown the practicality of this system (Fig. 119). Styrofoam logs are covered with a 3 cm thick layer of ferro cement with appropriate grooves and holes for cross-member fastenings. This method of floatation is particularly useful in exposed locations, for the waves wash over the raft instead of lifting it, thus preventing excessive action on the oyster strings.

The log raft is the simplest form of floatation. Discarded boom-sticks may be purchased at relatively low cost and these may last for several years. In compliance with log salvage regulations and for a reasonable price, it is possible to obtain beached logs that are unsuitable either as saw or pulp logs, but adequate for rafts.

Cross-members for such rafts may be of any type of material, even small logs. If the cross-members are long enough, more than two logs may be incorporated into a single raft as long as the distance between inside edges of the logs is not less than 1 metre. In addition to draping the strings over the floatation logs, they may also be placed on a series of cross-members placed across the logs (Fig. 120).

FIG. 119. Cement float logs.

FIG. 120. Raft with cross-members suspension.

If possible, the rafts, whatever the style, should be removed from the water or stored as high as possible on the beach as soon as the oysters are harvested. This permits a certain amount of drying to maintain buoyancy and also destroys marine borers and other fouling organisms.

Long Lines

Although log rafts are a basic means of floatation and have the advantage of serving as a working platform, the increasing cost of logs has caused greater use of long lines (Fig. 121). The ability of long lines to withstand greater exposure to wave action is a definite advantage and they normally do not need quite the anchoring system required for rafts. However, an important feature of suspended culture is the jerking action of the floating system, whether it be rafts or long lines because of the disturbing effect on the oysters as well as on the suspension, ropes or wires, particularly the latter. There are several types of wave shock absorbers such as horizontal plates (Fig. 122) or rubber snubs such as sections of inner tire tubes. Weight on each anchor line will stabilize the line as well as assisting in keeping it taut (Fig. 123).

Size of the longline components depend on the length, amount of wave action as well as potential weight of the culture apparatus such as trays, strings or sticks. The main line should be not less than 15 mm in diameter and for long life should be synthetic rather than manila which has a relatively short life.

Longline Floats

There is available a wide assortment of floats, ranging from short sections of a cedar fir log to plastic types specially designed for long lines. Oil barrels, used plastic shipping containers, styrofoam or urethane filled automobile tires, and styrofoam log sections are a few of the many alternatives.

The supporting power of a float is roughly equivalent to the weight of water displaced i.e., 28 cubic decimetre (1 cubic foot) weighs 28 kg (62.5 pounds). The weight in water of most oyster culture items such as shell strings is approximately one tenth of the weight in air. Only experience will determine the floating power required for a specific operation, but it should be designed in a way that will permit floats to be added as needed when the weight increases from oyster growth and fouling.

Long lines can be constructed as single lines or as doubles with a line attached to each end of the float (Fig. 124). It is also possible to construct long lines in the form of a grid.

A simple form of a long line is where each individual item such as a shell string or tray has its own floats to which it is directly attached and with the floats attached to each other (Fig. 121, 123). A further plan is to have drop lines from the floats so the main line is placed at whatever depth desired below the surface (Fig. 125). To avoid having numerous surface floats, the long line with its floats may be sunk below the surface by weights (Fig. 126). With this plan the line may be placed below the level where wave action is a factor or if it is necessary to avoid barnacle or mussel settlement.

The rope on all metal to metal contacts should be thimbled and shackle pins should be wired for constant tugging causes the pin to turn and be lost. Too much care cannot be taken to ensure the safety of marine installations. What is considered adequate in terms of size, strength and fail-safe features should be at least doubled.

Anchors

It is not possible to lay down specific rules for anchoring. The 13 m × 3 m log rafts with 50 double shell strings used in a coast-wide experiment were held successfully in a variety of wind and wave situations by two 28 cubic decimetres [1 cubic foot cement blocks (65 kg)]. It is preferable to have a chain attached to the anchor to act as a shock absorber on the anchor line which should be approximately three times the depth of water.

FIG. 121. Surface long line with styrofoam floats each suspending a unit of shell cultch. Two types.

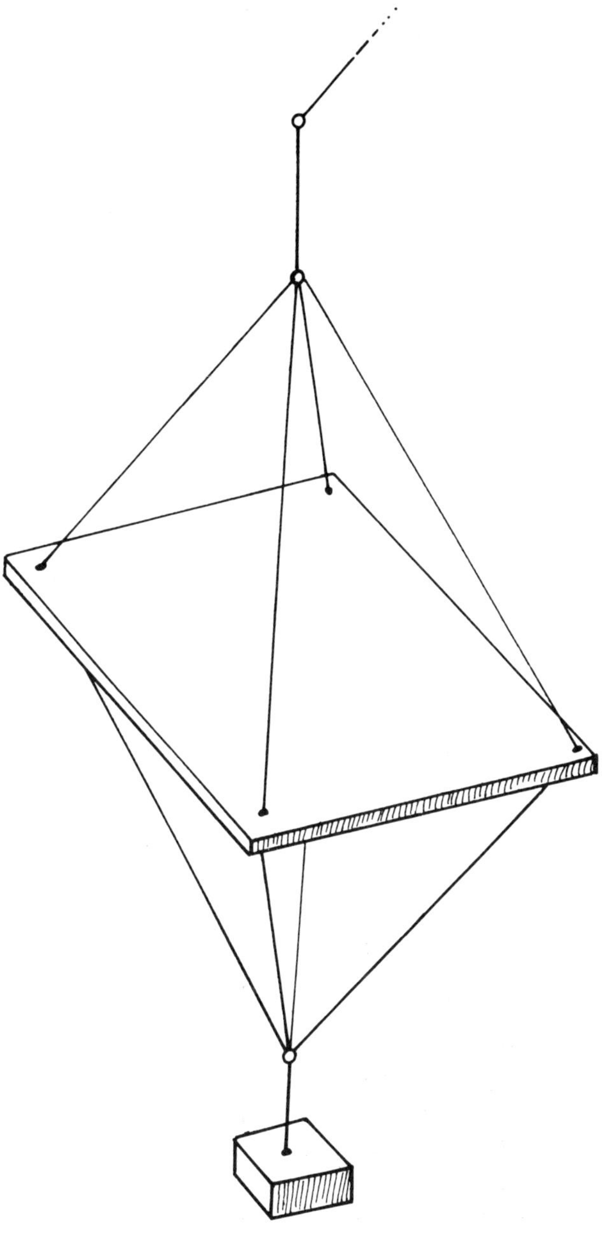

FIG. 122. Wave action damper of plywood. (× 1/9).

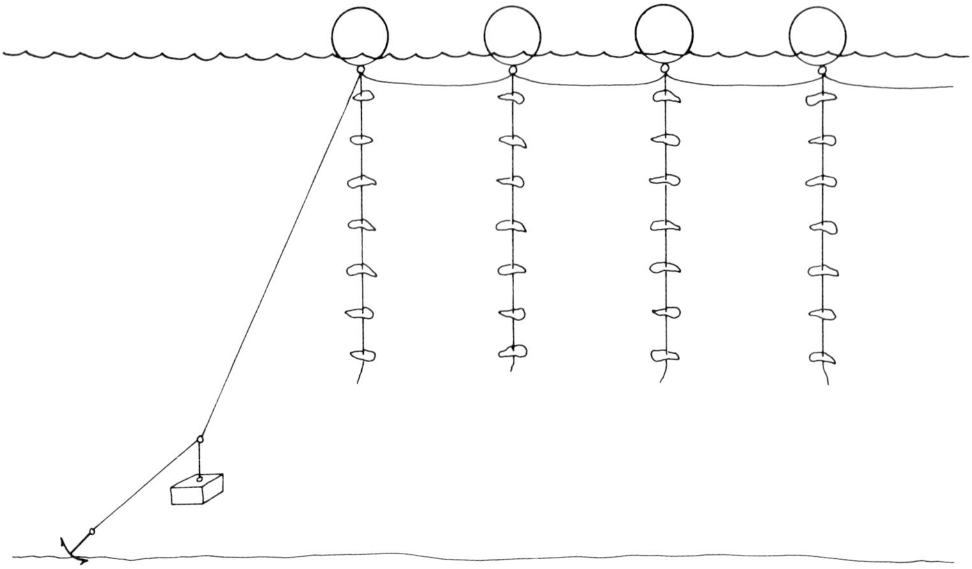

FIG. 123. Single float surface long line with weight for keeping the line taut and damping wave action.

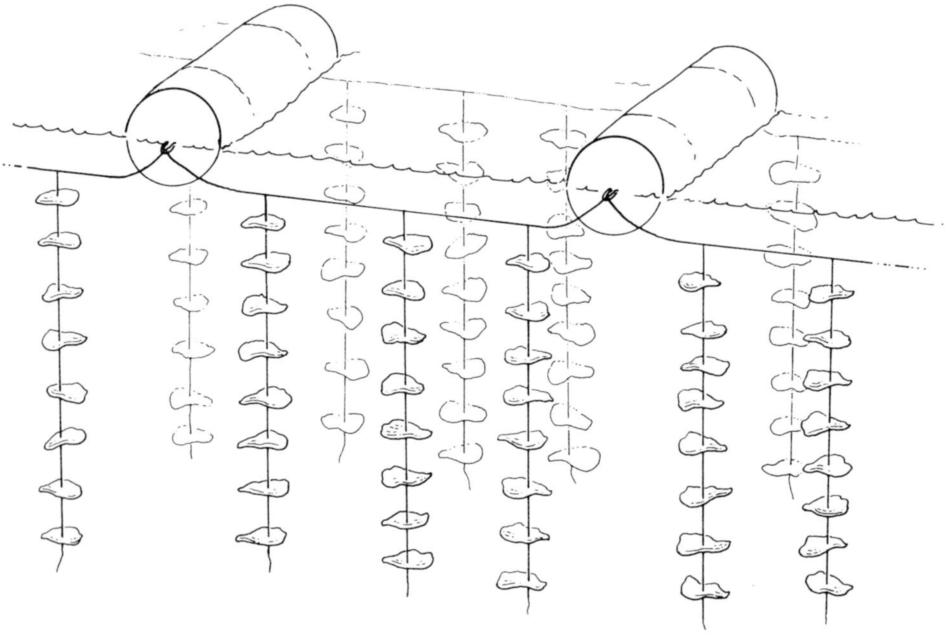

FIG. 124. Double long line. Surface type.

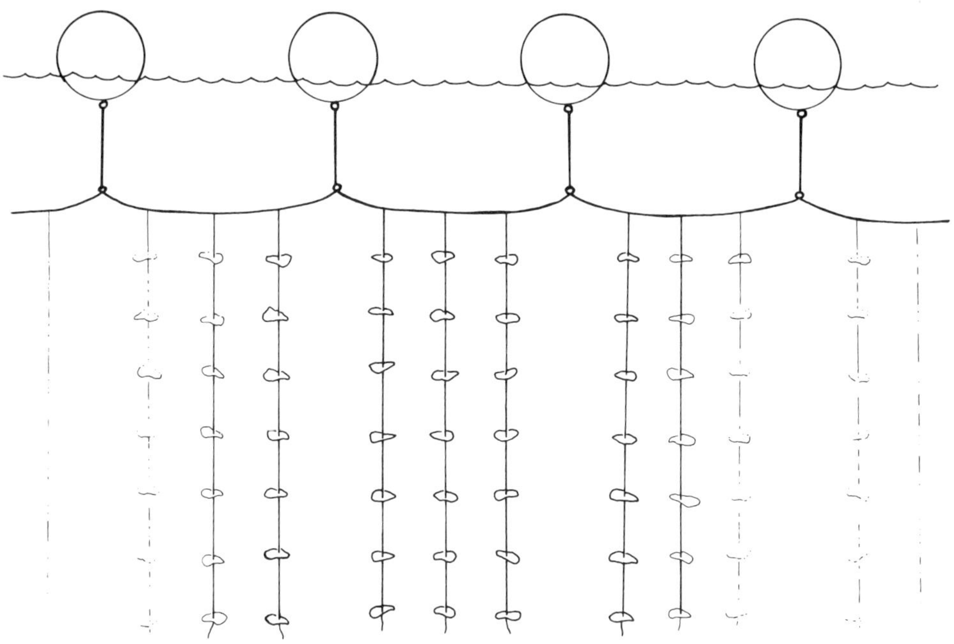

FIG. 125. Sub-surface long line.

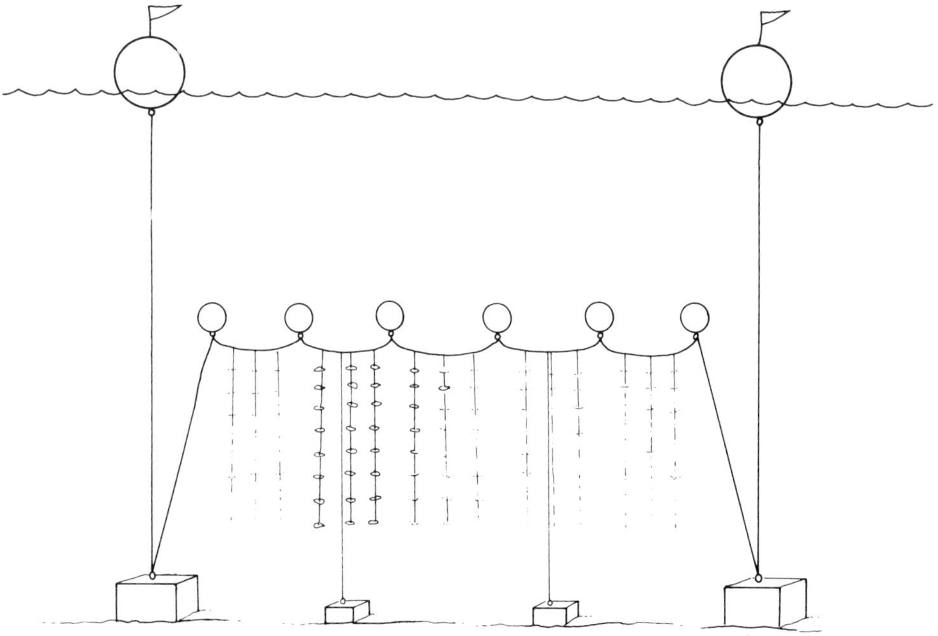

FIG. 126. Sunken long line.

The long line should be anchored at the time of low water slack to provide maximum strain. In setting out the anchors, one end is set first. Through the bolt on the other anchor a double tow line is reaved after the floated line has been stretched out. This anchor is lowered from the tow boat with the tow line to a point above or just touching bottom. It is then towed as far as possible from the other anchor for maximum extension of the longline and other anchor line before it is dropped to the bottom. The tow line can be drawn back through the anchor bolt.

The longline anchors may be deployed as singles from each end of the line. Double anchors of lesser weight may be placed at the ends, each one with its own line. Alternatively double anchors may be bridled together with a single anchor line (Fig. 127). Depending on length of the long line and current strength, lateral anchors may be necessary to maintain position. It is advisable to maintain a floated retrieval line from each anchor to allow replacement if and when the long line becomes too slack. There are many types of manufactured anchors. However, cement blocks, old tractors treads, disused engine blocks and similar materials are less costly and just as effective.

Long lines are most easily made up on shore.

String Culture

Local shell, because of its relatively large size and solidity, provides an ideal cultch for suspended culture. In most years local seed does not reach more than 1 cm in diameter by time growth ceases in the fall of the settlement year, and this is too small for it to reach a marketable size in one year of suspension. This means an extra year of occupation of valuable floation space, when growth is somewhat reduced and fouling becomes a significant factor. To confine floatation time to one summer it is necessary to hold the seed on intertidal ground for a year to allow it to reach a length of 25 to 40 mm. This is less costly than floatation. Further, holding seed intertidally permits the process of hardening to take place.

In Japan and Korea, where the Pacific oyster is grown almost entirely on rafts or long lines, it has been found, both through practical observation and specific research that there are significant benefits from a process called hardening. This was the process of holding seed on intertidal racks for stunting to prepare it for trans-Pacific shipment to North America. The exposure caused thickening and consolidation of the shell edges and presumably adductor muscle adjustment enabling the seed to withstand long periods of exposure. Seed hardened before suspension showed better survival than untreated seed placed directly in suspension.

There is also evidence that hardened suspended oysters recover and fatten after spawning more rapidly than unhardened seed and this difference may be the cause of mass mortalities in the latter. These are good reasons, apart for obtaining a crop from only one year of floatation, for one year of growth on the beach.

In some years, when the spatfall in Pendrell Sound occurs early in July, the set may attain a length of 3–4 cm by the time growth ceases in November. In this case seed may be used the following March and the oysters harvested at an age of 16 months in contrast to 40 months for bottom culture. If the Pendrell spatfall occurs in August, the size attained by November is insufficient to warrant using it the following March for raft culture and must be held on the beach for a growing season. The total age at harvesting will then be 30 months with two summers of growth instead of three, about a year earlier than possible with bottom culture. This is the normal situation.

The British Columbia industry should seriously consider marketing oysters at a size much smaller than has been the practice. If this occurs, whatever the size, seed may be placed on the raft in the year succeeding collection and will produce a large proportion of these smaller, but market-sized oysters at the end of that year.

Holding seed for 1 year on shore will probably be the most frequent practice to produce large-sized oysters, as shown in the oyster culture flow diagram (Table 29). It may be planted at a relatively high tidal level (up to 2 m above zero), to prevent it growing too

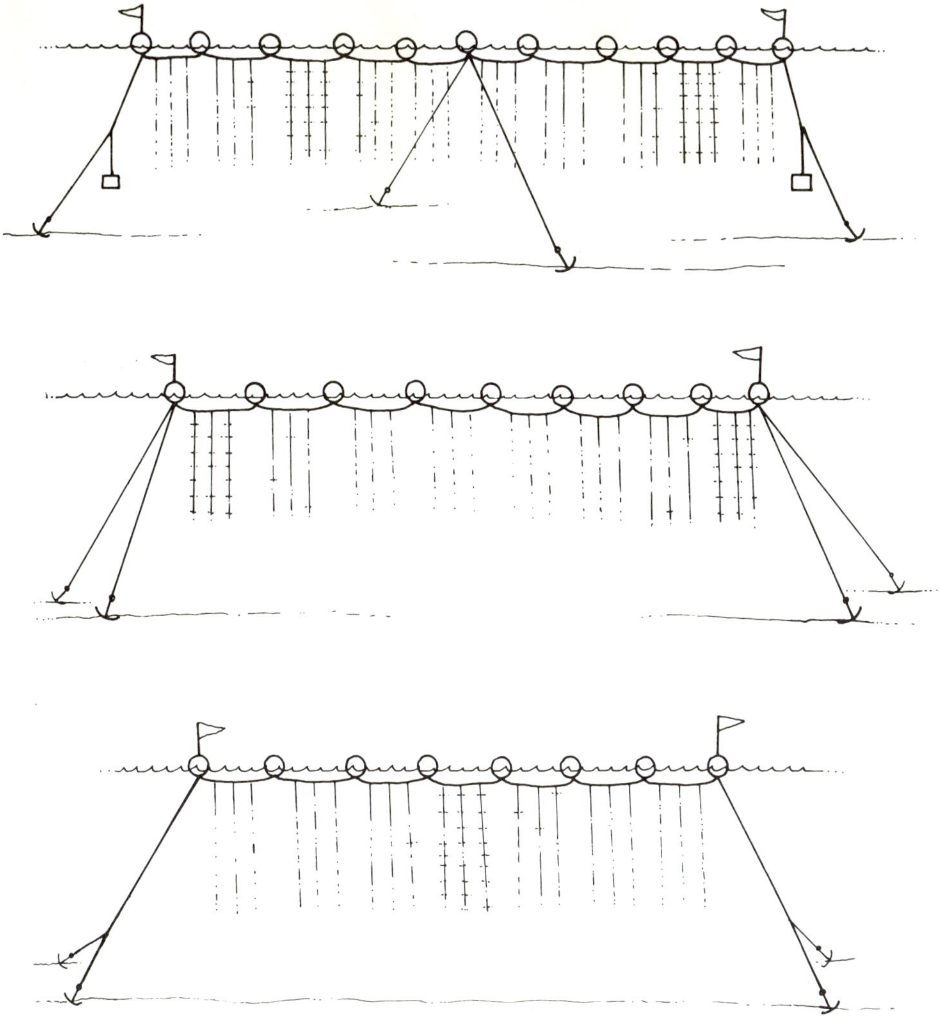

FIG. 127. Long line anchoring systems.

large. Also, it may be planted at a high density, up to the equivalent of hundreds of bags per hectare. Thus, only a small amount of what may be termed marginal oyster ground is required for holding seed, but this should be quite firm to prevent young oysters being killed by silting.

TABLE 29. Oyster culture flow diagram.

	Raft culture	
Bottom culture	A	B
Pendrell (year 1) Seed bed (Oct.-yr 1) Growing bed (yr 2) Fattening bed (yr 3) Harvest (Nov.-yr 4)	Pendrell (year 1) Seed bed (Oct.-yr 1) Raft (Mar.-yr 2) Harvest (Nov.-yr 2)	Pendrell (year 1) Seed bed (Oct.-yr 1) Raft (Mar.-yr 3) Harvest (Nov.-yr 3)

Of some importance is the number of spat per shell that will produce optimum returns. In Table 30 data are given from a raft culture experiment, in which strings held shells with varying numbers of small spat about 1.5 cm in diameter. Although the percentage survival decreased with greater original numbers of spat per shell, the absolute number of survivors and consequently weight and volume of oysters produced improved with increased original numbers of spat per shell. In Table 31 data are given from an experiment where the spat were larger than those described in Table 30, having been held on the beach for a summer's growth. In this case the drop in survival rate is not as pronounced, for a large part of the early spat mortality had already taken place. Early mortality of oyster seed held on the bottom in British Columbia is due largely to silting and to competition for space on the shell. In the first experiment (Table 30), virtually all mortality was due to competition for space, since only a limited number of oysters can live in a given area.

Extremely high spat count seed is of no more value than seed with moderate counts (Table 32). Survival of a maximum of 15–20 mature oysters per shell from Pendrell Sound seed may be expected and this is obtained from shells (as placed on the raft) with not less than 20–25 spat per shell. Although successful ground culture may be conducted with spat counts as low as five per shell, it may be uneconomic for raft culture. Because of the cost of rafts, it is necessary to obtain the maximum annual return possible. With small numbers of spat per shell, growth rate is somewhat higher but not significantly so to warrant its use.

TABLE 30. Floating culture production and survival of oysters from cultch with various intensity of set. Ladysmith Harbour, December 1, 1952 to November 9, 1953.

String no.	Original spatfall per shell	Original size spat (diameter in mm)	Total survival	Survival per shell	% survival	Weight (kg)	Volume (litres)
1	9.8	5–15	83	6.0	61.3	8	12
2	14.0	—	123	7.7	55.0	10	—
3	14.0	—	156	10.4	74.3	11	18
4	14.0	—	82	5.4	38.6	9	12
5	14.0	—	92	5.4	38.6	7	—
6	19.0	3–9	69	5.7	30.0	6	9
7	32.0	12–25	251	16.7	50.0	22	36
8	33.0	—	245	17.5	53.0	19	36
9	33.0	—	211	14.0	42.4	17	24
10	35.0	—	206	16.0	45.7	12	18
11	35.0	—	138	10.0	28.6	9	15
12	36.0	3–6	165	10.8	30.0	10	18
13	39.0	—	195	15.0	38.4	18	36
14	44.0	—	154	11.0	25.0	12	18
15	50.0	—	347	23.1	46.2	25	43
16	65.0	—	291	19.4	30.0	20	36
17	83.0	—	365	24.3	30.0	29	45
18	87.0	—	282	18.8	21.6	26	40
19	87.0	—	359	18.9	21.7	30	45
20	87.0	—	257	18.4	21.2	18	36

String No. 7 — Mean length 11 cm
Mean width 6 cm
50% of oysters shucked out at 38/L
Condition factor 130.2
Condition factor of adjacent shore oysters 71.6

String No. 8 — Mean length 10 cm
Mean width 5 cm
50% of oysters shucked out at 40/L

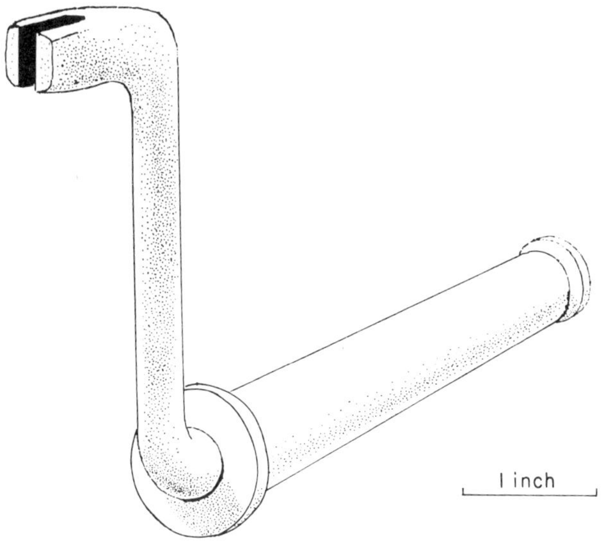

FIG. 129. Wire twisting tool.

FIG. 130. Twist in the wire to separate clusters.

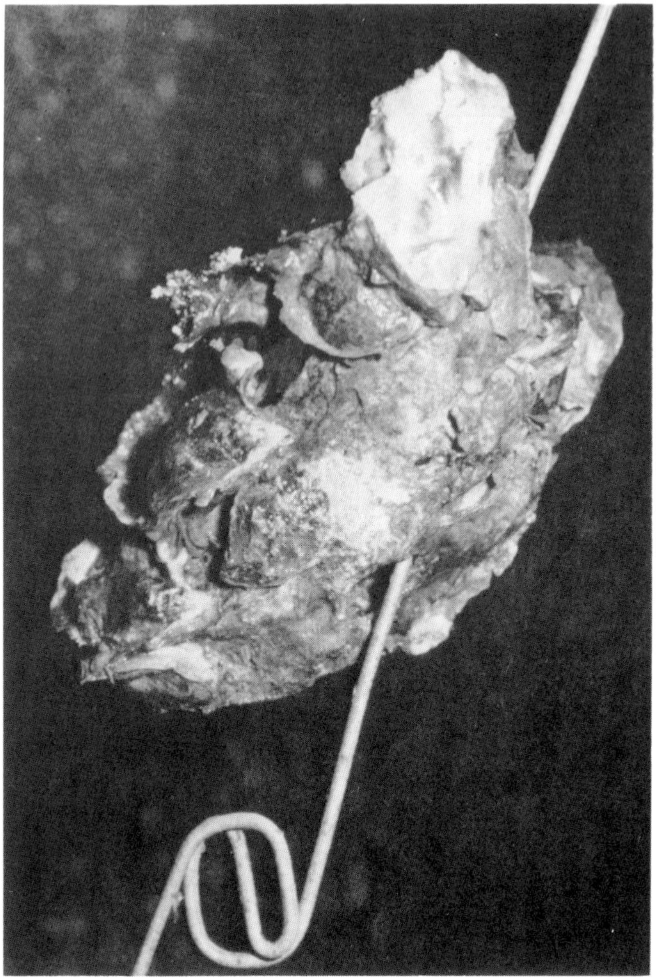

FIG. 131. Example of shell cultch.

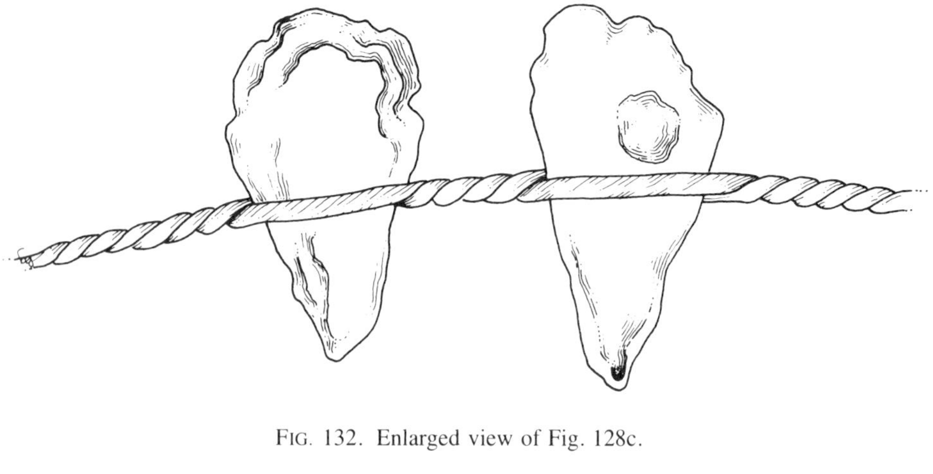

FIG. 132. Enlarged view of Fig. 128c.

Shell can be punched before packaging in the plastic bag containers so the wire system of stringing may still be pursued.

Harvesting Methods

Two methods of harvesting raft culture oysters are suggested, but others may be developed. Moveable rafts may be placed over an intertidal area at high water and the strings allowed to drop to the bottom where they may be picked up at low tide. The other method is to erect an A-frame on a top float or scow (Fig. 133), which is moored alongside the raft to be harvested. A spreader bar is placed under the wire of the double string to keep the two parts away from the logs so oysters will not be scraped off clusters as the string is lifted. The A-frame is swung over the side, and the string lifted by the spreader bar with a small winch. When the strings are free of the water the A-frame is swung aboard and the strings dropped to the deck.

Harvesting oysters from longlines, whether they be strings or trays, is best done from a catamaran type vessel or a log raft which can straddle the line, although standard vessels or scows are satisfactory. Powered line pullers such as used by crab and oyster fisherman

FIG. 133. A-Frame for harvesting.

are available. A swinging davit is useful and its arrangements depends upon what is being lifted. The height of the davit or boom can be reduced by lifting very long shell strings in sections with a snubbing device for holding the lifted section in place. An apparatus called a star wheel is also available. This is attached to the gunnel of the harvesting vessel and holds the long line without interfering with its attached lines.

Cluster Removal

If strings have been made up with spacers, the clusters will slide off the wire or rope by removal of the bottom one. This method is labour saving, particularly suitable for large-scale operations, but must be weighed against the labour of stringing with spacers and their cost. With the two strand rope system, the rope must be cut between each cluster or the oysters removed from the mother shell with the rope intact.

With the wire twist system the wire between the clusters must be clipped. An excellent time to wash the clusters is immediately after harvesting, while any adhering silt is still soft and loose. Since growth virtually ceases by the end of November, harvesting may be done at any time thereafter. Strings may be held on the raft throughout winter as there will be little further fouling. The sooner oysters are harvested the earlier the raft or long line may be removed from the water to dry out and be protected from winter storms.

After removal from strings the clusters are ready to be separated into single oysters. This should be done before they are sent to the shucking table, to avoid reducing the efficiency of the shucker. If separation is done just before shucking, the oysters with broken shells will still be useable, as the meats will not have had time to dry out.

Most separation may be done by hand simply by pulling the oysters apart. Some may require force with a flat breaking iron. If the cultch shell is not covered by oyster growth, a sharp blow with the breaking iron will shatter it and allow easy separation of the oysters. It will be satisfactory for the shuckers if the clusters can be reduced to not more than three or four oysters.

Because of the soft, thin shell of raft cultured oysters resulting from rapid growth, the jabbing technique of opening will be most effective. The oyster knife, suitably ground and sharpened, is driven into the centre of the upper valve about an inch from the bill end and the oyster opened from there. In the side-opening method normally used in British Columbia, the soft shell is often broken in the region of the adductor muscle before it can be cut and this makes the remaining operation awkward and slow. Initially, shuckers may dislike opening raft culture oysters, but it takes only a short time to learn a modified technique.

Replants

On separation of the clusters, varying numbers, up to 25% of the oysters, may be less than the desired market size. At present the only way in which these may be utilized other than for the half-shell trade, is to place them on a growing bed for another year to attain more growth. They may be sold as shell stock (oysters in the shell) to growers with growing ground, if the raft oyster culturist has none himself. The market size of Pacific oysters is determined by the number of shucked oysters required to make a litre and this varied between 25 and 35 depending on the particular market being supplied. It may be that raft culture will promote marketing oysters of a smaller size, which might be more acceptable to the customer than the larger ones generally sold at present. It is also possible that raft culture oysters may fetch a better price, but this remains to be seen.

Fouling

Fouling of strings by marine organisms is a major drawback to raft culture in many parts of the world for it can inhibit growth of the oyster and may even cause mortality.

Unfortunately, there are few ways to combat fouling. In British Columbia, the problem has not been too serious in most areas. The extent and kind of fouling varies considerably according to area and to the length of time strings are held on the raft. Fouling is also dealt with on page 99.

The chief fouling organisms in Ladysmith Harbour are barnacles (*Balanus glandula*), hydroids (*Obelia* sp.), and mussels (*Mytilus edulis*). There are also a number of other fouling organisms but so small in size or volume that they are not a problem. Hydroids and mussels are loosely attached and may be readily removed, as can barnacles if the set has been heavy. The oysters are harvested in November or later when many barnacles are dead, or the attachment is so feeble that most will be knocked off in handling and in removal of the hydroids and mussels.

In most of the Strait of Georgia, the major barnacle set occurs quite regularly within about 10 days before or after April 1. If shell strings are exposed after this time the heavy barnacle set will be missed but several weeks of oyster growth will also be lost. If strings are placed on the raft in early March, the initial spring growth of the oysters is obtained and the barnacle set tends to inhibit to a considerable extent other fouling organisms, while not interfering greatly with growth of the oysters. This, however, depends on the density of the barnacle set. In some areas it has been dense enough to overpower the ability of the oysters to complete, and it would be necessary in such cases to delay exposure until after the barnacle set.

That fouling may be a problem is evident from Table 33. In some areas the organisms may amount to 50% of the weight of oysters produced and where oysters have been held on the rafts through two summers in Ladysmith Harbour, fouling, particularly by the anemone (*Metridium senile*) becomes fairly extensive. This, together with possible limited additional growth attained in the second summer, as well as the probability of sea star predation, may prohibit using a raft for two summers to produce one crop in the Strait of Georgia. However, this is generally necessary in the more northern areas of the coast where growth is slower and where, in most bays studied, the fouling in a 2-year period is no more than in the Strait of Georgia.

The 1967–68 experiments showed barnacles and mussels were the chief fouling organism in respect to volume and frequency of occurrence, although each location had its own characteristic group of organisms. In Hecate Bay the white tube worms (*Serpula*) were most apparent (Fig. 96), and here too there were more sea urchins than anywhere else, all three species that occur in British Columbia waters were present. Ethelda Bay

TABLE 33. Weight and volume of fouling organisms from 40 clusters of oysters at varying floating culture stations.

	Fouling	
	Weight (lb)	Volume in standardized-sized pails
Fatty Basin	216	10
Quait Bay	227	12
Boca del Infierno	104	6
McKay Cove	51	4
Queen's Cove	65	4
Hecate Bay	41	2
Echo Bay	123	6
Pruth Bay	50	3
Ethelda Bay	51	3

was unique owing to the occurrence of fairly large numbers of the relatively rare gastropods *Velutina* and *Lamellaria*. Purple-hinged rock scallops (*Crassadoma*) were also abundant. Quait Bay and Annette Creek were notable for dense barnacle sets and Fatty Basin for heavy mussel fouling. At Boca del Infierno there was moderate fouling while at Kyuquot-McKay Cove little fouling was experienced. Quantities of fouling organisms are shown for the 1968 experiments in Table 33.

If a raft is left too long (3–5 years) in one place there may be a decline in productivity, particularly in clusters at the bottom of the string. This is believed due to the deposition and accumulation of oyster faeces and pseudofaeces that amount to 0.6–1.0 t (dry weight) of material per year from a raft 10 m × 6 m, holding between 60 000 and 100 000 oysters. This deposition is rich in organic material and when stirred by wave action, may create a suspension of silt accompanied by a release of hydrogen sulphide gas, both of which are detrimental to the well-being of oysters.

There are several remedies which include rotation of sites, selection of areas with some tidal action so that deposits are swept away, or actual removal of deposits by drag or suction dredging.

One advantage of suspended culture is that, provided the strings are not allowed to touch bottom, there are no known predators, except the possibility of sea stars in a two summer cycle.

Condition

In experiments so far, raft cultured oysters have shown a consistently high condition factor. In the Ladysmith experiment, raft culture oysters had a consistent condition factor of about 130 compared to 70 for those grown on the bottom only 50 m away.

Although quality is difficult to measure quantitatively, it is the general opinion that raft culture oysters are superior in colour and in flavour to those grown on the bottom.

Productivity

Using initial data from Ladysmith Harbour, seed from Pendrell Sound spatted in early September and considered equivalent in amount to 6 bags produced 180 L by the following November. A similar amount of seed held on the bottom for one summer, followed by one summer on the raft, produced 320 L of oysters. On an area basis with 65 rafts per hectare, occupying 25% of the area and 100 shell strings, each with 15 shells, the production would be 9 000 L or about 9 000 kg of meat. Results from more recent studies indicate these values are quite conservative. This is in contrast to 50–75 L per 6 bags of seed in 3 years of ground culture, equivalent to about 2 250 kg of meat per hectare per year. The per hectare yields for British Columbia are only estimates, for no large-scale operations have been carried out.

A raft will produce twice as many oysters as bottom culture from one unit of seed and about 10 times as many oysters as from the bottom in equivalent areas. The basic production cost is approximately the same for both types of culture.

It is estimated there are at least 600 ha of prime area suitable for raft culture in the Strait of Georgia. Assuming a period of one summer on the raft, this area is capable of producing about 4 000–12 000 t of meat annually, far more than any foreseeable market. In addition, it would be possible, with properly constructed rafts, or long lines to utilize more exposed sites that could well be more productive than protected inshore areas.

Culture Areas

The original raft culture studies in British Columbia were carried out in Ladysmith Harbour, but it was not known if this area was typical even for the Strait of Georgia. Information from a short-lived commercial raft culture operation in the Prince Rupert area showed that the system was feasible that far north. Bottom culture experiments in the Queen Charlotte Islands produced excellent oysters, but it required 5–6 years to attain

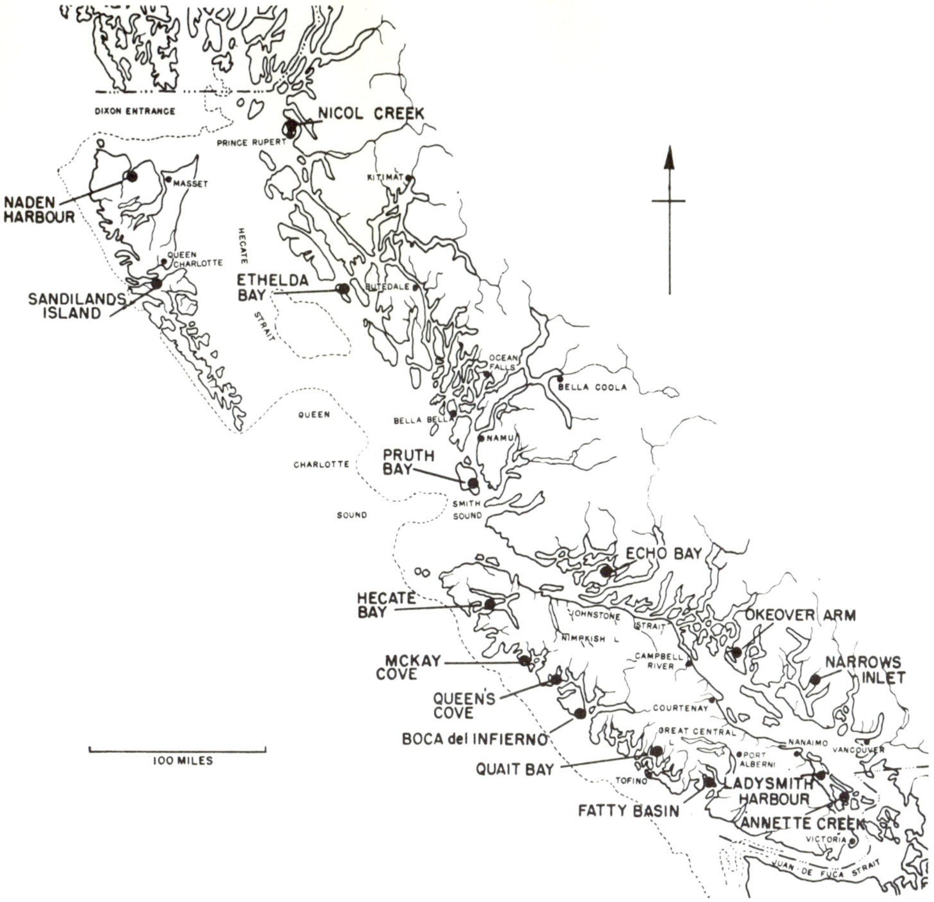

FIG. 135. British Columbia showing raft culture sites.

11. Sandilands Island – Skidegate Inlet, Queen Charlotte Islands
12. Naden Harbour – Queen Charlotte Islands
13. Nicol Creek – Prince Rupert Harbour
14. Echo Bay – Cramer Passage – Queen Charlotte Strait
15. Okeover Arm – Malaspina Inlet
16. Narrows Inlet – Sechelt Inlet

Seed was acquired from Pendrell Sound in October 1965 and placed on the Biological Station experimental area (lot 164) in Ladysmith Harbour. It was allowed to grow there (one summer) until January of 1967 when it had reached an average length of slightly over 2.5 cm with a range of 1.0–5 cm. In January and February the seed was taken to the raft sites. Log rafts were made on the spot and anchored at each end with 20 m of 12 mm diameter polypropylene rope attached with the necessary thimbles, shackles, and eyebolts to cement anchors weighing about 60 kg each. The rafts had an average length of 12 m and two logs were held 3 m apart by 30 × 4 cm fir cross-members secured by two 1 cm × 30 cm black iron drift bolts with clinch rings (Fig. 117).

Fifty double strings, each with 10 shells on each side, or a total of 1 000 shells, were placed on the rafts. The rafts were visited in November of 1967 and 1968 when several strings were harvested. Various production figures are given in Table 35. Of the 15 rafts, only those in Naden Harbour and in Narrows Arm were lost. Observations were possible only at the end of the first year on rafts at Okeover Arm, Skidegate, Prevost Island, and Ladysmith Harbour. Analyses are confined to the first nine locations where production in 1967 may be compared with that in 1968. These will be generally representative of conditions outside the Strait of Georgia. Ladysmith Harbour appears to represent those for optimum productivity.

TABLE 35. Production data from 1967 to 1968 Pacific oyster raft culture tests at various sites along the British Columbia coast.

	No. of clusters		Weight (kg)		Volume (litres)	
Station	1967	1968	1967	1968	1967	1968
Fatty Basin	40	40	35	92	60	150
Quait Bay	40	40	29	70	60	105
Boca del Infierno	40	40	31	88	60	165
McKay Cove	40	40	46	74	75	120
Queen's Cove	40	40	40	106	60	165
Hecate Bay	40	40	21	40	45	60
Echo Bay	40	40	18	54	30	90
Pruth Bay	40	40	32	53	60	105
Ethelda Bay	40	40	31	56	60	90
Nicol Creek	40		17		30	
Sandilands Island	40		41		60	
Ladysmith Harbour	40		73		150	
Annette Creek	40		1		15	
Okeover Arm	40		45		75	

	No. of oysters		No. of meats per litre		Percent longer than 10 cm	
	1967	1968	1967	1968	1967	1968
Fatty Basin	541	600	36	23	30	73
Quait Bay	491	432		19	30	70
Boca del Infierno	645	675		25	22	78
McKay Cove	787	580	36	18	25	62
Queen's Cove	654	730		23	30	78
Hecate Bay	723	630	62	32	12	38
Echo Bay	425	673		48	15	45
Pruth Bay	682	460	38	25	34	80
Ethelda Bay	754	592	53	40	30	75
Nicol Creek	465		53		12	
Sandilands Island	650				40	
Ladysmith Harbour	654		28		77	
Annette Creek					20	
Okeover Arm	724		44		70	

In 1967, on the basis of extrapolation from 40 clusters equivalent to one and one-half double strings of 30 clusters, the Ladysmith raft (two-log, 12 m long) with 50 double strings (1 500 clusters) produced 24 500 oysters weighing 3 t of which 18 800 were shuckable (over 10 cm long). They shucked out at 28/L giving a yield of 680 L together with 5 700 replants.

The nine outer stations on a raft equivalent to that described above for Ladysmith Harbour produced an average of 24 600 oysters weighing 1 200 kg. Of these only 25% were shucking size and 44 oysters/L a relatively small production of 136 L resulted.

These nine rafts were sampled again in November 1968. There was a significant improvement which indicates that in most areas outside the Strait of Georgia, 2 years on the raft may be necessary. In 1968, the average production was 27 500 oysters indicating no significant mortality had occurred (Table 36). These weighed 3 t, equivalent to the production in 1 year in Ladysmith Harbour. About 60% of shucking size and at 28 oysters/L the production would be 580 L together with 13 000 replants, which with another year on the beach, should produce another 450 L.

Growth in oysters may be measured by either length, weight, or volume and these data for the 1967–68 tests are given in Tables 35 and 37.

TABLE 36. Growth rate of raft culture Pacific oysters at various sites along the British Columbia coast, 1967–68.

Station	Mean oyster length in mm (25.4 mm) Jan. 1967	Nov. 1967	Oct. 1968	1967 increment	1968 increment
Fatty Basin	25.5	93.1	126.4	67.6	33.3
Quait Bay	29.0	90.6	127.1	61.6	36.5
Boca del Infierno	35.5	89.6	132.0	54.1	42.4
McKay Cove	29.1	90.8	116.8	61.7	26.0
Queen's Cove	29.0	88.8	131.0	59.8	42.2
Hecate Bay	33.3	70.2	87.6	36.9	17.4
Echo Bay	34.7	69.5	97.0	34.8	27.5
Pruth Bay	23.5	87.6	117.0	64.1	29.4
Ethelda Bay	30.0	84.1	112.0	54.1	27.9
Nicol Creek	25.1	73.7		48.6	
Sandilands Is.	30.7	100.1		69.4	
Ladysmith Harbour	30.5	115.3		84.9	
Annette Creek	29.0	78.5		49.5	
Okeover Arm	20.5	101.2		80.7	

TABLE 37. Measurements of Pacific oysters grown at various depths on rafts, Boca del Infierno, B.C., October 1968.

Depth (m)	Cluster no.	Weight of clusters (g)	Weight of oysters (g)	Weight of fouling (g)	No. of oysters	Mean weight of oysters (g)	Mean length of oysters (mm)	Condition factor (visual assessment)
0	1	1 180	445	735	3	148	145	120
0.3	2	5 120	3 012	2 108	20	151	138	125
0.6	3	3 890	2 185	705	17	128	128	115
0.9	4	1 375	695	680	4	174	146	115
1.2	5	3 685	2 626	1 059	20	136	133	100
1.5	6	5 325	2 795	2 530	20	140	136	105
2	7	4 570	2 587	1 983	24	107	125	100
2.3	8	2 686	1 612	1 074	12	134	126	95
2.6	9	4 406	2 776	1 630	17	163	131	95
3.0	10	5 442	3 515	1 927	28	126	136	115
3.3	11	5 105	2 594	2 511	20	130	132	120

Each location is a separate entity and comparisons should be made between individual locations or with the optimum production area such as Ladysmith Harbour. It may be seen at a glance (Table 35) that Hecate Bay, Echo Bay, Annette Creek, and Nicol Creek were well below the others in 1967. In the 1968 results, Hecate Bay and Echo Bay had similar ratings. This points out that care must be taken in selecting an area for raft culture and more than one test should be made before beginning a full-scale operation. For instance, it was anticipated that Annette Creek would be an excellent area. However, the barnacle problem was too great for the oysters to surmount, and it may be that this is one area where exposure of strings should be delayed until after the barnacle set. Hecate Bay and Echo Bay, both small bays off a main channel, possibly did not have enough circulation to provide sufficient food. Temperature has considerable influence on shell growth and this factor is demonstrated in the case of Ladysmith Harbour that had relatively high summer water temperature compared with the other areas tested.

Although one test in Ladysmith Harbour suggested a slight difference in growth rate or condition of oysters with depth (Table 34), a subjective assessment of all the areas indicated no significant differences. To determine if this were true, a single string of 11 clusters was studied (Table 37). These limited figures support the conclusion that within the depth range studied there are no real differences due to depth.

Judging from satisfactory results obtained from sites on the west coast of Vancouver Island, the Queen Charlotte Islands, and the mainland Hecate Strait areas, there is little doubt many other suitable areas exist.

Advantages of Raft Culture

Advantages of raft culture over bottom culture are:

1. More rapid growth
2. Better condition
3. Better flavour
4. No predators
5. Tidal independence
6. Higher production by per unit area
7. Better survival
8. More available area
9. Cultch flexibility

At present most oysters in British Columbia are produced in the Strait of Georgia. This is the area of present high population density and as this increases oyster production may in time decline, for high population density and oyster production are seldom compatible because of land drainage, sewage and industrial pollution, and priorities in intertidal land use. Oyster growing may therefore be forced away from centres of population, notwithstanding the use of purification plants. Outside the Strait of Georgia there is relatively little good oyster ground and growth on the bottom is slow, but there are numerous areas suitable for raft culture. Therefore, it is most likely that future oyster growing in British Columbia will consist mainly of raft culture, first inside and then later outside the confines of the Strait of Georgia.

Rack Culture

Rack culture is generally carried out in the intertidal area where there is a modest tidal range. However, racks may also be placed partially subtidally at a level which allows work to be done by wading.

The advantages of rack culture are:

1. Independent of bottom consistency.
2. Growth rate more rapid than on the bottom.
3. Few predator problems.
4. No silting mortality.

Racks may be erected either on firm or very soft bottom and at present the latter is usually more available and extensive than the former. Since the oysters are off bottom there is no danger from silting mortality unless water turbidity is excessive. Also they grow more rapidly than those cultured on the bottom, presumably because food is more available and less energy is expended in separating out food from silt which is always more prevalent along the bottom. Danger from predators is not completely removed but control is possible. Drills are the main culprits but they must scale the posts of the racks before they can reach the oyster and are amenable to control.

Rack Construction

There are four main types of rack construction (Fig. 136a, b, c, d): (1) Single beam, (2) Tripod, (3) Cross beam, (4) Parallel beam.

Single beam (Fig. 136a)

This consists of a single beam placed and secured to the top of a series of posts driven into the bottom. Rack construction material should be protected with wood preservatives such as creosote. The diameter of posts and beams should be in the 8–10 cm range. The posts vary in height depending on the type of culture material to be used. For trays, sticks and horizontal strings the posts may be no more than 60–70 cm above the bottom, and driven into the ground nearly as far. For string culture, the posts should be longer than the strings. One end should be pointed and a piece of flat wood placed on the other to take the force of the hammer blows to prevent splintering.

Length of the beams will also depend on the type of oyster culture material, but 3 or 4 m will be satisfactory for most. It must be remembered that as oysters grow the weight increases and it is a common error to use materials only strong enough to hold initial weights.

Tripod

Another form of beam support, useful if strong enough posts are not available or if the bottom is too hard to drive posts, is the tripod system. For the best support three poles are spread an equal distance apart on the ground and brought together near the top and tied (Fig. 136b) at the appropriate height. The second and next supports require only

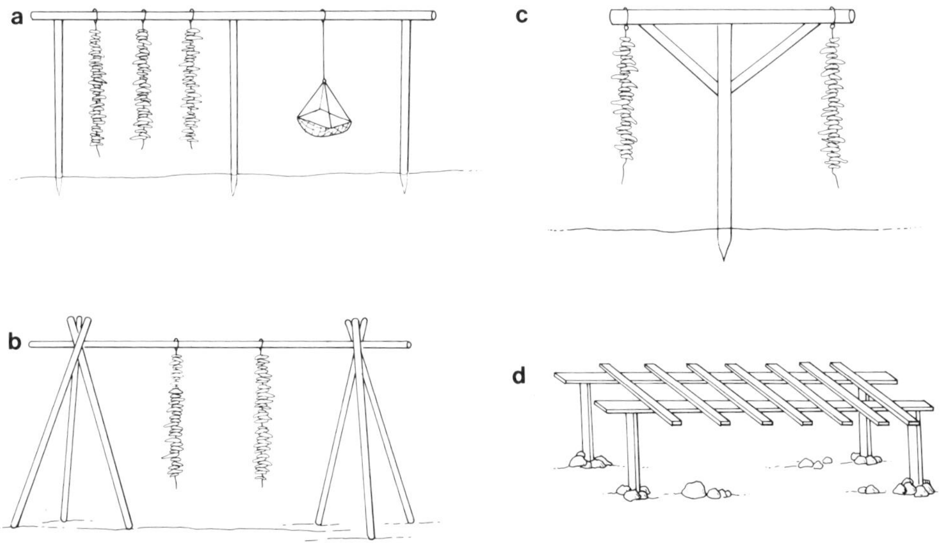

FIG. 136. Illustrations of various types of rack construction. a. single beam. b. tripod. c. cross beam. d. parallel or double beam.

two poles until the final one when again three poles are required. This rack is usually used for string culture where a longer distance from the bottom is needed.

Cross Beam

This is a slightly more complicated rack system because of the number of connections. On top of single posts a cross bar is placed and supported at each end by angular braces attached to the post (Fig. 136c). The long beams are placed on the end of cross beams.

Parallel Beam

This commonly used method is simply doubling of the single beam system with two lines of posts and beams connected by cross beams (Fig. 136d). Construction methods are similar to that for the single beam.

String Culture

String construction for seed with shell has already been dealt with (page 60). The strings for growing are constructed as described on page 159.

Strings may be hung vertically from single beam racks and the tripod form is particularly useful. Racks should be placed at a level to evade the dense barnacle zone if there is one and this may dictate length of the strings. For ease in handling, strings not much longer than a metre or so are satisfactory. The strings may also be laid horizontally on parallel beam racks and in this case the strings may be somewhat longer. Care is taken to see the strings do not touch the bottom and are placed about 25 cm apart.

With strings the seed may be of any size for full growth may be attained on the rack.

Drills or crabs may crawl up the posts so regular surveillance may be necessary and if the problem is severe both a picking and trapping programme may be required.

Stick Culture

This is a simple and direct form of oyster culture where the cultch is the grow out platform. The system has been in use for centuries although distinctions between stick and stake culture were not always clear. Stick culture has been developed to a high level in Australia and New Zealand with *Saccostrea*, a somewhat smaller version of *Crassostrea*. The basic process is collection of oyster seed on some form of stick or tube in the seed area and removing it to a growing area where the sticks are held on intertidal racks or suspended from floats. The seed grows to maturity on sticks from which they are harvested by chipping.

The Sticks

In Japan the original sticks were of bamboo. In Australia they were dry mangrove branches with a progression to 2 cm square, 1 or 2 m long sawn wood sticks. Eventually these were tarred partly as a protection against shipworms, but also to provide a more acceptable setting surface. In time wooden sticks were replaced by fibro-cement laths which are now in common use as are bars of ferro cement triangular in cross section. The fibro-cement laths measure 5 cm × 0.6 cm × 120 cm while the ferro cement sticks are 2.5 cm × 2.5 cm × 120 cm. The lengths of all of these may be changed according to circumstances. Aluminum as well as PVC moulding are useable but costly. More recently finely grooved PVC pipe with an outside diameter of 20 mm (approximately 1″) provides a good substitute stick (Fig. 56b). A modified PVC tube is being manufactured in B.C.

The tubes require at least a week or 10 days of leaching. Cedar construction laths manufactured in B.C. make low costs culture sticks providing they are coated with cement

FIG. 139. Stick culture. New Zealand.

FIG. 140. Stick culture. Spatted laths growing out on racks. New Zealand.

A rough approximation indicates that 1 ha (2.5 acres) can hold up to 10 000 sticks of 120 cm (4 000 per acre) although lesser numbers should be used initially to gain experience.

Suspended Stick Culture

Two forms of stick culture are possible on floats, either rafts or long lines. The first, and most simple, is to suspend each stick vertically and independently by a line through a drilled hole at one end. They may be placed at intervals of about 20 cm (8 inches) on a long line. On a raft, each stick should have something more than 90 square cm (1 sq. ft) of space, with a bit more those in the centre.

The alternative is to suspend the sticks horizontally one below the other. The sticks may be separated by knots in the end suspension lines, or by separators like those used on shell strings. The sticks arranged in this manner should be spaced no closer than about 20 cm, although this method requires more labour.

As with floating string culture optimum use must be made of the float space and seed that has attained a length adequate to produce marketable oysters from one summer's growth on the float should be considered.

Harvesting

If oysters on sticks have been conditioned by one summer on the beach, harvesting from suspended culture will take place during the fall, winter and following spring of the year in suspension (Fig. 141).

Oysters are removed from sticks by chipping them off with a breaking iron or with a pneumatic chisel. To do this the sticks are held vertically with one end on the ground. In France, machines are used to sliver oyster spat from the limed roofing tiles and a similar apparatus could be designed for the various types of sticks. With tubes, flexure usually removes most of the oysters if they have not attained a high degree of attachment

FIG. 141. Stick culture with PVC pipes.

FIG. 142. Tray growing of less than market size oysters that have been removed from sticks.

to each other. A proportion of oysters will be less than market size and of these a number will be of the size and shape suitable for the half shell market. This remainder may be grown to market size either on bottom plots or in trays as done in Australia and New Zealand (Fig. 142). A proportion of the non-shuckable oysters may have cracked or broken shells as a result of the chipping process. If the size of the breaks are not great the oyster will repair this damage by secreting new shell, but to prevent predation by crabs those oysters should be held on trays until healing is complete.

Productivity

It is difficult to determine productivity since it depends so much on the number and spacing of spat on the stick and how they are handled between spatting and setting out, as well as the site. Presuming an adequate spatfall of at least 100 spat per 120 cm stick, and a return of 20 to 30 marketable oysters per stick, a production of approximately 80–100 t may be expected from 1 ha on a 2-year cycle.

In addition there would be an equivalent amount of sub-market sized oysters.

It can be seen that stick culture can involve use of intertidal bottom, racks and trays as well as floats thus providing a measure of flexibility relative to growth rate, fouling, storage and type of market.

The system described above may be described as the classical method of stick culture for in practice many variations are possible and some of these are in use.

Tray Culture

Trays have been used to culture oysters for a very long time, although not extensively, probably because of the large investment in equipment and manpower required for significant levels of production. Only a small fraction of the world's oysters are grown on trays. Norway cultures oysters in this way partly because of temperature requirements, but also in the polls (small enclosed embayments) there is no suitable bottom and the European flat oysters (*Ostrea edulis*), does not lend itself well to string culture. Trays and pouches are used in Australia and New Zealand as a minor part of basic stick culture to finalize growth of sub-market size oysters from sticks at harvest. Pearl culture in Japan also uses trays to some extent. Modified trays in the shape of pouches are used in France as well as regular trays.

Trays were used quite extensively at one time in eastern Canada to hold spat until it was large enough to withstand sea star attack when planted out on the bottom, but eventually this sytem became uneconomic. Trays are used by shellfish biologists for comparative studies since they provide a uniform platform and can be placed in identical situations.

Tray culture is not well suited to mass production and tray oysters are deemed to be a specialty item eaten raw from the half shell. The industry places great emphasis on shell shape although the classical shape of each oyster species is different. Although this is a consideration, it would be expected the main requirement would be meat quality. As noted elsewhere, the period of prime meat condition occurs between October and May, which except for April and May is the time of minimum or no growth. By March a large proportion of oysters will have been harvested and the trays freed for reseeding. The remainder may be removed from trays for storage in larger, more economical containers and in high density.

The relationship between shell shape and meat quality is a tenuous one even in the prime season. A well shaped oysters has a somewhat better appearance than an elongated one since there is less mantle posterior to the adductor muscle. This is equally true for oysters destined for shucking and production; in the context of economics, meat quality and reasonably shaped oysters should be an aim of the culturist, regardless of the growing technique. In eastern Canada, oysters for the half shell are graded officially according to shell shape in four categories: Fancy, Choice, Standard, and Commercial. It is to be hoped that the B.C. industry does not opt for this system.

FIG. 143. Nestier tray with the number of spat that will eventually fill the tray.

The basic tray culture is simple but intensive, utilizing either an intertidal rack system for holding the trays or suspension from rafts or long lines. The disadvantages are a high labour content, high capital investment, seed requirements of a specific type as well as a specialized market.

Equipment

The basic equipment are trays plus either intertidal racks or floats. At present there are many types of trays available but the two main types used in B.C. are the plastic Nestier and McNichol. The Nestier tray (Fig. 143) is 55 cm × 55 cm and 4.5 cm deep, a suitable enough size for small-scale culture but it appears the perforation pattern is not able to provide adequate circulation. Holes on the bottom are 10 mm in diameter and those on the side 6 cm. The trays nest one within the other so they may be readily stacked.

The McNichol tray (Fig. 144) is circular with a diameter of 47 cm and a rim of 5 cm high. It is divided into three sections thus limiting the movement of oysters within the

FIG. 144. McNicol oyster culture tray.

tray. There is a central hole 15 cm in diameter to permit better circulation, but even so adequate circulation is also a problem with this tray.

Culture trays may also be handmade with a wooden frame and wire mesh. Formerly the wire mesh required a tar coating but the vinyl covered wire and plastic meshes such as Vexar, now available obviates this step, although the wooden frame requires protection from shipworms (*Bankia*) and gribbles (*Limnoria*). Custom made trays allow more flexibility in tray size.

In addition to Nestier, McNichol and home-made trays, people culturing the European flat oyster (*Ostrea edulis*) on the Atlantic coast use Japanese manufactured lantern (Fig. 145) and pearl nets (Fig. 146) for tray culture. These are made of metal hoops and fine monofilament nylon mesh which reduces the area for attachment of fouling organisms. The Komplett system of stacks of trays and a Mexican nesting tray are other designs.

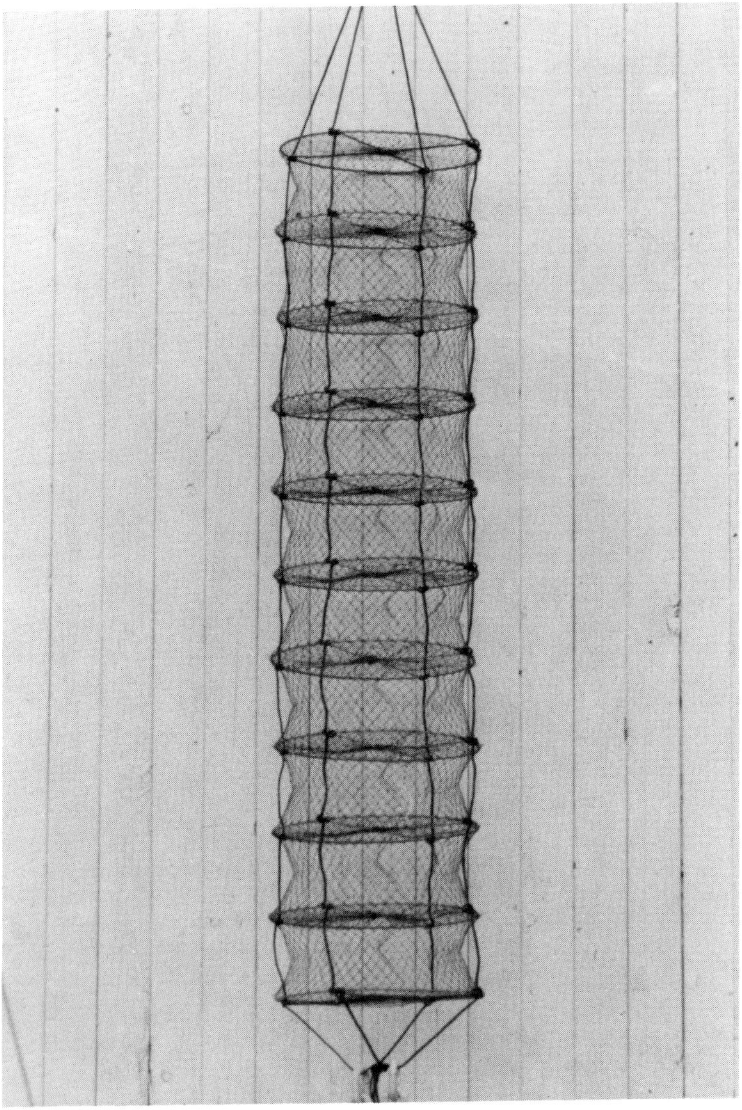

FIG. 145. Japanese style lantern net. (× 1/5).

FIG. 146. Japanese style pearl net. (× 1/5).

Suspension

Although trays may be placed on firm bottom, this negates the advantage of off bottom culture. The alternatives are intertidal racks or floats. Intertidal racks simplify operations since trays are more easily worked during low tide periods than floating trays and fouling is less than on floats. The trays may be either hung from or placed on racks, or both. Intertidal bottom too soft for normal bottom culture may be used for tray culture with racks.

The tidal level placement of racks is governed by the rate of growth desired and level of the barnacle zone. Fortunately in most areas the barnacle zone is high and growth rate is best at a low tidal level.

Floating suspension requires either rafts or long lines. Although growth rate and condition is better on suspended oysters, the fouling problem is more severe, necessitating constant attention, for it causes reduction of circulation.

Trays may be stacked one upon the other and held together by plastic strapping in layers approximately five deep. Attachment of bundles of trays to rafts or long lines should be of a nature to allow easy removal. Split links, now available in plastic and halibut snaps are used for this purpose. A clove hitch knot is satisfactory but difficult to undo. The bundles of trays may be placed adjacent to each other but each alternative one can be sunk to a greater depth to allow greater water circulation around each one.

Seed

Since tray culture implies specialty oysters, this in turn implies single, well shaped oysters. This requires seed that will produce these. The ideal seed for this purpose is a single or cultchless spat which is normally only available from a hatchery. Since the size of this spat is in the region of 3–5 mm the trays initially require small mesh liners. As mentioned previously, small seed in trays tend to concentrate in the corners or attach to each other or the tray, thus requiring constant attention. Rapid growth at this stage may also require frequent thinning, depending on initial loading (Fig. 39).

An alternative is a small sized cultch such as the diamond cement chip cultch now available (Fig. 59). These require some manual manipulation to separate the spat, either from the chip or each other when some growth has been added.

Single spat may also be obtained from cement coated wood veneer, either in the form of panels or circles (Fig. 57). Spat of an appropriate size can be removed by flexure of the cultch at a size before the spat begin to attach to each other. A much better surface is thin PVC sheeting which does not need a cement coating and from which the spat can be removed more easily than from the cement coating by flexure. PVC sheeting can be reused repeatedly. Single spat are obtainable from PVC tubing, also by flexure.

Seed from normal shell cultch may also be used. In this case the spat are held on the bottom for one summer after which the clusters are separated when the spat are from 25 to 40 mm in diameter. Another system is to crush the spatted shell cultch when the seed is no larger than abut 5 mm. This will result in some mortality, normally not significant. However the spat will be either single or attached in small numbers to small shell fragments. Three to four thousand spat on crushed shell has a volume of about 4½ L. As with single or cultchless spat, this requires lined trays for the initial growth period.

Culture procedure

One-Year Cycle

In trays, as with bottom culture, the obvious and initial plan is to put spat of the year in trays, assuming a natural source such as Pendrell Sound. These are normally not larger than about 1 cm in diameter when growth ceases in the fall. Spat of this small size requires two summers on the tray to achieve market size. This means a tray space occupied for two years as well as space on the float or rack. To obtain optimum use of the costly equipment (reducing the number of trays by half) it is necessary to reduce the 2-year cycle to one. This is done by growing the seed to a diameter of 25–50 mm by less costly means to a size where one summer in the tray and on the float or rack will produce a marketable half shell oyster of about 80 mm.

To do this the seed of the year is held (1985 seed until February 1987) for one summer on the bottom, thus requiring the possession or rental of an intertidal lease. Half a hectare (approximately 1 acre) will hold at least a million oysters in the 29–50 mm size range. Such a lease can be relatively high in the intertidal zone as long as it is reasonably well protected from wave action and most high intertidal ground is usually firm. How the seed is held on the bottom depends on the type of cultch and size of seed. Very small seed, up to 1 cm, should be held in special enclosures and possibly with a small mesh base until it reaches about 2 cm in diameter after which it requires little special attention.

Two-Year Cycle

This plan involves use of small seed, as small as 2 or 3 mm and up to 10–15 mm. With the smallest seed, fine mesh liners are required until the seed is larger than the meshes or holes in the tray. At this small size the seed tends to grow together and to attach to the meshes of the tray, thus requiring frequent stirring. If the trays are subject to even small wave action the seed tend to concentrate in corners of the tray again requiring frequent attention. Also rapid growth of the small seed, unless planted thinly, creates a need for frequent thinning.

Every attempt should be made to obtain seed not less than 10–15 mm in diameter.

As in the 1-year cycle, seeding density may be light at a level to eliminate sorting, or heavy to concentrate the seed in a few trays with periodic thinning. The particular plan to be used depends to some extent on the time of year when the seed becomes available and on the marketing time when trays become free.

With natural seed, the time of delivery is usually late summer or early fall, limiting the growth period for that year and consequently determining within fairly narrow limits the harvest period. Hatchery seed is available over a longer period of time usually spring to fall, so planting at intervals throughout the summer provides a longer harvesting period, keeping in mind the growing period is from April to November.

Seeding

For seed, there are two alternatives; the first is to seed trays at or near the number that may be expected to be grown to harvest size 3–5 cm in length. In this case the seeding rate will be 60–70 per Nestier tray or 180–210 seed per square metre (Fig. 143). In this way, repeated sorting is not necessary as the seed grows and fills up the space. This means most of the trays are continuously occupied.

The alternative is to load the trays to the maximum, with the seed one layer deep and touching or overlapping somewhat. With 3–5 cm oysters a concentration of about 200 per Nestier tray (600 per square metre) is possible (Fig. 147). Oysters this large usually

FIG. 147. Nestier tray with 200 spat.

do not attach to each other or the tray. With this plan it is necessary to sort and reduce the number per tray as the oysters increase in size.

Planting density is a perennial problem in oyster culture and forces compromise between time and space. There is the proposition that growth deceases with increasing planting or hanging density. This is so up to a point, but the difference in growth rates between a medium and heavy planting is often not significant. It is possibly less costly to hold oysters an additional month on the tray or float to obtain the required size, rather than use extra space (i.e., additional trays) with a low density. High density planting is generally more economical than low density.

Growth

Growth of tray cultured oysters is superior to those grown on the bottom and generally they grow more rapidly in suspension than on intertidal racks. Regardless of the method, it must be recognized the growth of oysters is variable from site to site and this is so even within one general site, as well as with individual oysters.

The data given in Table 12 gives a general indication of growth that may be expected in suspended culture in the one summer system. Cessation of growth during winter months should be noted and taken into consideration when planning a sequence of culture events.

Fouling

Since tray oysters are destined for the half shell trade, fouling should be minimal, particularly on the left valve, for the diner does not see the right valve, which has been removed before serving.

The main fouling organisms that disfigure the appearance of half shell oysters are barnacles and folliculinids. Although barnacles can be removed a scar remains. Folliculinids are small (1 mm) flask shaped protozons dark green in colour, of no significance individually, but most apparent in concentrations.

Oysters within a tray, particularly with a cover, are seldom seriously affected by fouling. The major difficulty occurs on the outside of the trays where pores can be covered over by fouling organisms such as hydroids, mussels, anemonies or barnacles, thus reducing the water flow to the interior of the tray. Periodic cleaning is necessary.

If the fouling is known, by season and by depth, it may be possible to culture around it. For instance the main barnacle set in the Strait of Georgia occurs about April first, plus or minus 10 days. Exact timing of settlement may be determined by examination of the interior of the barnacle. If the material at the base of the barnacle is yellow, the larvae have not yet developed; when this becomes dark coloured, nearly black, the larvae are ready to be released. Another system is to place a few barnacles in salt water in a small dish indoors every few days starting in mid March. When the time is right the larvae will be released and can be seen swimming in the dish. At this time the trays can be lowered to a level at four to six meters deep if in suspension. This may not be the correct level for a particular site which may be determined by experience. When the rate of settlement of barnacles decreases, determined by exposing clean oyster shells every week, the trays can be returned to the surface layers.

If this system is not successful, trays can be removed from the water as soon as the barnacle set has taken place before the individuals are no larger than a millimetre or so, and exposed to the air for 24–48 h. This should kill most of the barnacles without harming the oysters. Failing this, the barnacles may be removed from the trays with a stiff brush when no more than 2–3 mm in diameter.

The ecological method of avoiding fouling may also be applied to other organisms such as mussels, where there is usually a spring and a fall peak breeding time.

Shipping

Half shell tray oysters, particularly those grown suspended, require careful handling for the shell edges are generally fragile in comparison to bottom grown oysters. It is possible to thicken the shell edges somewhat by holding the oysters intertidally for some time and this will aid in extending the shell life since it will assist in preventing leakage. This type of oyster should be stored with the cup valve down and shipped in waxed cartons and in polyethylene bags if shipped by air.

Stake Culture

Stake culture, as the name implies, is where stakes driven into the ground carry oysters either directly attached to the stake by spatting there or on cultch attached to the stake.

Stake culture in one form or another has a long history in oyster growing countries. In many tropical countries with mangroves, the aerial roots of that tree act as stakes wherever oysters occur naturally.

A modified form of stake culture uses flat rocks such as shale or sandstone which are used both as spat collectors and growing platforms such as stick culture. Further development is to use bricks or cement slabs especially made for the purpose. These forms are found extensively in China and at one time in Australia. Stake culture is a common practice in the Phillipines where cultch such as shell or even tin cans are attached to bamboo stakes.

Stake culture utilizes oyster ground too soft for bottom culture and lends itself well to growers unconcerned with high production or with limited ground space. There appears little opportunity for mechanization.

Equipment

Direct equipment needs are small, simply stakes of some form. The obvious stake material is wood usually 2.5 cm square, 45 to 60 cm long. These may be treated chemically to prevent shipworm (*Bankia*) attack. Light treatment with antifouling materials (i.e. 500 g of copper oleate in 2 L of kerosene added to 4 L of tar) is unlikely to create problems of uptake by the oysters. A better solution for stakes is materials impervious to shipworm attack. PVC pipe is one possibility but entails a special cultch attachment device, whereas the wooden stake requires only a galvanized nail driven into the top of the stake, on which is placed a punched shell (Fig. 148). A wooden peg in the end of PVC piping will take a nail.

Seed

The cultch material normally available in British Columbia is oyster shell. A flatter cultch such as a 10 cm × 10 cm × 3 mm fibro cement square would be suitable since it produces a high proportion of well shaped oysters, many of half shell quality, but the cost of such a cultch is quite high and not readily packaged for spat collection, although it could be used over and over again. An alternative, although probably not too practical, is the right valve (upper flat) only of shell cultch.

For optimum production the spat count should be not less than about 25–30 spat per Pacific oyster shell, of average size. This is about the largest number that such a piece of cultch may produce under normal conditions whether it be bottom culture, suspended string culture or stake culture. Loss of spat more than 25 per shell should not be calculated as true mortality in the sense of potential returns.

As with other forms of culture it would be preferable and less costly to grow the seed for one year on the bottom when they will attain a length of up to 5 cm. This will ensure a marketable oyster after one summer on the stake. The main problem here is that the hole in punched shell will be most often covered over by growth of the spat. Re-drilling the hole would cause some mortality but not of significance, but would be an additional labour cost. The shell, should be placed with the cupped left valve down to minimize silt build up. The stakes are placed about 30 cm apart.

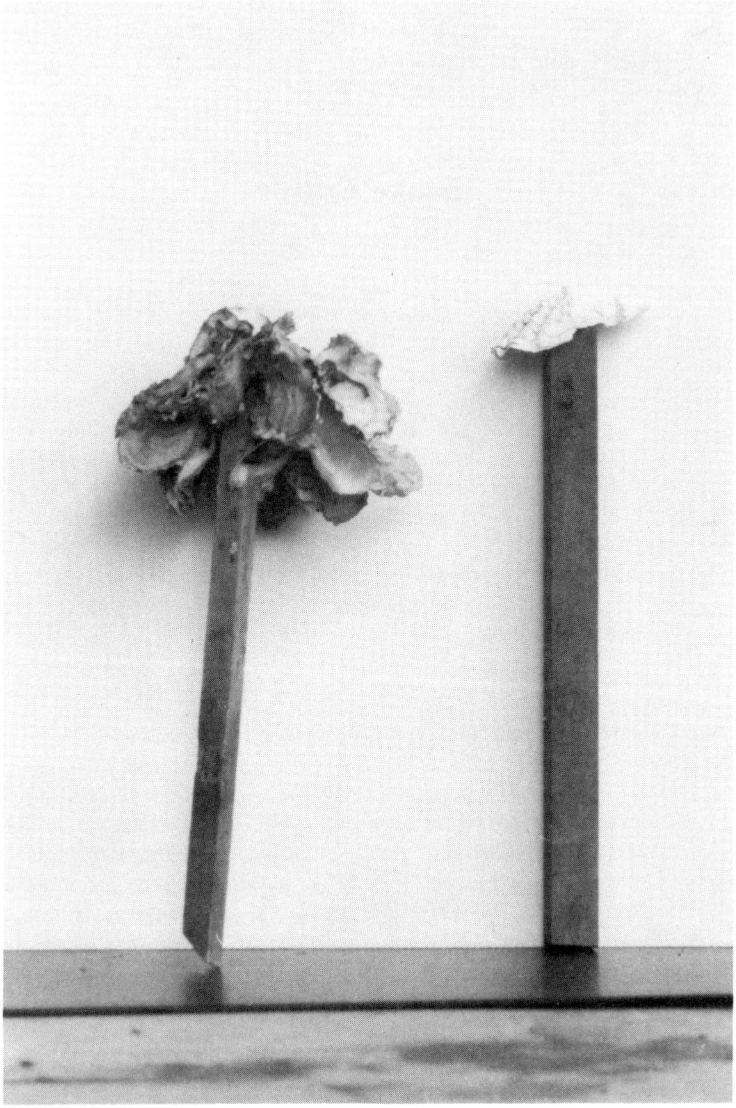

FIG. 148. Stake culture showing position of shell cultch.

Site Selection

Stake culture may be conducted on any type of beach except rock and requires somewhat less wave protection than bottom culture (Fig. 149). The main consideration is tidal level and this is important relative to fouling, particularly with barnacles and to a lesser extent with mussels. The site should be well below the barnacle zone: first to avoid this area and to take advantage of good growth usually at the lower tidal levels. Algal fouling is usually transitory.

Potential difficulty from drift logs should be considered. The amount and age of driftwood already on the beach may provide an indication. Boating is also a consideration and may require the site to be well marked.

FIG. 149. Stake arrangement.

Growth

As in several other forms of culture the alternatives are 1–2-year cycle. In a 1-year cycle, one summer on the bottom provides seed up to 35–40 mm in diameter and when placed on the stake, most will attain market size at the end of the second year (Fig. 150). In the 2-year cycle, the seed of the year is placed on the stake either in the fall (natural seed), or early spring after the barnacle set in April, and at the end of the first summer should attain a length of about 50 mm and market size at the end of the second summer. Either method requires two summer's growth, but the 1-year cycle requires only half the number of stakes for the same production, with the possible exception of some silting mortality if the storage area is not completely suitable. With hatchery seed, the timing of the operation may be somewhat more flexible to provide a supply of market oysters on a continuous basis, since the seed will be available over most of the summer instead of only in the fall with natural seed but the small size will require special treatment. These estimates apply to the Strait of Georgia only.

Predation

The main predators to be expected in stake culture are sea stars. On the softer ground normally used for stake culture the 20 rayed sun star (*Pycnopodia helianthoides*) is the species of concern and possibly the green and grey sea star *Evasterias troschellii*. The purple star (*Pisaster ochraceus*) is normally associated with rocky shores. An occasional patrol should provide adequate control.

Productivity

There are insufficient data to provide definitive production figures. What follows are estimates derived from small-scale plantings and the experience from a similar cluster type culture such as suspended string culture or bottom culture.

FIG. 150. Stake with cluster of matured oysters. (× 1/3).

With 25 plus spat per stake a maximum of 10 marketable size oysters with a number of smaller size, may be realized. A hectare of bottom can hold about 350 columns of 3 stakes 30 cm apart with a 60 cm separation between the columns, totalling about 75 000 stakes. This theoretically results in 24 tonnes of market sized oysters (33/L), providing a reasonable condition factor. The time period is either two years on the stake or 1 year on the bottom and one on the stake as compared to at least 3 years on the bottom.

Harvesting

The harvesting method depends partly on the type of stake, whether it be wood, metal or plastic and its degree of possible reuse. Picking the stakes into a floating shallow draft scow such as a herring skiff is one method. If the ground is firm enough a front end loader may be used as long as the columns and between paths were pre-arranged to take the width of the loader. There are numerous possibilities.

As with all cluster type culture, clusters should be separated before reaching the shucking table. This enables shuckers to work without interruption and permits separation of oysters destined for shucking from those suitable for half shell and the smaller ones to be planted for further growth.

Storage

Storage is a most important facet of producing oysters particularly with bottom culture, and there are a number of problems. First, there is shell stock storage, where oysters are held in the shell until ready for shucking. Since there are tidal cycles to consider, there must be a stock on hand during periods when oysters cannot be readily obtained from the beds. A second problem is to hold shucked oysters until, and during the time of shipping.

Shell Stock

Because of the alternation of spring and neap tides, there will be periods of 7 or 8 days when oyster beds are not exposed and harvesting by picking or raking is not possible. To compensate for this, it is necessary to obtain a supply of shell stock during the spring tide period adequate to carry over until the next spring series.

These oysters may be stored either wet or dry but attention must be paid to the pertinent section of the sanitary regulations. Wet storage is the term given to a method of holding oysters in water during the period between harvesting and shucking. The regulations state:

"1. Floating and wet storage. Floating and wet storage shall not be practiced unless written approval is given each year by the Minister. A sketch drawn to scale shall be shown on the reverse side of the certificate indicating the fixed location of the float or structures and all the potential hazards to which shellfish in the designated areas may be exposed. The presence of usable sink floats in the water shall be deemed to be evidence that floating is practiced.

2. Dry storage. Shell stock in dry storage shall be adequately protected from contamination at all times."

Wet Storage

The problem with wet storage is that oyster shucking houses are often close to concentrations of population, which may give rise to pollution. If a shucking house is adjacent to an approved area, and many shucking houses in British Columbia are so placed, it is permissible, as stated in the regulation, to practice wet storage. This may be done simply by placing shell stock in a receiving area which should be a beach, floored with clean firm gravel, or a cement platform relatively high in the tidal zone. Oysters should be accessible, at least to waders, during the neap tide period. That they will be uncovered for much longer periods than at the lower tide levels from which they came will have no significant effect over a short period of up to 2 or 3 weeks. An alternative is to use a sink float (which is merely a floating platform with the floor submerged to a depth of 30–40 cm (12–16 inches). This is anchored near the shucking house so that oysters are available at all times, whether the float is floating or is dry on the beach. In either sink floats or beach storage, oysters may be piled to a depth of 30 cm (12 inches) or more without harm. A clean beach storage area or float storage also allows oysters to cleanse themselves of accumulated mud within the shell and the movement in and out of the water enables much of the mud adhering to the outside of the shell to be cleaned off. The cleaner an oyster is when it reaches the shucking table the cleaner and more sanitary will be the end product. Shuckers tend to do a better job with clean oysters.

Dry Storage

This must be used when there is pollution or possible pollution of wet storage areas in the neighbourhood of the shucking house. An obvious disadvantage is that the oysters are out of water completely and storage life then depends on air temperature and humidity, as well as on how badly the shell edges have been broken during handling. Shell edges are more fragile and more easily broken in spring and summer than during winter. The lower the air temperature and the higher the humidity, the less oyster meats will be desiccated. It is difficult to place exact time limits on dry storage, because of these variables, but, during most of the winter, they should hold quite easily for a week or more. Freezing should be avoided, although, under certain conditions, oysters may be frozen and yet live, if carefully thawed without jarring. This often occurs on the beach. Ice is often used to keep dry-stored oysters cool during warm weather.

On the Atlantic coast of Canada, where oyster grounds freeze over during winter, the supply of oysters for that period must be harvested during the fall, so a good deal of storage space is required. Dry storage is always used in this case, and it has been found that well-shaped oysters with fat meats store best. Optimum storage temperature appears to be about 1°C (34°F) or just above freezing, but some storage is carried on as high as 5°C (40°F). Storage life at these low temperatures may be 4 months or longer.

Shucked Oyster Storage

There are specific points in the sanitary regulations of British Columbia to cover this aspect of processing oysters.

Sanitary Regulations

Refrigeration Rooms or Ice-boxes. Where a refrigerator or ice-box is used for retention of shellfish, it shall be so constructed as to permit easy and thorough cleaning and have adequate capacity to store all stock which may be shucked or packed in the plant in any one day. The refrigerator or ice-box shall be well insulated and have an impervious lining: the floor shall be graded to drain quickly; and an accurate thermometer shall be kept in each refrigerator at a point predetermined to have the highest temperatureé. No drain from a refrigerator or ice-box shall be connected directly to a sewer or other waste disposal system.

Ice. Any ice used shall be obtained from an approved source and shall be stored and handled in a clean manner. No ice or other foreign substances shall be allowed to come into contact with shucked stock.

Refrigeration. Shucked shellfish shall be cooled to a temprature of 7°C (45°F) or less within 5 h after the shellfish are shucked, and kept at or below 7°C (45°F) until subjected to a cooking process or delivered to a consumer. If shucked shellfish are frozen, they shall be kept in a frozen condition until delivered to the consumer. If subjected to a quick freezing process, the temperature of shucked shellfish shall be held at − 12°C (10°F) or lower until delivered to the consumer.

Refrigeration

Two factors involved in storage of freshly shucked oysters are the speed of cooling after shucking, and storage temperature. Even though an oyster originates from non-polluted water, it still harbours, quite large quantities of bacteria which, in normal numbers, cause no harm. Under suitable conditions, however, they can multiply rapidly in the body of the oyster, or in the container, and so cause spoilage. The purpose of refrigeration is to hold numbers of bacteria in the oyster, or any other food, at as low a level as possible. There are several methods of measuring spoilage and one of these is by bacterial counts, where samples of oysters are incubated in suitable chemicals for certain periods of time. Another method utilizes the pH or acidity of the oyster after a period of time, but this has limited use, for the initial pH of oysters varies according to area of origin. A third is the chromate method which is a colour test. Here potassium dichromate and sulphuric acid (1 mL of each) are added to to liquor (2 mL) from the oyster sample. In the oyster liquor are organic substances capable of reducing the chromate ion to chromium (111), the concentration of which may be determined, either by spectrophotometry or by comparison of its colour with visual standards. The reducing substances have been shown to increase with time and temperature, and by applying concentrations of chromium (111) found to standard curves, the temperature or age of storage may be determined and acceptability limits established. Finally there is an organoleptic or taste and smell test and, of course, this is an important one. Usually a panel of tasters is used, and quite consistent results may be obtained.

The speed of cooling of oysters in containers of various sizes has been measured by recording the temperature in the centre of the container and these results are shown in Table 38.

TABLE 38. Rates of cooling of Pacific oysters in containers.

Container		Initial temperature (°C)	Final temperature (°C)	Time (min)
	Litres		Cooling in crushed ice (−0.5–1°C)	
1 gallon	4.5	17	6	285
1 gallon	2.2	26	7	345
½ gallon	2.2	18	4	200
½ gallon	0.6	29	5	275
1 pint	0.6	18	5	100
1 pint	0.4	27	5	125
12 ounces	0.4	18	5	92
12 ounces	0.3	26	5	105
½ pint	0.3	19	5	60
½ pint		28	5	80
			Cooling in dry refrigeration (−0.5–1°C)	
1 gallon	4.5	27	6	400
1 gallon	4.5	16	4	400
½ gallon	2.2	28	5	350
½ gallon	2.2	18	5	350
1 pint	0.6	29	5	190
1 pint	0.6	19	5	170
12 ounces	0.4	29	5	160
½ pint	0.3	29	5	110
½ pint	0.3	18	5	90

It may be seen from this table that it requires a considerable time to cool oysters, particularly in larger containers, and that dry cooling requires more time than cooling with crushed ice. The greater the delay, after shucking, washing, and packing, in beginning the cooling process, the greater an opportunity for spoilage to set in and the shorter the "shelf life". This is the term used to describe the length of time the oyster can remain on display by the retailer and still be a good product. The longer the shelf life, the greater is the opportunity for a sale.

After an oyster has been cooled as rapidly as possible, it should be held constantly at as low a temperature as possible without freezing. Permitting the temperature to rise for appreciable periods will destroy the effect of rapid cooling. Tests have shown that oysters held at 11.7°C were unacceptable after 3–5 days, while those held at 7.8°C were unacceptable after 7–8 days and those held at 1.7°C on ice were still satisfactory after 16 days. These are of course, approximate figures, for they will vary depending on the origin of the oysters, how well they were handled prior to storing, and, particularly, on cooling rate.

The importance of good refrigeration cannot be overemphasized. An example of what happens to naturally occurring bacteria in oysters at different temperatures is given in Table 39. This is very convincing evidence of the need for adequate refrigeration.

TABLE 39. Relationship between temperature and number of bacteria on stored shucked oyster meats.

Original number of bacteria per gram	Final number of bacteria per gram	Temperature (°C)	Time (days)
1 600	1 800	−5	24
1 600	8 900	0	24
1 600	1 600 000	5	24

Processing

Hand Shucking

As indicated previously, most Pacific oysters are marketed in the shucked or opened state so that the art of removing an oyster from its shell is of some importance.

Shucking skill comes only with practice and each opener develops his own individual opening technique. However, there are some basic points that are fairly standard practice for a right-handed shucker.

First, the shucking table should be the correct height for the individual; tall people usually have additional platforms placed on a table of standard height. Shucking knives with a fairly long blade are now the standard type used for Pacific oysters (Fig. 151). Gloves and finger stalls or cots (short rubber coverings for the finger tips, much like long sewing thimbles) are needed, for shell edges are very sharp. Each opener has his own shucking container which is most often a stainless steel vessel, usually with a capacity of 4 L for ease of tallying, for shuckers are normally paid on a volume basis. The shucking container may or may not be perforated.

The opening process begins by placing the oyster on a table with the cupped or left valve down and with the hinge pointed toward the opener's left. With the oyster in this position, the single adductor muscle, which must be cut to allow the shell to open, is located about two thirds of the distance from the hinge toward the right. The point of the knife is inserted between the valves at this point with a slight twisting motion of the knife. The handle of the knife is elevated slightly, for most often the upper flat valve is slightly inside the lower valve. After the knife point has entered, the blade is forced into the oyster to about 3 cm (1½ inches), when a movement of the knife to the right and to the left will sever the muscle. The knife is then turned until the blade is vertical and a prying motion will break the hold of the hinge and the two valves will separate. The cutting of the adductor muscle by a right-left motion must be done against either the upper or lower valve. This should be decided before the knife is forced into the shells so that the knife blade will be against either one valve or the other, so as not to injure the oyster meat and also to ensure the adductor muscle is cut right against the shell. The valve with the oyster still attached to its is then picked up, held at the outside edge of the shucking container and the remaining attachment of the adductor muscle is cut flush with the valve surface and the oyster flipped into the pail.

In an alternative method the shucking knife is driven through the shell like a dagger an inch or so from the posterior or "bill" end of the shell. After entry, which may require more than one stab, the knife is worked forward to sever the adductor muscle.

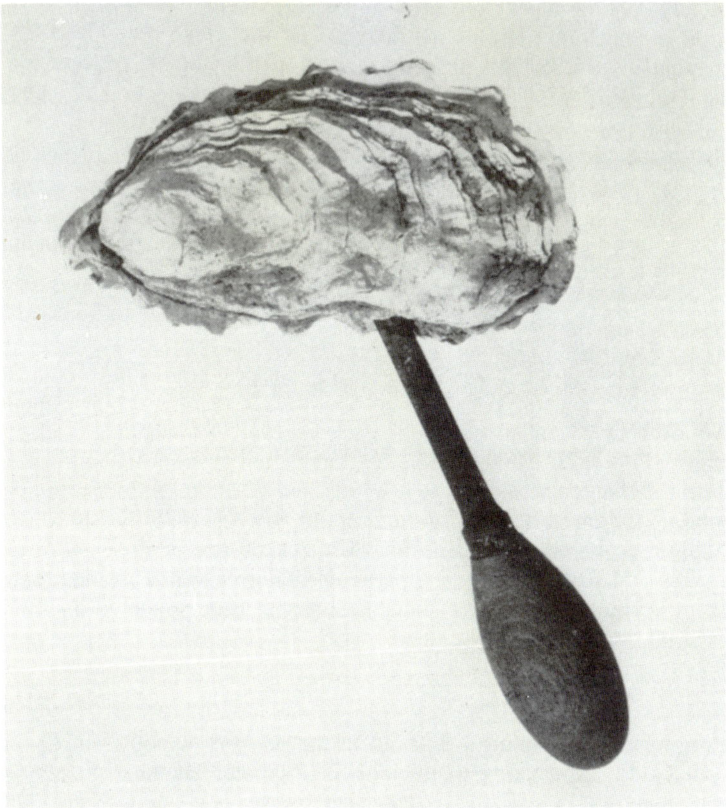

FIG. 151. Oyster knife typical of those used for shucking Pacific oysters.

It can be seen that up to this point the oyster has come into contact only with the oyster knife, and this is the way it should be. Also it can be recognized that if the oyster shell is well washed and clean, there will be little opportunity for mud to get into the shucking pail or to touch the oyster meat.

Mechanical Shucking

Many attempts have been made to develop a mechanical method for shucking oysters. Unfortunately, lack of uniformity in size and shape of oysters makes a purely mechanical system difficult and so far no suitable method has been devised.

Chemicals can cause the adductor muscle to relax and so enables the shucking knife to be introduced more easily; but from then on, essentially all operations used in normal opening must be carried out. There is also the necessity of providing tanks and chemicals; and, moreover, the time required for anaesthesia can be fairly long. Thus there is little advantage in using this method.

Steam is used to open oysters when they are to be processed by canning as whole or smoked oysters or as stew. The oysters are carried on an endless belt through a steam tunnel where the temperature is high enough to cause them to gape. Some drop out of the shell but others may require removal, so, while not an entirely mechanical method, it is more rapid than raw shucking. Microwaves will no doubt eventually replace steam in this process.

What is called the "heat shock" method for opening clustered and irregularly shaped South Carolina oysters has been developed. Oysters are given a brief dip of 2–3 minutes in water at a temperature of 63–66°C. This causes relaxation of the adductor muscle without actually cooking the oyster meat, and is followed by an immediate chill. Again, all of the normal shucking motions still have to be carried out.

An interesting discovery unlikely to be of practical benefit is that oysters will open quietly when subjected to a pressure of 700 atmospheres or 5 000 kg (11 000 lb) per 6.255 cm (square inch), equivalent to a water depth of 7 km (23 000 feet).

It may be that, in the future, most oysters will be sold fresh in the blanched condition. The blanching process consists simply of a brief dip of the oyster meat in boiling water to solidify the outer layers of tissue. This prevents slime formation during storage and it firms the oyster just enough to make it more attractive to handle. The blanching process is essential prior to freezing oysters.

Shell Disposal

Disposal of shell is a problem and some oyster shucking houses have holes in the table leading to chutes which carry the shells to a suitable collecting point. Small operators usually have a tub or wheelbarrow beside each shucker. The shells are taken to a shell pile for storage (Fig. 152). They may be used as cultch or sold to shell processors who grind them up for chicken scratch. It is also possible to use them for building or firming up oyster bottom. More soft oyster ground could be reclaimed in this way and it is a more profitable use of shell than disposing of it to shell processors.

Flies are often attached to shell piles, particularly during warm weather, and this leads to difficulty in keeping them out of shucking houses. Biodegradable insecticide sprays might be used on the shell pile when necessary.

Washing

After the shucking pail is full it is taken to the delivery window since, according to regulation, shucking and packing rooms must be separate. The next process is washing, the ideal way being to "blow" the oysters. The blower is a stainless steel tank with air pipes in the bottom; the tank is filled with water, oysters placed in it, and the air turned on. The air bubbles churn the water and oysters, providing a cleansing action without damaging the meats. Smaller operators, however, normally use stainless steel deep sinks

FIG. 152. Shell pile adjacent to a shucking house.

with a colander insert and stir the oysters gently by hand or with paddles. Rubber gloves are to be worn during washing and packing, according to regulation.

After washing, the oysters are poured onto a shallow perforated stainless steel tray called a skimmer. Here the oysters receive a final wash from an overhead hand held spray. Bits of shell as well as any reject oysters which are badly cut, of poor colour or condition are removed. Here, too, the oysters may be graded. The skimmer usually has several pour lips under which the final containers are placed.

Sanitary regulations state that all shucked stock shall be thoroughly washed with cold water for a period not exceeding 3 min. Also, the *Canada Food and Drug Act* states (B.23.017) that "No person shall sell shucked oysters that contain more than 10 per cent of fluid separable by draining for 5 minutes through a 10-mesh sieve".

The reason for these regulations is to prevent adulteration, for when a shucked oyster is placed in freshwater there is an exchange of fluids, with freshwater entering the cells of the oyster body and salts with minerals leaving those cells. In this way the fluid content of an oyster is increased as well as the whole volume and weight of the oyster, and the process is known as "plumping" or "floating". In this way volume yield can be unjustifiably increased in favour of the oyster producer while, at the same time, quality is impaired, for the consumer is then purchasing more water and less of the minerals and salts that contribute to the flavour and nutritive value of oysters.

The problem of oyster standards based on fluid content is a controversial one, for when an oyster is opened some blood vessels are usually severed and bleeding may occur. The amount and time of bleeding and indeed the amount of mineral salts–water exchange is variable from area to area and, very likely, from species to species as well as from season to season. At any rate, to comply with present regulations and to provide a quality product, washing and holding in freshwater should be kept to an absolute minimum. More than a 3-min wash may increase the weight of an oyster by 1–12%.

Packaging

The basic sanitation regulation regarding packaging states: "Shucked shellfish shall be shipped and sold, retail or otherwise, either in single service containers made of clean impervious materials positively sealed or in such containers so sealed that tampering with the container can be detected. After sealing, the container must be water tight." The purpose in using non-returnable containers is to ensure a minimum of handling for each additional step increases the hazard of contamination. In this way, the grower is assured that consumers receive a product in the same sanitary condition as it left the plant. Further, containers must be marked by the packer's certificate number, and it is thus possible to trace the source of any shipment. This is important to the producer, the governmental agency concerned, and the industry as a whole, for any problem such as contaminated oysters will reflect on every producer. In this way the source of contaminated oysters can be pinpointed immediately and steps taken to correct the difficulty. Containers are many and varied in size, shape, and material such as waxed paper, plastic, or both. Gallon containers are usually of metal. Many are of United States manufacture and are consequently in American measurements rather than Imperial. The relative number of each size of container used varies with sales, but the following gives an approximate idea of the manner in which oysters are marketed in British Columbia: 0.25 L, 75%; 0.5 L, 5%; L, 11%; 4.5 L, 9%.

Processed Oyster Products

In addition to standard fresh and half-shell oysters, there are several ways of processing oysters; these include smoking, canning whole oysters, canning oyster stew, and freezing. Only brief outlines will be given here, for anyone considering these processes will require a considerable amount of detail beyond the scope of this bulletin. So far in British

Columbia, the main processed pack of oysters has been the smoked variety, with lesser amounts of others. Assistance may be obtained from manufacturers of can and canning equipment.

Canned Smoked Oysters

This method requires oysters that have been steamed open or which have been shucked fresh and partially cooked. They are rinsed for 5 min in a 2.5% brine solution to salt the product slightly and to wash off bits of grit. They are then spread in a single layer on a 1 cm-mesh galvanized-wire tray which has been previously oiled with cooking oil to prevent adherence of oysters to the wire. At this time the oysters should be left in the open air as briefly as possible otherwise oxidation will cause them to become dark. For this reason, the smokehouse should also be ready at a smoking temperature of 50°C. Smoking time is variable but about 4 h at this temperature gives a light brown colour without causing the meat to become shrivelled or tough. Birch sawdust gives a sweeter flavour than alder, and does not darken the oyster as much; crab apple is also often used. The wood, time and temperature of smoking are largely a matter of individual taste and opinion.

Smoked oysters are usually packed in 225 g flat cans filled with 2 cm^3 of salad oil and exhausted for 15 min at 4.5 kg pressure in the retort. After double-seaming, the cans are then processed in a retort at 115°C for 60 min and immediately cooled.

Canned Oysters

This is the pack in which oysters are canned whole in brine. Either cold (fresh) shucked oysters or steam-opened oysters are used but in any event they are blanched before being placed in cans. Thereafter, standard canning procedures are used.

Oyster Stew

This product forms an outlet for oysters which are too large for the fresh oyster trade; however, such oysters bring a lower price to the grower. As with canned oysters, they are usually precooked, and chopped into small pieces. Milk, spices, and finally a pat of butter are added followed by standard canning techniques.

Oyster Sauce

An important cooking item in South-east Asia is oyster sauce. The present popularity of Oriental cooking in North America should provide an appreciable market for this product. Oysters less than table quality such as those cut in the shucking process or oversize could be utilized for this purpose.

Frozen Oysters

This method of preserving oysters is relatively new and as yet an ideal process has not been perfected. Just freezing a container of oysters results in a product that leaves much to be desired because of so-called "drip", a slimy secretion produced when they are thawed. This may be avoided by freezing individual oysters (I.Q.F.) and by preliminary blanching.

Precooked, breaded, frozen oysters have also been marketed with some success.

Half Shell Oysters

On the west coast of North America, in both the United States and Canada, virtually all oysters are marketed in the shucked state and, except for a small proportion used in oyster or seafood cocktails, they are consumed in a cooked state.

This is completely different from Europe and the North American Atlantic coast where oysters are consumed raw or, as it is often described, "on the half shell". Oysters are sold to restaurants in the shell and they are opened on the premises when ordered by customers. They are served in the cupped left valve, usually six on a plate.

Almost all oysters produced in the Canadian Maritime provinces are eaten raw on the half shell, and the cultural practices are designed almost entirely for this particular market.

In the past, small numbers of British Columbia oysters were used for the half-shell trade but this is now being expanded. Two problems are involved; the first is to make the oyster consumer aware that such a product is available and the second to have it available. This is not difficult, for it merely involves culling out small-sized oysters which should not be more than 60–70 mm in length and of good shape, nearly circular in outline and with a deep cup. Indeed, it is possible to grow nearly perfect half-shell oysters by growing single seed on hard gravel bottom. Raft culture oysters also make ideal half-shell oysters because of the short mantle and attractive colour. The value of half-shell oysters is, of course, considerably more than that of shucked oysters.

Since about 1970 there has been an upsurge in the production of half-shell oysters, associated with an increase in tray culture activity, and production has reached about 170 000 dozen, the unit of measurement for this type of product.

Nutritive Value and Chemical Composition

The seasonal change in chemical composition of oysters in British Columbia has been studied and the results are given in Table 40.

TABLE 40. Chemical composition of Pacific oysters in British Columbia.

			Solids calculated on moisture-free flesh				
	Average weight (g)	Moisture (%)	Protein (%)	Glycogen (%)	Fat (%)	Ash (%)	Balance
Feb. 9	18.90	78.20	47.80	20.50	10.68	8.66	12.36
Apr. 12	18.30	79.84	46.65	24.95	12.94	7.62	7.84
May 30	16.72	77.50	47.70	23.80	13.31	6.87	10.32
Aug. 3	11.07	81.41	54.60	11.85	15.75	7.80	10.00
Oct. 3	17.40	80.00	52.20	14.25	11.72	5.78	16.05
Dec. 7	20.10	83.37	49.90	20.05	13.08	8.42	9.45
Feb. 6	20.00	79.80	49.60	19.00	15.27	6.81	9.32

The balance represents material not accounted for, and is probably the result of destruction of some of the carbohydrate by chemical action in the analytical procedure. As previously observed, oysters are usually in best condition during spring, and this is reflected in the table with high values for glycogen which gives a prime oyster its characteristic colour and flavour. In early summer, as spawn is developed, apparently at the expense of glycogen, the value for this material drops but there is a corresponding increase in the protein value. Presumably in the experiment few, if any, of the oysters spawned. After the August low, glycogen content began to rise and the December value is indicative of the winter condition with a value somewhat lower than the high level in spring.

In addition to organic chemicals just mentioned, oyster meats also contain most of the numerous inorganic elements that occur in sea water such as sodium, potassium, calcium, magnesium, iodine, and phosphorus in small concentrations, and the following in very low concentrations: iron, copper, zinc, aluminium, silicon, strontium, lithium, rubidium, nickel, silver, titanium, vanadium, gold, zirconium, platinum, and manganese. An analysis of Pacific oyster meats and shells was made by the Washington Pollution Control Commission in 1959 and the results are given in Table 41.

Heavy metals such as copper, iron, zinc, and manganese can be concentrated and stored by oysters through direct absorption if there is a concentration of these metals in the sea water in which they are growing. Oysters stored in galvanized containers or on galvanized trays will concentrate relatively large amounts of zinc.

TABLE 41. Metallic element content of Pacific oysters, 1959 — Mean of eight samples.[a]

	Shell	Meat
Total solids		11.36 g
Ignition residue (% by weight)	93.66	—
Calcium — % ash	10+	—
Sodium — % ash	1+	—
Elements in ash (ppm)		
Aluminum	1 490	7 762
Silicon	4 562	9 500
Iron	1 024	1 070
Strontium	300	412
Zinc	—	250
Titanium	69	381
Copper	34	89
Managanese	—	128
Chromium	9	49
Vanadium	—	62
Boron	40	84
Silver	—	8
Lead	180	150
Nickel	—	72
Magnesium	4 162	—

[a] Data condensed from: "On Oysters and Sulphite Waste Liquor" by G. Gunter and J. McKee. 1960. Washington Pollution Control Commission.

Strangely enough, arsenic is one of the heavy metals that is stored only in modest amounts.

Copper is picked up readily, as is shown by the bright green oysters that grow near slag piles from former copper smelters in Ladysmith Harbour and in Howe Sound. Heavy metal analyses of British Columbia oysters from various areas are shown in Table 42. The highest concentrations were in oysters from near the slag pile of the former copper smelter in Ladysmith Harbour and from Howe Sound near Britannia Beach. In the case of copper, the colour is diffused throughout the body, especially where there is an ample supply of blood. Such oysters should not be eaten. This should not be confused with the green-coloured gills which result from the oyster feeding on certain phytoplankton. A well-known practice at Marennes, France, is to place oysters in ponds or claires in which a particular diatom flourishes in order to colours the gills, thus producing an oyster which fetches a premium price.

Oysters have been shown to be a good source of vitamins A, B, and D, as well as niacin, riboflavin, and thiamine. Recent studies have shown that some molluscs such as the eastern oyster and abalones contain antibacterial and antiviral substances.

TABLE 42. Heavy metal analysis of Pacific oysters from various areas in parts per million.

Area	Zinc (ppm)	Copper (ppm)	Lead (ppm)
Crofton	11 000	590	10
Ladysmith 1	13 000	700	10
Ladysmith 2	13 000	1 200	10
Ladysmith 3	21 000	940	10
Slagpile Ladysmith 1	28 000	9 600	10
Slagpile Ladysmith 2	34 000	20 000	10
Porteau	2 300–5 000	50–1 600	—
Commercial	330–480	5–27	—

Food Value

Dr. J. P. Tully has calculated the food value of Pacific oysters and his data are shown in Table 43. The indication here, also, is that oysters are in the best condition during winter and early spring.

TABLE 43. Energy content of Pacific oysters.

	Calories per 100 g (3.5 oz)			
Date	Protein	Glycogen	Fat	Total
Feb.	42.7	18.3	23.4	84.4
Apr. 12	38.4	20.6	24.2	83.2
May 30	42.0	21.8	27.9	91.7
Aug. 3	41.6	9.0	27.2	77.8
Oct. 3	42.9	11.6	21.8	76.3
Dec. 7	33.4	14.9	20.2	68.5
Feb. 6	40.9	15.7	28.8	85.4

Marketing

Oysters are marketed in a number of ways. These are (a) direct to the consumer (at the plant, to the public, or to restaurants); (b) to retailers (fish markets and supermarkets); (c) to wholesale fish dealers; (d) to other oyster producers; (e) to processors (canners).

Regardless of type of culture in use, growers should consider the range of potential markets. Though an operation may be primarily engaged in producing shucked oysters, a proportion of the production may be of half shell quality which fetch a higher price than shucked oysters and the harvesting process should take this into consideration by having well-shaped singles set aside.

Until recently, oyster marketing could only have been described as a ''free-for-all'' with considerable price-cutting and exchange of markets. For a number of years, the price of oysters increased very little in proportion to relatively large increases in production costs. However, on December 17, 1964, the British Columbia Oyster Marketing Board came into being and virtual chaos was changed into order. A plan was instituted, voted for by a majority of oyster growers, for the effective promotion, control, and regulation, within the province, of the transporting, packing, storing, and marketing of oysters.

The Board consists of three oyster growers who are elected by the British Columbia Oyster Growers Association at its annual convention. The Board licenses and collects fees from all persons handling oysters and fixes minimum prices at which the regulated product may be bought or sold in the province. The Board is also required to promote the oyster industry by advertising and other means.

Economics

Although economic implications played a large part in the design of many studies in this bulletin a thorough analysis of this aspect of oyster culture is beyond the scope of this account and would be soon out of date.

Oyster culture is required to be a profitable venture and indeed it is, as demonstrated by the considerable extent of worldwide production and in British Columbia by the continued participation of growers. There are three major studies on the economics of oyster culture in British Columbia including computer simulation models. These indicate the various factors that must be considered but rapid fluctuations in the value and cost make estimates quickly obsolete. Much depends on expectations of the individual and whether only production without processing is the approach, as well as the type of the culture.

The publications listed in the section ''Literature'' will provide a guide.

Oyster Production

Statistics

The earliest official record for oyster production goes back to 1884 when 220 barrels with a landed value of $1,250 were produced. These were, of course, native oysters. By the turn of the century production had reached 283 t (in the shell) and was valued at $15,000. Production fluctuated widely (Table 44) and analysis is made difficult because published statistics did not differentiate between native, Atlantic, and later Pacific oysters. However, it was not until 1930, when Pacific oysters were first marketed, that the landed value reached the $50,000 mark, and not until 1944 that production was valued over $100,000 (Fig. 153, 154). It is remarkable that production was maintained as high as it was during the later years of World War II, for there was no importation of Japanese seed from 1942 to 1947. The 1941 Japanese seed reached market size in 1944 and the 1947 seed in 1950 so there was a period of 6 years of production during which seed requirements were filled by local seed fortuitously produced in 1942 by general breeding in most of the Strait of Georgia growing areas.

During the postwar period there was a gradual year-by-year increase in production with minor fluctuations until 1962 and 1963, when two significant increases placed the landed value at $635,000 in 1963. This was due to the availability of large quantities of wild seed and oysters from the extensive and widespread breeding in the Strait of Georgia in 1958. Oysters resulting from this breeding were available to the industry in fairly significant quantitites until 1965 and in reduced numbers until 1969. Between 1939 and 1961, inclusive, 324 000 bags of Japanese seed were planted on British Columbia oyster beds. At an estimated yield of 113 L per 6 bags this should have produced 49 090 t during that period. The actual production was approximately 63 630 t giving an estimated 14 540 t from local seed, equivalent to 100 000 bags of seed.

TABLE 44. Total production of all species of oysters in British Columbia, 1884-1985.

	Production value			Production value			Production value			Production half-shell value		
Year	(t)	($)	Year	(t)	($)	Year	(t)	($)	Year	(t)	(dozens)	($)
1884	20	1,250	1912–13	252	11,282	1940	789	36,874	1968	3 508		562,000
1885	23	1,250	1913–14	243	9,380	1941	1 326	80,100	1969	4 427		720,000
1886	27	2,100	1914–15	161	13,840	1942	1 362	56,749	1970	2 898		530,000
1887		3,500	1915–16	99	14,337	1943	1 372	76,514	1971	3 067		536,000
1888	109	2,400	1916–17	141	20,788	1944	1 807	129,644	1972	4 000		702,000
1889	136	5,250	1917	163	32,202	1945	1 370	99,835	1973	4 787		875,000
1890	68	7,000	1918	132	26,926	1946	2 427	231,523	1974	3 930	22 000	880,000
1891	91	3,000	1919	216	38,569	1947	2 207	197,600	1975	3 246	35 000	883,000
1892	145	4,000	1920	156	36,830	1948	1 638	187,600	1976	3 245	35 000	887,000
1893	145	8,000	1921	144	21,136	1949	2 172	233,900	1977	2 998	40 000	981,000
1894	145	8,000	1922	186	30,406	1950	2 280	287,800	1978	2 793	32 005	1,021,000
1895	145	4,800	1923	234	23,625	1951	2 143	233,500	1979	2 231	36 000	893,000
1896	109	8,000	1924	151	26,492	1952	2 952	368,600	1980	1 922	41 000	1,134,000
1897	145	4,800	1925	134	22,905	1953	2 417	276,000	1981[b]	1 415	64 000	1,030,000
1898	218	8,000	1926	214	34,122	1954	3 122	307,000	1982[b]	1 579	101 000	981,000
1899	218	12,000	1927	199	30,792	1955	3 084	316,000	1983[b]	2 453	132 000	1,554,000
1900	270	12,000	1928	219	40,334	1956	3 272	340,000	1984[b]	2 897	172 000	2,109,000
1901	270	15,000	1929	384	49,952	1957	2 450	265,000	1985[b]	3 420	269 000	2,613,000
1902	270	16,000	1930	290	56,825	1958	2 282	334,000				
1903		18,000	1931	323	61,247	1959	2 715	407,000				
1904	363	13,000	1932	183	28,638	1960	2 850	339,000				
1905	230	7,190	1933	203	25,670	1961	3 097	369,000				
1906	132	5,075	1934	299	33,886	1962	3 679	466,000				
1907–08		10,000	1935	305	43,173	1963	6 195	635,000				
1908–09	145	7,263	1936	563	51,978	1964	5 530	444,503				
1909–10	363	30,935	1937	212	36,199	1965	5 479	337,643				
1910–11	247	22,362	1938	271	36,258	1966	4 885	672,627				
1911–12	489	39,053	1939	1 028	48,719	1967	4 667	870,981				

[a] 1884–1980 data from Fishery Statistics of Canada.
[b] 1981–1985 data was provided by the Shellfish Unit, Marine Resources Section.
[c] Includes a small amount of wild oyster harvest from picking permits.
Conversion: Shell oysters — 1 barrel = 200 lb = 90 kg = 2.5 gallons = 11.3 L.

During the 10-year period from 1951 to 1960, average annual production was approximately 3 400 t. During the period 1948 to 1957, inclusive, when seed which produced oysters between 1951 and 1960 was on the beds, approximately the equivalent of 200 000 bags were planted. This shows a return of 18 L per bag.

In 1963 there were 1 400 ha with about 160 lots under lease, including both Provincial government, and Esquimalt and Nanaimo Railway leases, held by 78 leaseholders. The total return for that year was 6 290 t and on the basis of an area of 960 ha held by the 38 growers who produced oysters in that year, the average return was 820 L/ha. When it is considered that an hectare of oyster ground should produce between 2 300 and 3 000 L/ha annually, it is evident that British Columbia oyster ground is not producing at full potential.

In mid-1966 there was a total of 80 leaseholders with 1 100 ha of which 210 (20 leaseholders) were granted by the Esquimalt and Nanaimo Railway on its Crown-grant foreshore in Baynes Sound. The remaining 2 211 (60 leaseholders) were granted by the Provincial Department of Lands. Some leaseholders may have had a number of separate leases. The leaseholder at that time held an average of 14 ha. More than half the leases were less than 8 ha, and 40% were less than 4 ha, indicating either that a considerable proportion of leaseholders did not use their leases or that oystering was a part-time occupation. Later the province acquired jurisdiction over the E & N crown grant foreshore.

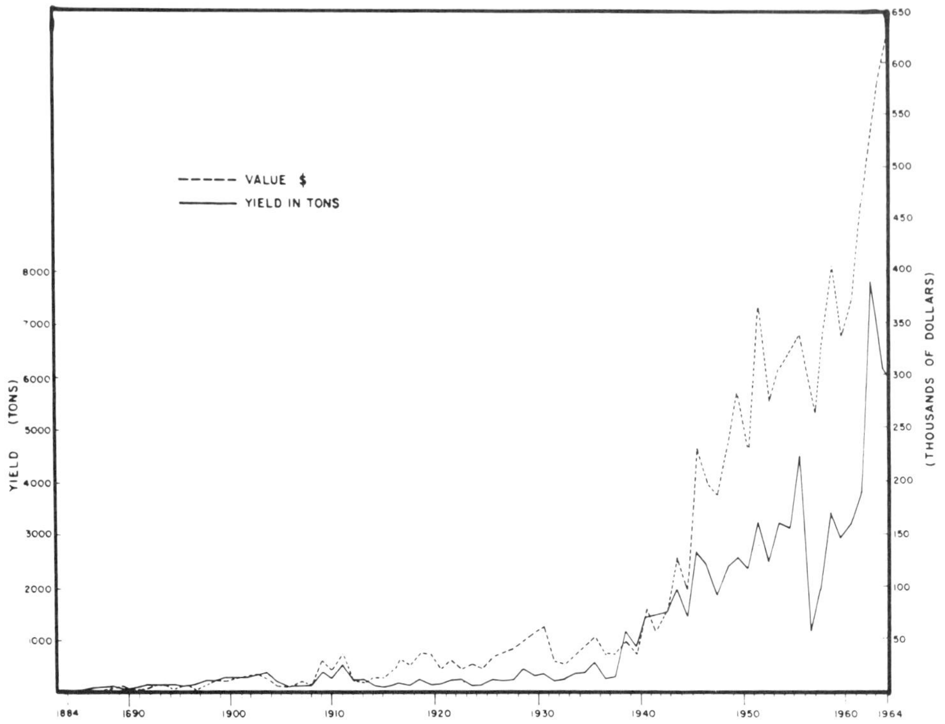

FIG. 153. Oyster production in British Columbia, 1884–1964.

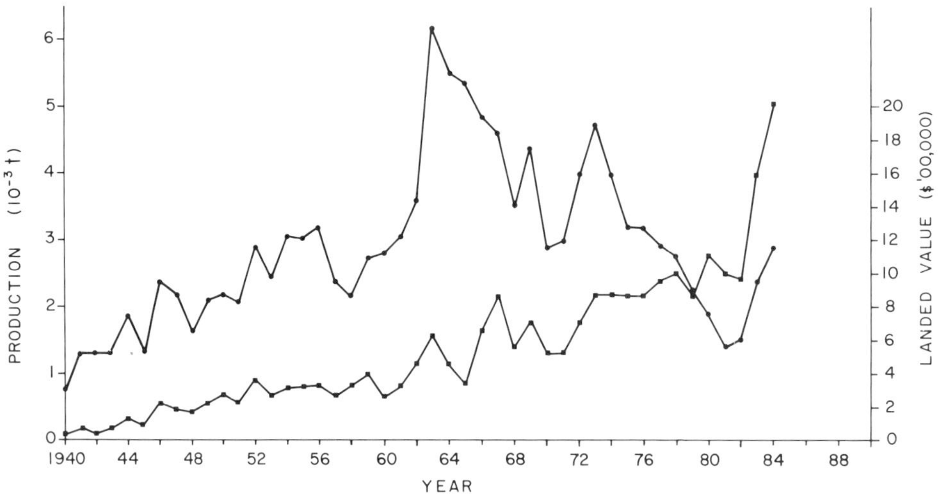

FIG. 154. Pacific oyster production in British Columbia (in gallons), 1940–84. Upper — production, lower — value.

It is estimated that probably not more than 85% of oyster ground under lease is actually usable, for often a proportion of unusable or marginal ground is included to simplify survey. On this basis, the total potential yielding area (1966) was 1 080 ha less 15%, or 920. A 3-year cycle may produce 8 500 L or 2 700 L/ha per year, so the estimated potential annual production should approximate 21 000 t.

By 1977 there were only 65 growers, but by 1983 the number had grown to 164 and is increasing rapidly. Because of time lag in production the effect of this dramatic increase will not be felt immediately. Production rose to another peak in 1973 with 6 000 t and then dropped to a low of 1 300 t in 1981, after which an increase appears to be developing with a production of 2 600 t in 1984. In 1985 there were 138 growers leasing 1 495 ha of oyster ground.

It may be seen that the British Columbia oyster industry can expand its production considerably with bottom culture alone. There is a natural desire to expand as the potential market increases and there is an automatic expansion in markets due simply to population increase but rate of expansion could be accelerated considerably by advertising. Whether or not advertising is economically feasible is problematical considering the size of the industry. Also, most growers are single-family producers who are content to maintain a production sufficient to provide a modest income.

British Columbia is poorly endowed with good oyster ground for bottom culture and in the past much leased ground has lain fallow or underutilized as shown above. Now there is effort by provincial authorities to pursue the concept of diligent use of leased oyster ground. In contrast to the paucity of bottom ground, the islands and protected bays of the coast line provide a wealth of area suitable for suspended culture. The potential of these techniques, in spite of long time availability, has not been realized. This may change with the new surge of interest in oyster culture.

Production Factors

There are, however, other techniques that will permit expansion of oyster production e.g. suspended culture. There is also modest amount of what may be termed marginal oyster ground. This type of ground lacks one or more of the three basic requisites of oyster ground: (a) correct tidal height, (b) correct bottom consistency, (c) adequate protection from wave action. As oyster ground becomes more valuable, as it will, it becomes economically feasible to change the level of ground that lacks the correct tidal cover either by excavation if it is too high, or by filling if it is too low. Bottom consistency can be changed by filling with appropriate material such as gravel or shell on soft ground. Unprotected ground maybe guarded by appropriately designed breakwaters. However, even now, some marginal ground could be integrated in a system of oyster culture as outlined on page 125.

In addition, as mentioned previously, the grower should maintain careful records of everything he does with his oysters from time, place, and density of planting of seed to the time it is marketed. Source of seed should be noted as well as an average spat count and the average spat size. Size of every planting of seed at the end of each year should be noted, as well as the estimated mortality. Every transplant should be recorded.

When oysters are harvested, some record of the volume of oysters harvested, and their condition of fatness, should be maintained as suggested on page 142. Only in this way can the grower make an intelligent appraisal of what he is doing and whether he is doing it correctly. The expert can only suggest broad principles on how to grow oysters, for each oyster bed is different from every other and there can be considerable variation within a single bed. It is up to the oysterman to be aware of the details of his own ground, points that can mean the difference between only adequate and excellent production.

An example of a typical record form is shown below. Individual growers can easily devise a form to suit particular needs.

PRODUCTION REPORT

Oysters transplanted to or harvested from: ____________________ Date ____________ , 19 ______
(Bed name)

Oysters transplanted to ____ Date ____________ , 19 ______
(Bed name)

Year:					
Type:	Hatchery seed	Collected seed	Wild seed	Wild oysters	Other

Harvesting method ____

Gross Tonnes	Av. No. Oysters per Gross Tonne	Gross No. of oysters

Mortality: Estimated % ______________________ Reason ______________________

Comments: ____

____

Pollution

This is probably the most serious of the problems of the shellfish industry today, not only in British Columbia, but throughout the world. In the Canadian Maritime provinces of Nova Scotia, New Brunswick, and Prince Edward Island, there are about 150 shellfish growing areas which are unproductive because of pollution. British Columbia is relatively newly settled, with a small population and it would be expected there would be few, if any, shellfish pollution problems as yet. However, sewage pollution has already closed or limited a significant proportion of oyster producing ground, and industrial pollution is a problem in several oystering areas.

The two major types of pollution which may affect oysters are industrial and sewage.

Industrial Pollution

As the name implies, this type of pollution emanates largely from industrial wastes. In addition to manufacturing plants, such as pulp mills, oyster grounds may be polluted by washings from ore dumps or by adjacent log booming grounds. A single pollutant may not create significant harm, but often the single source provides a nucleus around which further industrial development takes place and the resultant combination of pollutant sources may be sufficient to cause difficulty.

So far, the two main industrial pollutants of concern to the oyster industry are the waste liquor from pulp mills and the effects of log booming operations.

Pulp Mills

The two types of pulp mills are sulphite and sulphate (kraft). In British Columbia, most mills are of the latter type, whose effluent as a whole is considered to be less toxic than that of the sulphite type. In pulp processing large quantities of water and a variety of chemicals are used. After the water has been used, it contains some of these chemicals and it must be discharged somewhere as effluent. The obvious place to discharge waste water is into a body of water such as a river or the sea. Naturally, mills attempt to recover as much of the material in waste water as possible as a matter of economics. However, some chemical waste and some wood fibre are lost in the effluent, which may amount to more than 113 million L/day or the equivalent of a small stream. Mill effluent is termed sulphite waste liquor or spent sulphite waste liquor (S.W.L. or S.S.L.), and kraft mill effluent (K.M.E.).

These effluents, when discharged into a body of water, either salt or fresh, have certain characteristics which are able to affect living organisms. These characteristics are (1) Toxicicy, (2) Oxygen demand, and (3) Particulate material.

Toxicity

The source of toxicity in effluent is some of its chemical constituents. These are generally more toxic in kraft mill effluent than in sulphite effluent, although their total effect is less because of the very low concentration. Two types of effect may result from toxins in the effluent. A direct effect is the immediate toxic action on the oyster, whose tissues may be harmed or whose physiological activities may be affected. Indirect effects influence the environment of the oyster, particularly food organisms on which it depends.

Oxygen Demand

Pulp mill wastes contain quantities of organic materials which, upon decomposition, require large amounts of oxygen. This oxygen requirement is called biochemical oxygen demand (B.O.D.) and can denude surrounding waters of this material upon which living organisms depend. If the oxygen concentration of water surrounding an organism is lowered sufficiently, the animal will virtually suffocate.

Particulate Material

This consists of wood fibres, bark, and chips which are often sufficiently heavy and concentrated to sink and form a mat on the bottom. This prevents circulation of water to the bottom and animals living in or on the bottom will perish. In time, these begin to decompose and an oxygen demand is set up with attendant production of hydrogen sulphide gas. This can often be seen bubbling to the surface near log dumps. This again is an indirect effect. A direct effect of suspended material, such as fibre, results when there is enough present to clog up the gills of filter-feeding organisms such as oysters.

The effect of these factors on oysters in the vicinity of a pulp mill effluent may be felt in several ways:

(1) Mortality
(2) Reduced growth rate
(3) Reduced fatness
(4) Effect on breeding.

However, measurement of these factors, while not particularly difficult in itself, becomes a complex problem when effects of the effluent must be separated from the wide variations that occur normally where there is no pollution. Of these, mortality, because it is definite, is the simplest of factors to measure, but only rarely are effluent concentrations high enough to cause significant mortality in the hardy Pacific oyster.

Through the years, particularly in the United States, a considerable amount of research effort has failed to produce conclusive results regarding the effect of pulp mill pollution or the generally acceptable tolerance limits above or below which there will be or will not be significant effects.

The effects of pollution on oysters pose a particular set of problems. Immobility forces it to accept whatever environmental stresses are placed on it. However, ability to close its valves and the sensitivity of the mantle edge to chemical changes make it possible for the oyster to exclude the external environment for a time.

The Pacific oyster is an extremely hardy and adaptable animal. The marine environment of oysters, usually an estuarine one, is highly dynamic and the temperature, salinity, currents, and other water characteristics are continually undergoing large daily, seasonal, and annual changes. Detailed measurements of any one or all of these characteristics relative to its total effect on the oyster is difficult. As the environment is in continual flux, so are the various physiological activities of the oysters in response to these changes, so the total effect of physiological activity is difficult to measure.

An obvious solution to these difficulties appears to lie in the laboratory where the environment can be controlled. Much laboratory work has been done, measuring mortality, growth, and various physiological activities of oysters, sometimes for quite long periods, against various "set" concentrations of mill effluent. Interesting results and correlations of value have been derived, but parallel conditions do not exist in nature and it is virtually impossible to duplicate in a laboratory the dynamic aspect of the natural environment.

It is necessary to turn to field studies, and required is some single measureable factor which integrates or combines the various physiological activities of the animal. The condition factor nearly fits this requirement. Most activities of the oyster such as heart beat, ciliary motion, adductor muscles action, mantle movements, and sensitivity as well as others are directly or indirectly involved in the fattening process since they control the intake of food. Fatness is, together with growth and mortality, a major factor in determining whether or not an oyster-growing operation is profitable. Every grower can observe and appreciate changes in fatness, while this is not so for such a thing as rate of ciliary action. Fatness can also be readily measured by the grower (page 137) and he should do so regularly, particularly if there is the slightest possibility his beds may be affected by industrial pollution, now or in the foreseeable future.

Kraft Pulp Mill Pollution

To demonstrate how pollution problems may be studied an account of an experiment with pulp mull effluent and its effect on oysters is given. A kraft pulp mill located at Crofton which began operations in 1958 created a problem for the oyster industry there. The mill is located directly above oyster leases and the effluent line passes over the oyster beds to be discharged in Stuart Channel in a depth of 20 m just off the edge of the tidal flat drop-off. In connection with this situation a number of studies of condition factor, growth, and spat survival were made but with inconclusive results. However, to demonstrate the effect of kraft mill effluent on an oyster, an experiment was conducted at the Harmac kraft pulp mill near Nanaimo.

Experimental Design

The experiment was designed to test the effect of varying concentrations of kraft mill effluent (K.M.E.) on the condition of oysters. The assumption made was that the concentration of effluent affecting oysters would decrease with distance from the outfall discharge at the Harmac pulp mill near Nanaimo.

A randomized block design of three blocks with six replicates at each station (block) was selected. Each replicate group of experimental oysters was held on 1 m × 1m × 15 cm wire trays coated with neoprene. The trays in this instance were held on floats and suspended 1 m below the surface. One site was located on a float above 50 m from the outfall which was just above the high tide level from where it simply poured down the beach. At this station there was little marine life except for slime algae on the float logs and the barnacle (*Balanus glandula*) on the adjacent rocky shore. Because of discolouration of the water by the effluent, it was not possible to see the trays at any time.

Another station was located on the end of a standing boom in the log pond about 300 yards from the outfall. The fauna and flora on the float logs at this point were similar to that in the general area. The control, was located at the Biological Station in Departure Bay, about 6 miles from Harmac, and where K.M.E. was presumably absent.

Distribution of oysters, was randomized in the trays with 150 oysters per tray. Samples of 15 oysters from each tray were used for condition factor determinations.

The experiment was begun in May 1962 and completed in May 1963 with samples taken at intervals of about 2 months.

Results

Results demonstrated a significant deleterious effect of effluent on condition factor with a mean difference over the year of 50 condition factor points between the mill and control in Departure Bay. With growth rate there was a difference of 20 cm^3 in volume and 13 g in the weight of shell added during the year of the experiment.

Log Booming

Sites for log booming leases are most often chosen on tidal flats and frequently occur close to oyster leases. There are two main problems. The first is deposition on the bottom of bits of bark broken from logs. Most of it drops directly below the log booms but some pieces are just light enough to float for a time. Some of this inevitably land on the surrounding area and in time enough collects to cause anaerobic conditions. The bottom becomes foul and hydrogen sulphide is produced causing mortality of any animals, including oysters, living there. The effect on oysters of toxins leached out of log booms is not known. Consequently log booming ground and oyster leases should be kept as far apart as possible. However most of the suitable intertidal ground, particularly in the Strait of Georgia has already been taken up by either log booming or oyster leases and there is now full awareness of the conflict between the two. The main danger in the future lies in the possible competition between deep water log booming, suspended oyster culture and fish farm sites, all of which have quite similar requirements. The propellor wash of tugboats towing logs in the vicinity of oyster beds may create furrows in the bottom and move oysters, if the boats are off course.

Heavy Metal Pollution

Shellfish are capable of accumulating fairly high concentrations of a number of heavy metals such as zinc and copper and this has already been described. While the animal itself is not harmed by the metal, its consumption by humans can result in illness if the metal concentration is high enough. High concentrations of copper in oysters results in a grass-green colour of the meats. This occurs particularly near copper slag piles in salt water such as at Ladysmith or near Britannia Beach in Howe Sound. Light tinges of green colour may also be found in other areas.

In British Columbia, a solution of an arsenic salt is sometimes used to spray log booms to control shipworm attack. Oysters extract and accumulate arsenic from this water but fortunately in quite low concentrations.

Organotin

Tributyl tin antifouling paint is causing some concern where many vessels occur in the neighbourhood of oyster operations. Antifouling paints operate on the principal of leaching of the toxic elements from the paint coating to form a thin protective layer of poisonous solution. Turbulence can disturb this layer and cause distribution of the toxic solution beyond the object being protected. Both experimental laboratory and field studies have shown that organotin compounds such as tributyl tin oxide in antifouling paints affect shell deposition in Pacific oysters as well as larval survival. It leads to thickening and layering particularly of the upper (right) valve, sometimes forming inclusions of gelatinous material between the layers (Fig. 155a, b, c). Growth and condition factor is also reduced in the presence of organotin compounds. An index of shell thickening is given by the ratio between the length and thickness of the upper valve. Low values (circa 3–5) indicate malformation while normal valves have a ratio of about 10–15. The high degree of correlation between the phenomenon and its occurrence in areas of vessel

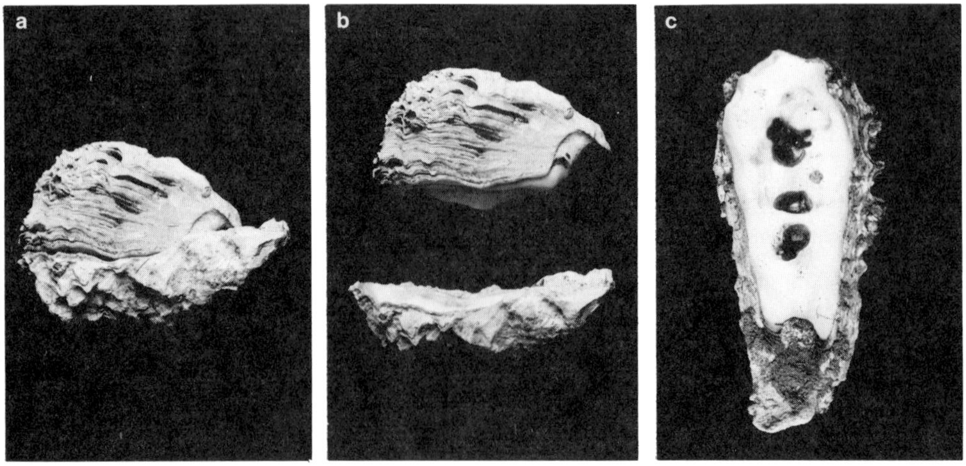

FIG. 155. Pacific oyster from the Nanaimo yacht basin. a. Lateral view of whole shell. b. lateral view of single valves showing extreme thickening of the right (upper) valve. c. black markings associated with the valve thickening phenomenon. (× 1).

concentrations has been established and in France, antifouling paints containing organotin compounds are banned on vessels less than 25 m in length. Shell thickening in Pacific oysters has been observed in Nanoose Bay and the yacht basin in Newcastle Island Passage, Nanaimo, but whether this is the sole result of exposure to tributyl tin is not known at this time. The black areas on the internal surface of the shell presumably of a conchiolin base occurs in a few specimens of the malformed oyster from the above two sites. It has not been observed elsewhere in oysters of normal shape.

Shellfish Sanitation

Sanitation is an extremely important part of oyster culture for public health is dependent on the integrity of the grower and purity of his product. Past history of outbreaks of hepatitis and cholera have shown how these events can cause devastation in the industry for a considerable period of time. It is therefore necessary the grower has full knowledge of this aspect of his industry. Although he will not be conducting bacteriological tests himself, he should have an appreciation of the complexities of the procedure, hence the detail given here.

Sewage Pollution

Oysters and other shellfish have no disease of their own which they can transmit to man. However, they have frequently been agents by which diseases such as typhoid fever and infectious hepatitis have been transmitted. Since the oyster is a filter-feeding organism, it collects and concentrates the most minute particles from the water and, among these are bacteria and viruses. The filtering mechanism has limited selective power, so both harmless and harmful particles are accumulated.

It has been shown that bacteria, once collected by shellfish, are able to survive for long periods even under refrigeration. In addition, under normal storage conditions they are able to multiply within the bodies of shellfish. The usual cooking methods may destroy some bacteria, but some are resistant to heat and among these are some harmful (pathogenic) types. Processing methods for fresh oysters are such that destruction of bacteria is not possible, and, indeed, they are often a means by which shellfish may be further contaminated. Therefore, it is necessary that shellfish be bacteria-free, or nearly so, before they reach the consumer.

Contamination may originate from two main sources. The first is from waters in which oysters are grown and the second is the subsequent treatment at and between the shucking house and the consumer.

Sanitary Survey

Sanitary control of bivalve shellfish is based on bacteriological examinations supported initially by sanitary surveys. The intent and plan of the bacteriological phase will depend to a considerable extent on findings of a sanitary survey which studies sources of pollution in the area, how pollution is distributed, and examines such factors as:

1. Topography, geology and oceanography of the area.
2. Natural and artificial drainage systems.
3. Rainfall and seasonal rates of stream discharge.
4. Shipping channels and moorages.
5. Population distribution in place and time.
6. Industrial, institutional and recreational areas.
7. Economic status of the community.
8. Sewage treatment and disposal.

With these data a bacteriological survey may be undertaken and attention given to:

1. Bacteriological quality of the discharge from streams and stream waters.
2. Statistical validity of possible sampling techniques.
3. Seasonal factors.
4. Effects of hydrographic factors.
5. Levels of pollution caused by the main sources.
6. Distribution of bacteriological levels in the shellfish of the area.

Pollution in Growing Waters

Waters where shellfish are grown may be polluted by direct discharge of trunk sewers into the area or by drainage from individual, improperly installed, or improperly functioning septic tanks. Pollution may occur indirectly, by runoff from the land through seepage after rains, or river discharge. Boats discharging raw sewage may be a significant source of pollution. One population equivalent (fecal sewage production from one individual) is considered to require daily dilution by 25 000 m^3 (8 million cubic feet) of pure water to meet required sanitary standards for shellfish.

Sea water always carries a certain content of marine bacteria which is a normal part of life in the sea. However, from sewage sources are added bacteria which normally inhabit the digestive tract of man and other animals such as gulls and wild or domesticated animals. Most of these bacteria are not particularly harmful but others, such as the species causing typhoid fever or the virus causing infectious hepatitis, may also be found if there are carriers in the area. Since harmful types are generally quite rare, it is too difficult to search for them, so bacteriologists use the more common, harmless types originating in the colon portion of the digestive tract as indicators. These are normally divided into two groups which may be identified by bacteriological tests. These are the coliform group and a human faecal coliform type known as *Escherischia coli*.

By appropriate tests, quantitative estimates of the occurrence of these two groups can be made, and numbers found are assumed to represent the level of sewage contamination. In turn, this is an indication of a possibility of the occurrence of harmful organisms.

Shellfish growing areas are classified in three ways, partly on the basis of a bacteriological study of growing waters and partly on the basis of a sanitary survey, where a study is made of such factors as population density and distribution, sewage disposal systems, river discharge points and flow rates, geology, distribution of docks. The following classification is taken from "The Regulations for the Sanitary Control of the Shellfish Industry in British Columbia."

1. APPROVED AREAS — Shellfish may be taken for market purposes from approved areas at any time provided that any rules or regulations of the Department of Fisheries concerning conservation or the regulations concerning relaying are not violated.

2. RESTRICTED AREAS — Shellfish shall not be taken from any restricted area except under written permission of the Deputy Minister of Health and only under any specific conditions imposed by the Deputy Minister of Health.

3. CLOSED AREAS — Shellfish shall not be taken from closed areas, except for the purpose of depletion, and then only under written permission and strict control of the Deputy Minister of Health. Oyster leases shall not be granted in closed areas.

Following granting of a lease and classification, subsequent periodic checks are made of the bacteriological quality of the growing waters and oysters, with observations on general sanitation of the area. The grower should assist by keeping a careful watch on any deleterious sanitary condition that arises, for the time to correct a difficult situation is in its early stages.

As indicated, closure has been applied to growing areas and a number of leases in British Columbia have been placed in the restricted category. In other parts of the world, such as England and France, where dense populations occur close to shellfish growing areas, pollution has been a problem for many years. The problem has been solved by the use of various methods to purify oysters. Here in British Columbia, partly due to recent closures, a system based on the bactericidal properties of ultraviolet light has been designed for conditions in this province.

Purification of Molluscan Shellfish

Shellfish purification (sometimes called "depuration") is based on the knowledge that filter-feeding molluscs remove solid particles from the water around them, digest some, and pass out the rest enmeshed in mucus of faeces and pseudofaeces. In time, it is possible for them to remove enough bacteria from a small volume of water to make it pure enough for them to cleanse themselves in it. Naturally, cleansing action is most rapid when pure water is used initially. This system presumes adequate temperature and salinity. It is known, of course, that at low temperatures, most filter feeders do not filter as rapidly as at higher temperatures and this must be kept in mind in purification procedures, as the rate of purification will normally change with temperature.

Pure water for purification may be obtained from nonpolluted sources or slightly polluted water may be purified by treatment with small amounts of chlorine as is done in many cities with drinking water, or by treatment with ultraviolet light or ozone, which can destroy bacteria.

Chlorine has been used for many years, with apparent success, in England and France. It has been shown that chlorine has an inhibiting effect on the feeding rate of oysters and, consequently, it is not advocated for shellfish purification. Ultraviolet light treatment, on the other hand, does not have such a deleterious effect and, since proper dosage can destroy over 99% of the bacteria, this is the treatment advocated.

The "United States Public Health Service Manual of Recommended Practices for the Sanitary Control of the Shellfish Industry" stipulates that water used for the purification of shellfish, if treated, should meet Public Health Service drinking water standards.

Shellfish Purification Procedures

1. Restoration of polluted areas

Restoration of polluted waters to the former pure state is the most positive and, in the long run, possibly the most economical approach to shellfish purification. Realistically, this is not always possible, but even a reduction in the level of pollution would be desirable. Also, cost of restoration should not be charged entirely to shellfish purification, for there is also an aesthetic aspect, relative to use of these waters by the public.

2. Relaying

Fortunately, in British Columbia, there are still many areas in close proximity to oyster beds now under closure because of sewage contamination, which are still free of pollution. Such areas could be used for relaying oysters from polluted beds for purification. In some sites, natural bottom would be suitable for relaying: in others, tanks or embayments may be used. It may also be possible to transport and hold oysters for purification in a container which could be held either on the bottom or suspended from a raft. Such relaying sites might be governmental reserves set aside for this specific purpose.

Natural purification involves moving — called "re laying" — contaminated shellfish from a polluted growing area to an uncontaminated area. Most jurisdictions require a relaying time of two weeks but the actual purification is accomplished within two or three days.

In the feeding process in clean water, contents of the digestive system are removed as well as renewal of mucus on the gills, mantle and body. The main costs of relaying is due to double harvesting and transportation. This can be about 5% of the cost of producing oysters.

Costs may be reduced somewhat by relaying oysters in wide mesh baskets (Fig. 156). Studies on relaying with baskets 62 × 62 × 30 cm holding 55 kg of oysters (*Crassostrea gigas*) and easily carried by two people showed adequate purification within 48 h over a range of temperature. There is a slight increase in purification rates at higher temperature levels. Even grossly polluted oysters may reach equilibrium with the coliform counts in the purifying environment. The shellfish can be transported from purification to processing sites in the same container.

3. Artificial Purification

Artificial methods of shellfish purification are used extensively in other parts of the world, particularly in Europe. Apart from capital costs of an artificial purification plant,

FIG. 156. Relaying oysters for purification in baskets.

the high level of bacteriological control necessary cannot be provided by the small operator. In British Columbia a feasible approach to artificial purification would be a large central plant operated on a cooperative basis.

At present, the several suggested methods of relaying appear to hold more promise than the artificial methods described below.

Purification (Depuration) in British Columbia

For some years there has been a significant amount of oyster producing ground lost through closure or restriction caused by bacteriological pollution. Also there is a steady increase in population density in coastal B.C. particularly in the Strait of Georgia. This caused concern that in the not too distant future artificial purification of shellfish in B.C. might become a necessity, as it has been in most European countries. All of the 120 000 t (1983) of mussels produced annually in Spain are artificially purified.

To prepare for this eventuality, a small scale purification plant was established in Ladysmith Harbour in a joint Federal–Provincial project in 1971. A suitable protocol for the purification of oysters under British Columbia conditions was established.

Purified water was supplied through ultraviolet light units that provided a flow of 90 L/min through 4 tanks of 2 000 L capacity. Each tank held 1000 kg of shell stock in baskets. Temperature, dissolved oxygen and turbidity were monitored through the 48 h required for purification. Bacteriological quality of the water and of the shell stock was also monitored on a strict programme. A plant of this size is capable of processing shell stock producing about 5 000 L of shucked oyster meats per month.

Cost of purification is greatly dependent on scale. Apart from capital outlay, the difference between relaying and depuration is slight.

Relaying, where possible, is the economic solution to shellfish purification, but the maintenance of unpolluted growing areas is of prime importance.

After harvesting, shellfish are subject to storage, processing and shipping. Without suitable precautions the bacteriological quality can rapidly deteriorate to an unacceptable level.

Wet storage requires a level of refrigeration and speedy utilization. Shellfish transported and sold whole in the shell is termed shell stock. Producers using this method should be certified and each shipment should be labelled in case of problems so its origin may be traced. Otherwise shellfish are processed in a plant where, at least initially, shucking (opening) and packing is carried out. Certification of the plant is required for cleanliness and health of the personnel are factors in maintaining a satisfactory bacteriological level.

Sections of the British Columbia Fish Inspection Act covering essential requirements for mollusc processing plants are given in Appendix D. Growers should obtain a copy of both the Federal and Provincial Fish Inspection Regulations.

In summary, shellfish sanitation is an important phase of a shellfish industry requiring a level of technological competence which may only be reached by special training. Health problems created by eating unsanitary shellfish can harm or destroy an industry for a long period of time.

Bacteriological Testing

Water samples for bacteriological analysis are collected in clean sterile bottles with care taken to prevent contamination during and after sampling. They are stored at or below 10°C and testing is done preferably within one hour of sampling and not later than 30 h.

Presumptive tests are done in replicate test tubes (usually 5) containing laural sulphate lactose broth or lactose broth innoculated with three concentrations of the sample (usually 0.1 mL, 1 mL, and 10 mL). These tubes are incubated at 35°C, ±0.5°C and examined after 24 h and then at 48 h. Formation of gas within these time periods indicate a positive presumptive test for the coliform group of organisms. The test is now confirmed by innoculating fermentation tubes containing brilliant green lactose bile broth and incubating these for 48 h, (3 h at 35°C) 5°C. Formation of gas within this period confirms the presence of coliform organisms.

The density of coliforms is expressed as the Most Probable Number (MPN) per 100 mL of sample. This is a statistic based on the number of positive tubes in an array and is taken from a set of tables.

To test for fecal coliforms similar innoculation procedures are used with an E.C. broth medium but at a higher incubation temperature of 44.5°C ± 0.2°C.

If necessary various species of coliforms may be differentiated by a further series of tests.

Similar techniques are used for testing shellfish themselves by masceration and innoculation of the tubes by appropriate quantities of meats.

For testing quality of shellfish at the market level standard plate counts are used. Agar plates are innoculated with samples of mascerated meats and the colonies counted and converted into a quantitative measure of coliform count.

In addition to the tube fermentation method of bacteriological analysis it is also possible to estimate coliform density in sea water by means of a membrane filter method.

The water sample is filtered, with partial vacuum, through a specially manufactured membrane filter with a rated pore diameter for complete retention of coliform bacteria. It may be necessary to sterilize the filters. Absorbent pads are saturated with Endo type media in special petri dishes and the filter membrane placed on the pad. These are incubated at 35°C ± 0.5 for 24 h. Colonies of bacteria are produced and those with a golden green metallic sheen are considered to be the coliform group. The coliforms are estimated from the count of colonies which should be in the region of 20 to 100 per filter. There are also membrane filter-techniques for fecal coliforms.

Paralytic Shellfish Poison

Safe utilization of molluscan bivalve shellfish is affected by sewage pollution, sanitary handling, chemical contamination or a phenomenon known as paralytic shellfish poison.

Solutions for problems of sewage pollution are relatively simple for it is directly related to the bacterial content of the water in which the shellfish are grown. There is constant surveillance of shellfish entering the market, by public health authorities providing adequate sanitation control between the producer and the consumer.

Chemical contamination usually occurs when shellfish are grown in proximity to industrial plants with uncontrolled effluent discharge.

Paralytic shellfish poison, generally referred to as PSP or red tide is a more difficult problem for prevention of development is impossible and forecasting its occurrence is difficult. Oyster growers should have some background knowledge of it.

History

During Captain George Vancouver's voyage of discovery in British Columbia waters in 1793 one of his men was reported to have died and several others became ill after eating mussels at what is now known as Poison Cove in Matheson Channel near Butedale (Fig. 157). In 1942 several Indians in Barkley Sound on the west coast of Vancouver Island died as a result of eating clams and mussels. These were local incidents in a long record of illness or death after eating molluscan shellfish on the Pacific coast of North America, from Alaska to California. However, shellfish poisoning is also known from other parts of the world.

As a result of studies in California, it became known that paralytic shellfish poisoning, as it was called because of the symptoms, was caused by a microscopic dinoflagellate plankton organism which produces a virulent poison in its body. When filter feeders, such as clams and oysters, feed on a poisonous species, the toxin from it is released in

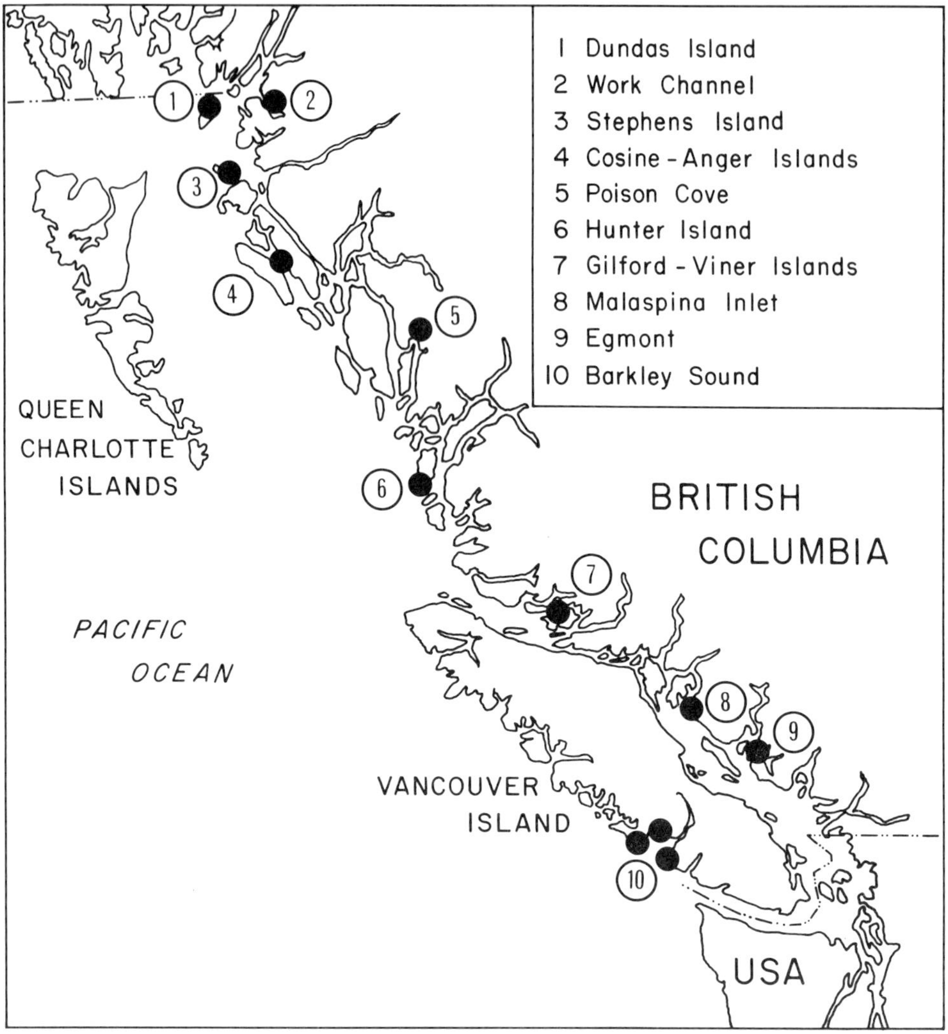

FIG. 157. Map of the British Columbia coast showing the location of some of the major outbreaks of paralytic shellfish poisoning.

the digestive process, accumulated, and stored in the shellfish body where the poison causes no harm.

It was also discovered that about the only means by which the poison could be detected in shellfish was by its reaction on mammals such as mice. Consequently, a bioassay technique was devised, whereby the poison is extracted from shellfish with a solution of weak acid. This solution is injected into a standard type laboratory mouse. Knowledge of the dosage and time required to kill mice made possible a quantitative determination of the amount of paralytic poison in the shellfish.

In California, and in eastern Canada, paralytic shellfish poisoning occurs mainly during summer, whereas some species of shellfish may be toxic at any time of the year in British Columbia and Alaska. However, initial toxification in these areas occurs mostly during summer months but there have been instances as late in the year as December.

A more recent occurrence of shellfish poisoning in British Columbia since 1942 occurred in October of 1957 when a number of persons became ill from eating shellfish in the Courtenay area of the east coast of Vancouver Island. This was the first time oysters were involved in a significant manner in paralytic shellfish poisoning. The next known outbreak occurred in mid-June of 1963 when high toxicity levels in butter clams were found near Namu on the central British Columbia coast. Subsequent testing showed toxic clams to occur along the whole coast between Cape Caution and the Alaskan border, including the Queen Charlotte Islands. The closure applied at that time (1963) is still in effect; as a result, commercial production of butter clams has been seriously curtailed.

On June 1, 1965, one death occurred and four other persons became ill from eating cockles taken in a small bay in Malaspina Inlet which lies in the northeast corner of the Strait of Georgia. Several other species of shellfish including clams, mussels, and oysters were found to be quite toxic.

Subsequent outbreaks resulting in illness of death are listed as follows:

Viner Sound	May 1970	butter clams	illness
Barkley Sound	November 1972	manila clams	illness
Work Channel	June 1975	butter clams	illness
Viner Sound	May 1977	butter clams	no illness
Gilford Island	May 1980	butter clams	death and illness
Church House	December 1981	butter clams	illness
Work Channel	May 1982	butter clams	illness
Simoom Sound	May 1985	butter clams	illness

Symptoms

Mild symptoms of PSP after eating toxic shellfish take the form of tingling or numbness in the lips and toes, usually with in one hour after consumption. More severe symptoms are incoherent speech, prickly feeling in arms and legs with lack of co-ordination in the limbs. Extreme symptoms are muscular paralysis, breathing difficulty and shaking sensations. Fatality is caused by respiratory failure and heart problems. Death times may vary between 3 and 12 h after severe poisoning.

So far there is no known antidote and treatment consists of emptying the stomach with an emetic or induced vomiting and use of a rapid laxative. Artificial respiration should be applied if breathing becomes difficult. Treatment may be given for primary shock. No alcohol should be given because it aids in the body's absorption of the poison. Warm water in quantity assists in elimination of the poison in urine.

Symptoms and Order of Appearance

(a) numbness of the face and lips
(b) difficulty with speech
(c) numbness of fingers and toes
(d) dizziness or giddiness
(e) numbness of arms and legs
(f) difficulty with breathing
(g) difficulty with standing or sitting
(h) stomach pain
(i) vomiting
(j) headache
(k) backache

Domestic animals such as dogs and cats and birds can die from eating toxic shellfish while cold-blooded vertebrates such as fishes and amphibians are generally not affected.

Gastropods (snails) are not filter feeders but may become poisonous by feeding on toxic bivalves.

Source

The source of the poison is a microscopic one celled plankton organism belonging to the order Dinoflagellata, which are motile and coccoid or filamentous. In British Columbia the responsible organism is called *Protogonyaulax*. When oceanographic and weather conditions occur in the appropriate combination these organisms can multiply rapidly and become so abundant the sea becomes coloured, usually red, brown, or green but other colours also occur.

In one form or another the organism, either as a spore which may lay dormant on the bottom, or as plankters, occurs throughout the year, but it is only when the right combination of light, temperature, salinity, nutrients, minerals and vitamins are present that a bloom may occur. In a broad sense these conditions may occur fairly frequently but the specific correct combination only rarely. A filter feeder, depending on size, may become somewhat toxic when the concentration of the toxic dinoflagellate approaches 200/mL. The poison itself, one of the most potent known, affects the transmission of nerve impulses.

Three thousand cells of a *Protogonyaulax* weighing 100 mg wet can yield 15 mg of dry extract of 1 mg of pure poison. The abundance may reach up to 50 000 organisms/mL but colouration can appear at about 20 000 cells/mL. Such developments are called blooms and small scale occurrences can develop in less than an hour and may disappear almost as rapidly. Because of colouration due to the xanthophyll peridinin, such blooms are often called red tides. There are about nine dinoflagellates that may cause PSP, most of which belong to the genus *Protogonyaulax*, formerly called (*Gonyaulax*. Figure 158 shows this species, as well as *Gymnodinium* and *Noctiluca*, which are the most common causes of coloured water in British Columbia.

Many blooms, depending on species, are innocuous, while others may cause invertebrate or fish kills or respiratory irritation in humans. PSP blooms occur throughout the world but the north east and north west coasts of North America are areas of persistent and high level concentrations. The toxin is known variously as saxitoxin (after the butter clam *Saxidomus* on which early studies on PSP were based) or mytilitoxin (from the mussel which is the source of many PSP difficulties in humans). The toxin, originally considered as a single entity is now known to be a complex of poisons.

Fish kills related to plankton blooms may result from oxygen depletion resulting from the decay of dying organisms rather than to the poison from toxic or non-toxic species.

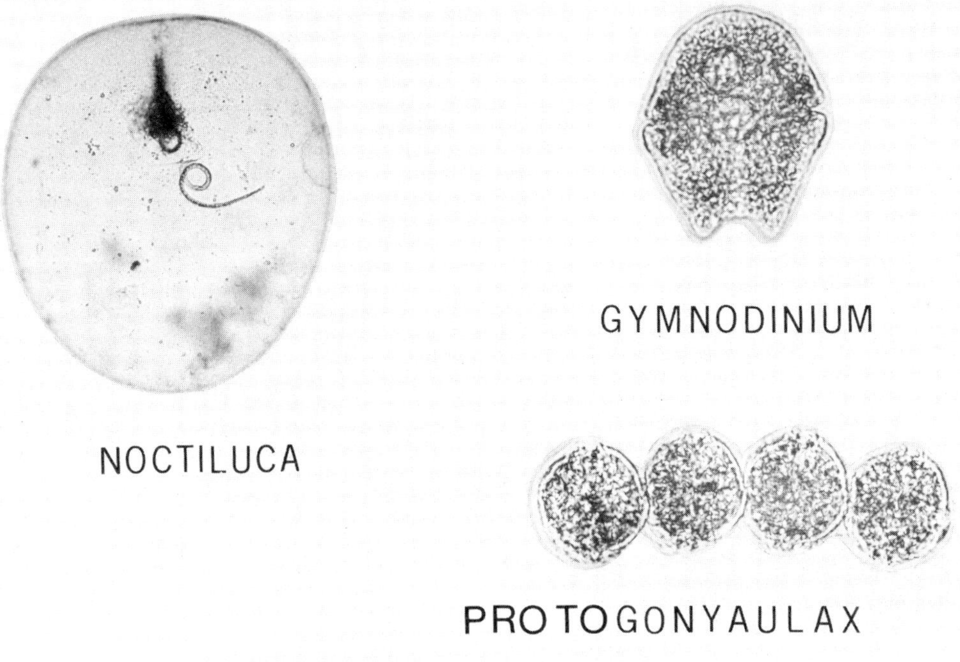

FIG. 158. Three types of organisms that may cause coloured plankton blooms. *Noctiluca scintillans* × 50; *Gymnodinium* **splendons** × 600; *Protogonyaulax acatenella* × 600.

Anatomical Distribution of PSP Toxin

Anatomical distribution of the toxin varies between molluscan species and in time after initial toxification. In most species the digestive gland, gills and mantles are the sites with high values. In some species, such as the butter clam *Saxidomus giganteus*, siphons contain a considerable portion of the toxin at the time of toxification and can retain it up to 2 years. Non siphonate forms, such as oysters and mussels in B.C. lose the toxin relatively rapidly in 2 months or less.

During initial stages of toxification, the body of *Saxidomus* may contain 60% of the toxin in the whole clam, but in 6 wk there can be a complete reversal with the body containing only 15% of toxin with a high proportion in the siphon–gill complex.

Measuring PSP

One of the greatest difficulties in dealing with PSP is the quantitative measurements of toxins, and after much effort and many years of research there is still no rapid method.

The standard procedure, first developed as recently as 1935 is to mince the meats of a sample (possible a dozen, depending on size) and extract the toxin from a known quantity of meat by boiling it in a weak hydrochloric acid. The volume and acidity are adjusted and the liquid clarified, after which a small quantity is injected peritoneally into the body cavity of a mouse. A lethal dose of 1 mL of extract, termed a "mouse unit", kills a 20 g mouse in 15 min.

If there are 200 mL of extract then the toxicity would be 200 mouse units per 100 g of meat. By using a purified poison extract as a standard reference, it was possible to express the toxicity in terms of weight (micrograms) rather than mouse units. Equivalence between mouse units and weight depends on the strain of mice used in the bioassay but is generally in the area 0.20 (conversion factor) equal to one mouse unit (0.17 in British Columbia). Accuracy of the test is estimated to be ±20% and low toxicity values (80/100 g) may be low by as much as 60%. The amount of toxin in 3 000 *Protogonyaulax* cells is equivalent to one mouse unit. Since an averaged size oyster can filter up to 20 L of water per hour and with the possible concentrations *Protogonyaulax* in the plankton (upto 50 000/mL) it is possible for an oyster to ingest a considerable quantity of toxin.

Several chemical methods of assay have also been developed but none has so far superceded the mouse bioassay for common use. The method has shortcomings, one being the necessity for a pure strain mouse.

Control Measures

So far no entirely safe method of PSP control has been devised beyond bioassay of filter feeding molluscs before they are marketed, but this is costly and time consuming.

Sampling plankton, reporting visual "red tide" occurrences, area and seasonal closures, and monitoring standard stations have been used, but each has its limitations.

Unless a plankton program is extensive in terms of sampling, frequency and number of stations there are too many gaps, in addition to high cost. Visual sitings are useful up to a point but missing crucial occurrences could be disastrous. Many toxic blooms do not cause coloration of the water. Closure of areas where PSP occurs with a degree of regularity is effective and this also applies to seasons and this method is in use in several countries. The most common control method is to monitor a set of standard stations, at least during the time of year when likelihood of PSP occurrences is high.

Each PSP area has its own peculiarities. In the PSP area on the Atlantic coast of Canada the occurrence with regard to area is annually quite regular, with only timing and level of toxicity varying somewhat. Monitoring a single station permits timing of a seasonal closure. In B.C. there is no such regularity in occurrence over much of the coast either in time or place. The north coast and part of Alaska is permanently closed owing to the persistence of PSP particularly in butter clams. California and Washington state have annual summer closures in certain areas. Monitoring specific stations is limited by dependence on frequency of sampling. High frequency is costly but low frequency permits the possibility of a toxicity occurrence between sampling times.

Processing and Toxicity

Boiling bivalves until the valves open may reduce toxicity by as much as 70%, but the liquid in which the shellfish were boiled contains about 50% of the toxin that was in the raw meats. Steaming and frying reduced toxicity more than boiling. Cooking for as short a period as five minutes reduces toxicity somewhat and longer periods decreases it further to the 70% level.

The canning process can reduce toxicity in clams as much as 90%.

Detoxification

After initial toxification, most species, except for *Saxidomus* lose the toxin in a few weeks. Siphonate forms take a little longer than non-siphonated forms such as mussels and oysters. Natural methods include detoxification of the live animal *in situ* and this is the normal method for in most areas repeated blooms are infrequent. Transplanting to non-toxic areas, and there may be no assurance they will remain so, is costly. Artificial methods include chemical treatment by adjustment of pH or salinity or temperature but are ineffective and costly. Removal of more toxic parts such as the digestive gland and siphon-gill complex is a successful and positive method for the siphonate forms, particularly in such species as *Saxidomus*.

To ensure only safe shellfish reach the public, permits must be obtained from the Department of Fisheries before harvesting molluscs from closed areas. The Department then samples these for PSP levels before and during harvesting.

Shellfish along the British Columbia coast are under constant surveillance for shellfish toxicity by means of sampling programs. Since development of paralytic shellfish poison is often associated with rapid multiplication of the causative organism into numbers sufficient to colour water reddish or brown, such occurrences are often taken as a sign of imminent toxicity. However, there are many types of organisms, nearly all nontoxic, which also discolour sea water. Growers should report occurrences to the Department Fisheries and Oceans.

Oceanography

Oceanographic factors of most concern to oyster culturists are temperature, salinity and currents. Of lesser importance are hydrogen ion concentration (pH), oxygen content and turbidity. Other chemical constituents such as nutrient salts (nitrogen compounds, phosphates and silicates) can influence production of potential oyster food but are difficult to measure or to apply to oyster culture activities. Except in special circumstances oxygen is not a limiting factor nor is pH. Currents are of considerable importance since they influence siting of culture structures such as rafts or racks, the movement of oyster larvae and of food.

Tidal information is obtainable from published tide tables.

Temperature

Temperature may be measured in a number of ways and for oyster work a high degree of accuracy is not necessary and readings to 0.5 degrees are more than adequate.

Standard Glass Thermometer

Standard glass thermometers should be protected by a metal container. For surface temperature the simplest way is to take a bucket of surface water, immerse the thermometer in it and wait until movement of the mercury column stabilizes (a minute or so), before taking the reading.

Recording Thermometer

One of the most useful temperature instruments for shellfish study is the recording thermometer. One type has an open clock face and one or two tails for placing at the depth at which the temperature is required. These are clock wound and operate for a week on one winding. There is another more compact type that is submersible which permits keeping it hidden from vandalism or theft. The clock work winding operates for up to several months. The continuous record shows all temperature fluctuations within the period at that depth and permits an analysis of the temperature (i.e., degree-days) to which the oysters in that particular area are subject.

Maximum–Minimum Thermometer

Another fairly useful instrument is the maximum–minimum thermometer which determines the maximum and minimum temperature for any given period of time. These are inexpensive.

Temperatures may be recorded as daily, weekly or monthly means and are usually so graphed. The high water slack of the day is an appropriate time if only a single reading is taken.

Salinity

Salinity measurements are based largely on determination of the chloride ion concentrations rather than on sodium chloride. Both salinity and chlorinity are expressed in grams per kilograms of sea-water or in other words, in parts per thousand, or "per mille", and the symbol ‰ is used.

Methods of determining chlorinity or salinity are:

1. Hydrometers

Salinity is determined by measuring its density where a hydrometer or float is used and the density measured by the weight of the hydrometer and the volume of water displaced. Hydrometers are accurate enough for most shellfish work. However, they are fragile, made of glass with a fine stem.

2. Refractometers

This is a small sturdy instrument requiring only a very small water sample and is probably the most suitable of all salinity measuring instruments for field work.

Oxygen

The oxygen content of oyster growing waters is important. In most natural circumstances there are generally adequate amounts of oxygen in seawater. Extraordinary circumstances include reductions caused by biological oxygen demand from pollutants or from decaying organisms such as plankton blooms.

If oysters live naturally in an area it may be assumed the oxygen supply is satisfactory. Temporary oxygen depletion for a number of days usually has no significant effect on oysters for they can close their valves and live for a time without an external source of oxygen.

There are electronic instruments for measuring the oxygen content of seawater.

Hydrogen Ion Concentration (pH)

As with oxygen, if shellfish are living in an area it may be assumed that the pH level is suitable. Changes may be caused by reduced salinity but fairly drastic differences over a period of time are necessary to cause difficulties for oysters. Unless abnormal conditions develop it is unnecessary to be concerned with pH in oyster culture operations. Hydrogen ion concentration is a measure of alkalinity or acidity and is measured on a logarithmic scale so a unit change in pH denotes a tenfold change in the acid and alkaline ions. A neutral solution, neither acid nor alkaline has a pH of 7; an acid solution has a pH less than 7 and an alkaline solution a pH greater than 7. Sea water is normally alkaline usually between 7.5 and 8.4.

Hydrogen ion concentration is measured with electrometric instruments or by colorimetric methods. In the main colorimetric technique, a controlled amount of an indicator solution such as cresol red or phenol red is added to the sea water sample and the colour developed is compared to that in a set of standardized tubes. This method is accurate enough for most shellfish field work.

Turbidity

Since many oyster areas are in estuaries, turbidity (transparency of water) exists in varying degrees and has a direct influence on oyster culture. Turbidity may be caused by the silt load, by the amount of detritus (suspended organic material), by plankton or by a combination of all three. Results of turbidity are shown in the deposition of silt which in sufficient concentration may smother bottom living oysters. It also affects the feeding efficiency of oysters when in high concentrations, since energy is expended separating out food and discharging unwanted particles.

Turbidity may be measured by testing the limits of visibility or by light measurements. For most oyster culture studies, limit of visibility is adequately and simply measured by the Secchi disc. This a circular plate, usually of metal, 20 cm in diameter. It may be all white or divided into 4 quadrants, alternately black and white on the upper surface with the lower surface black to prevent reflection of light. The Secchi disc is lowered into the water on a measured line. The depth at which the disc may no longer be seen is noted as well as the depth at which it reappears when lifted. The mean of the two readings is the limit of visibility and is a rough but useful measure of turbidity. The time of day, amount of cloud and wave action affect readings so these are noted at each reading. Readings may be standardized at mid-day on the shaded side of the boat and preferably with a water glass and observations made at one metre above the water surface.

Tides

Tides are a most important factor in determining the type of oyster culture in a given area. Tides are generated mainly by the gravitational effect of the moon and sun, particularly the former. However, wind and barometric pressure may exert a smaller influence in certain localities. For most areas in the world it is possible to predict fairly accurately the movement of tides and these are published for most of the shipping ports and a typical tide table is shown in Table 45.

Thus on June 1 at 0135 the height of the tide is 4.5 m above the chart datum which is the plane below which the tide will seldom fall and is the basis for marking on charts. Approximately 6 h later at 0850 the tide has fallen to 1.5 m above the chart datum. There is then a succeeding high and a succeeding low. Thus there are two complete tidal cycles or oscillations per day and this is termed a semi-diurnal tide. If there is only one complete cycle in 24 h the applied term is diurnal. In some areas there are no tides or a range of only a few cm.

In addition to complete data for reference ports on which the predictions are based, there is information where the time and height differences between the main reference port and a number of secondary ports. However oyster growing areas may be far from even secondary ports so it may be necessary to establish rough indications of the time and height differences from the nearest reference port. The times may be determined by observing the time of slack water which is the brief period when the tide begins to fall

TABLE 45.

Day	Time	Ht/ft	Ht/m
1	0135	14.8	4.5
	0850	5.0	1.5
	1535	12.5	3.8
	2020	9.5	2.9
2	0220	14.5	4.4
	0925	4.2	1.3
	1630	13.3	4.1
	2120	10.0	3.0
3	0250	14.2	4.3
	1005	3.5	1.1
	1720	13.9	4.2
	2225	10.3	3.1

after rising or to rise after falling. Levels or heights may be determined by installing a marked pole and observing the levels at the time of slack water. A satisfactory amount of data may be accumulated in a few months, and it will be sufficiently accurate. In protected embayments or estuaries where most oyster culture is carried on, strong winds may cause differences in height of 30 cm or more.

Currents

Currents are water movements developed by differences in tidal levels or are induced by wind. In open waters, currents generally have small velocities in the range of one or two kilometers per hour. In constricted areas such as estuaries or island complexes, currents may be quite rapid and speeds of 25 kilometers per hour are known. Currents are important to oyster culture relative to location of beds or positioning of racks or rafts and to distribution of oyster food and larvae. Current speed and direction should be investigated but the fact that these may vary with depth and time must be considered. There are several methods for measuring currents.

There is a wide range of complicated mechanical current meters but these are not necessary for most shellfish work. Drift poles which are simply lengths of wood or poles (bamboo) weighted at one end so they float upright have been used extensively (Fig. 159). Depth may be varied with the length and weight used. For surface currents plastic envelopes or other simple surface floating materials are satisfactory. Also in general use are drogues which are simply 4 plane surfaces set at right angles to a common centre (Fig. 160) weighted to float at a given depth. Either one or several may be set out at one or at several stations and at differents times to cover varying weather and tidal conditions. From direct observation and time rough current patterns may be developed. For more accuracy some means of taking bearings such as a sextant or bearing compass with or without range finders would be required.

Administration

History

Oyster growers soon realize they must deal with a number of governmental agencies, Municipal, Provincial and Federal. An understanding of the respective jurisdictions of these agencies may assist growers in their dealings with them.

The British North America Act provided that the legislative rights to all of Canada's inland coastal fisheries would be vested in the Federal Government providing jurisdiction

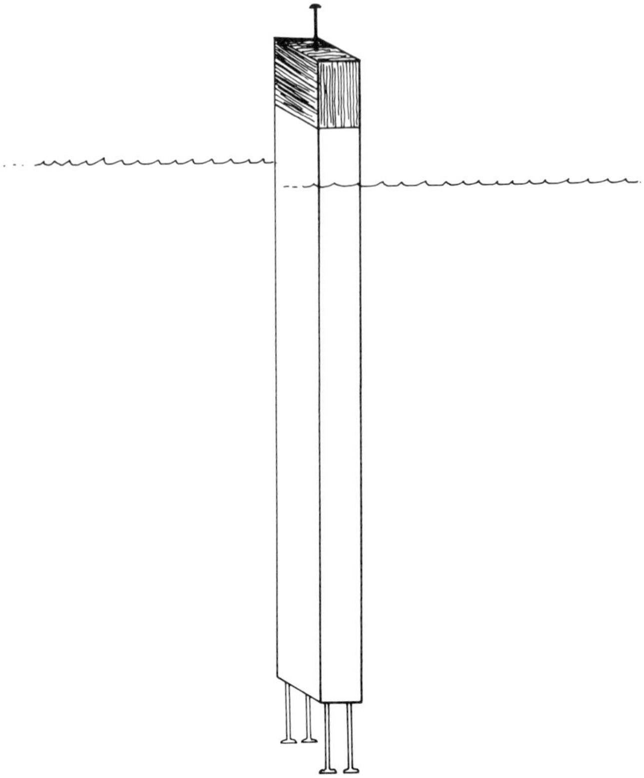

FIG. 159. Simple drift float. (× 1/2).

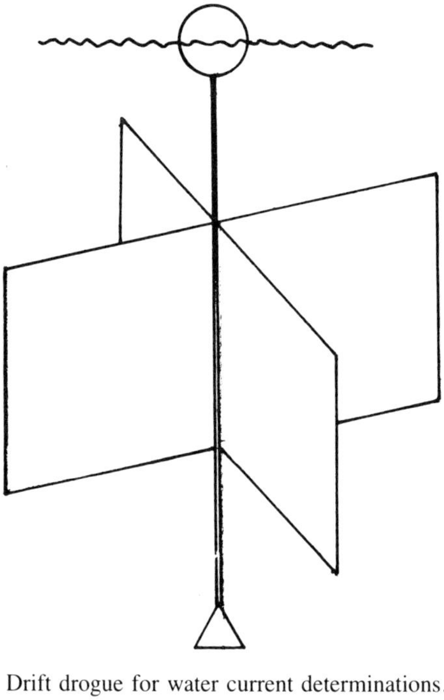

FIG. 160. Drift drogue for water current determinations. (× 1/6).

over all tidal and nontidal fisheries except those of Quebec. The nontidal (freshwater) fisheries are administered by the Province through the Fish and Wildlife Branches of the Ministry of Environment and Parks.

Through the years, since British Columbia entered Confederation, conflicts have arisen between Federal and Provincial authorities on various aspects of fisheries jurisdiction, notwithstanding the delegation of responsibilities in the terms of union. These problems were generally resolved in the courts of Privy Council and, as a result of a number of decisions, it now stands that regardless of ownership, insofar as tidal fisheries are concerned, the Federal authority controls and regulates the fisheries and, under its agreement with British Columbia, has the obligation to protect and develop them. The Provincial authority regulates the processing of fish within the Province and also taxes the fishery for Provincial revenue, although the exact management of this phase is Federal, through the Federal–Provincial tax-sharing agreement.

The British Columbia government has two ministries concerned with oyster culture.

The Ministry of Forests & Lands is responsible, under the Land Act, for the administration of Crown land, including aquatic land. This includes management, planning and allocation of aquatic Crown land for aquaculture purposes.

The Ministry of Agriculture & Fisheries has overall control of the oyster industry as it has been designated by Cabinet as the "lead agency" responsible for developing provincial policy for aquaculture. The Ministry also provides financial and marketing services and has a statutory mandate for the licensing and inspection of fish buyers and processors and the establishment of fish production and quality standards. The Ministry also has a formal agreement with the Ministry of Forests & Lands to ensure diligent use of aquaculture tenures and to encourage maximum productivity of the Province's aquaculture resources.

The Ministry's regulatory powers are based on the Fisheries Act and the Fish Inspection Act, from which pertinent excerpts of concern to oyster growers are given in Appendices D and E.

Various branches of the Federal Department of Fisheries and Oceans enforce the regulations emanating from the above two acts and exercise paralytic shellfish poison control. Sport fishing for shellfish also comes under this jurisdiction.

Aquatic land is that which is covered by water and is located below the high-water mark. It is divided into two zones, the foreshore zone, which lies between the high and low water marks and is commonly known as the beach or intertidal area, and the offshore zone located below low water mark and referred to as the deep water or subtidal area.

In British Columbia, all aquatic land is held by the Crown in the right of the province. This means that title to this land is held, with few exceptions, by the Province of British Columbia. Some of these exceptions are Federal harbours, such as Victoria, Vancouver, Nanaimo, and Port Alberni, where aquatic land is ministered by the Government of Canada. There are several other smaller areas held outright by private owners.

The British Columbia government has adhered to the policy of not disposing of any foreshore outright, but will lease or license it for various purposes, for various lengths of time.

British Columbia Crown lands are administered by the Ministry of Forests & Lands which will provide, upon request, information on tenure, acquisition for specific uses (e.g. aquaculture, private boat moorage, log booming, etc.) along with application forms.

Tenures for oyster culture are available in the form of a lease or as a license of occupation.

A *lease* is the most secure form of tenure and is issued for long term occupation (up to 20 years) within specific boundaries (a legal survey is required at the lessee's expense). The granting of a lease gives the lessee exclusive use of the site for the purpose specified in the lease contract. Conveyance of estate is implied and a mortgage can be registered against a lease. A lease is generally not cancelled except for breach of contract (e.g. noncompliance of Production Plan).

A *license* of occupation is issued strictly for use; no rights of ownership are implied. Consequently, a mortgage cannot be registered against a license. The term of a license may be up to 15 years, but it carries a 90-day cancellation clause which allows the government to cancel tenure at its discretion. The licensee is then allowed 90 days to remove stock and improvements. However, this does not generally occur except in cases of non-diligent use, or if the Ministry of Forests & Lands receives a request for use that it deems more beneficial than oyster culture.

Upland Owner Consent

Shoreland property owners have the right to unimpeded access to and from deep water from every point along the foreshore in front of their property. Therefore, written consent from upland owners is required for both leases and licenses of occupation when structures that will interfere with the upland owner's access to and from deep water may be placed.

Consent must be unconditional, except for purposes and tenure, and must be renewed with each tenure renewal. Should upland ownership change during the tenure period, consent must be re-established with the new upland owner at the time of ownership transfer.

The general procedure for acquisition of aquatic land for aquaculture is as follows:

1. Information and advice obtainable from the Ministry of Agriculture and Fisheries, Commercial Fisheries Branch regarding site criteria and management.
2. Application for aquatic land tenure, plus $25.00 application fee, submitted to the Ministry of Forests & Lands and site posted according to application requirements.
3. A Production Plan of the proposed operations must be presented for approval to the Ministry of Agriculture & Fisheries, Aquaculture and Commercial Fisheries Branch.
4. Adherence to local municipality zoning regulations and business license requirements.
5. Structures in, on or under navigable waters require approval from the Federal Ministry of Transport, Vancouver, B.C. Details of requirements are given in ''Navigable Waters Protection Act — Application Guide''.
6. Advertise application in a local paper and B.C. Gazette once per week for two consecutive weeks.
7. Obtain written upland owner's consent for tenure where structures are proposed that will interfere with the upland owner's access to and from deep water.
8. Registration of oyster growers is required by the Aquaculture and Commercial Fisheries Branch, Ministry of Agriculture & Fisheries, Victoria.
9. Oyster growers must register with and be licensed by the British Columbia Oyster Board. Regulations and requirements of this board are contained in ''The British Columbia Oyster Board General Orders (1973) Part 1''.
10. Participation with the non-governmental British Columbia Oyster Growers Association is optional. The Association was established to promote the interests of oyster growers.

Rental Fee for Leases and Licenses

As of 1987, rental fees for leases and licenses are paid annually and are designed to reflect the productivity of the site as rated by the Aquaculture and Commercial Fisheries Branch. This rating is based on the sum of ten environmental factors, each of which has a value of zero to five, giving a maximum score of fifty.

The productivity rating is converted to a land value per acre on ''Farmland Valuation Schedule''. For bottom culture, a minimum productivity rating of 20 gives a hectare land value of $625 and a perfect rating of 50 gives a land value of $2,500 per hectare. For off bottom culture, a minimal productivity rating gives a hectare land value of $500 and a maximum rating of 50 gives a value of $2,500 per hectare.

The fee for an initial lease or license of occupation is 1% of the land value from years one to five.

After the fifth year the annual rental fee for leases if 5% of the land value, and for licenses of occupation 4%.

In no case is the fee for a lease or license of occupation less than $200.00 per tenure.

Taxes on Leases and Licenses

Land tax on oyster leases and licenses of occupation is collected by local or municipal governments. This is in addition to the rental fees.

Tax Exemptions for Aquaculturists

Aquaculturists are assessed as farmers and are eligible for various tax exemptions. The Provincial Acts are:

1. Assessment Act
2. Social Services Act
3. Coloured Gasoline Tax Act

Federal Tax Exemptions Plan

1. Investment Tax Credit
2. Sales Tax
3. Tariffs

Reference: Taxation and the British Columbia Farmer, a laymans guide: B.C. Ministry of Agriculture and Food.

Administrative responsibilities of the oyster grower.

1. Annual registration with the Aquaculture and Commercial Fisheries Branch
2. Annual production returns
3. Payment of annual rental fees
4. Insurance policy coverning public liability and damage
5. Payment of annual property taxes
6. Licence from the British Columbia Oyster Board
7. Certificates of sanitary compliance for:
   (a) shipper of shell stock only
   (b) shucking and packing plant
8. Business licence where necessary
9. Transfer permit where necessary
10. Transplant application for importation of seed or larvae.

Acknowledgments

No literature references are given in the text for this would have enlarged the bulletin excessively. In any case oyster growers to whom this work is primarily directed would not have ready access to most of the literature. However, a list of the most important texts and reports are given and most of these contain extensive bibliographies.

The works of Drs. C. R. Elsey, P. S. Galtsoff, V. L. Loosanoff, Sir Maurice Yonge and many others have been drawn upon as well as those of the staff of the Washington State Shellfish Laboratory (Mr. C. E. Lindsay, Mr. R. Westley, Mr. C. Woelke, and Mr. C. Sayce) with whom many discussions on oyster culture were held. Most of the material has been taken from the bulletins of the British Columbia Provincial Shellfish Laboratory at Ladysmith which were produced by the author in mimeographed form in 55 numbers in 8 volumes between 1949 and 1957 while at that laboratory.

The late George G. Alexander, Deputy Minister of the Provincial Fisheries Department was responsible for the establishment of the Provincial Shellfish Laboratory at Ladysmith where much of the practical work on oyster culture was conducted with the capable assistance of E.P. Kent as boatman, carpenter and technician.

The revision of this bulletin was initiated by the British Columbia Oyster Growers Association whose cooperation and encouragement through many years is acknowledged.

The Aquaculture and Commercial Fisheries Branch of the Provincial Government has had considerable input through the informative British Columbia Shellfish Mariculture Newsletter and assistance of G. Chislett, R. Cox and K. Albrecht.

Appreciation is expressed for the assistance of Drs. F. Bernard, N. Bourne, and G. Jamieson of the Pacific Biological Station.

Illustrations are the work of the author, photographers C. Morely, W. Carolsfeld, and P. Fraser, and artists A. Boyden, A. Denbigh, and P. Franklin. Typing by Maureen Palmer is gratefully acknowledged.

Literature

There is a considerable amount of general literature on oysters, but much less on culture. A number of what may be considered major texts are now out of print and available only in major libraries, universities or fisheries institutes. Much of the literature, both in biology and culture is published as relatively short papers in periodicals and journals, again not generally available, although some libraries may make them available in copy form at cost. The general material is given in the reference list. To provide current information, subscriptions may be obtained for the following journals: *Aquaculture, Aquaculture Digest, Aquaculture Engineering, Veliger*, and *Journal of Shellfish Research*.

Details and addresses may be obtained from the Provincial Aquaculture and Commercial Fisheries Branch. Some publications may be obtained directly from the authors or from the originating institution.

BISSEL, G. E. 1970. Economic evaluation of the oyster industry in British Columbia, Part. I. The market area. Western Consultants. 40 p.

1970. Economic evaluation of the oyster industry in British Columbia. Part. II. The production area. 31 p. Western Consultants. 31 p.

CAHN, A. H. 1950. Oyster culture in Japan. U.S. Fish. and Wildlife Service. Fishery Leaflet 383. U.S. Government Printing Office, Washington, 25, DC.

CLAYTON, W. E. L. [ED.]. The B.C. Mariculture Newsletter, 6 volumes 1981–1986. B.C. Ministry of the Environment.

CLAYTON, W. E. L., and T. T. POBRAN. 1981. A Pacific oyster stake culture pilot project in British Columbia. Fisheries Development Report No. 33, B.C. Marine Resources Branch, 33 p.

DEVLIN, I. [ED.]. 1973. Operation report: oyster depuration plant-Ladysmith B.C. Report. Industrial Development Branch, Fisheries and Marine Services, Environment Canada and Commercial Fisheries Branch, Department of Recreation and Conservation, British Columbia. 103 p.

DORCEY, A. H. J. [ED.]. 1979. Coastal resources in the future of British Columbia. Vancouver Westwater Research Centre. 123 p.

GALTSOFF, P. S. 1964. The American oyster, *Crassostrea virginica* (Gmelin) U.S. Fish Wildl. Serv. Bull. 64: 430 p.

GRIFFITH, L. M. 1967. The intertidal univalves of British Columbia. Handbook No. 26. British Columbia Provincial Museum, Victoria. 101 p.

GUNN, R. C. 1984. A report on the development and use of an artificial cultch suitable for the bottom oyster culture in British Columbia. Fisheries Development Report No. 3. Marine Resources Branch, Ministry of the Environment, Province of British Columbia. 25 p.

HUMPHRIES, M. 1976. The production and marketing of tray-cultured raft oysters in British Columbia. Industrial Development Branch, Fisheries and Marine Service, Environment Canada and Marine Resources, Department of Recreation and Travel Industry, British Columbia, Technical Report No. 90, 74 p.

IMAI, T. 1978. Aquaculture in shallow seas. A. A. Balakemia, Rotterdam. 615 p.

JONES, C. and B. JONES. 1982. Methods for setting hatchery produced oyster larvae. Information Report No. 4. Marine Resources Branch, Ministry of the Environment, Province of British Columbia. 61 p.

JOURNAL OF SHELLFISH RESEARCH (formerly Proceedings of the National Shellfisheries Association). Publication of the National Shellfisheries Association. S. E. Siddall, Marine Sciences Research Centre, State University of New York, Stony Brook, N.Y. 11794–5000 USA.

KORRINGA, P. 1976. Farming the cupped oyster of the genus *Crassostrea*. Elsevier Sci. Publ. Co., Amsterdam. 224 p.

MAGOON, C. 1982. An introduction to shellfish aquaculture in the Puget Sound Region. Department of Natural Resources, Olympia, Washington. 68 p.

MEDCOF, J. C. 1961. Oyster farming in the Maritimes. Fish. Res. Board Can. Bull. 131.

PROBRAN, T. T. 1983. Oyster farming in British Columbia. A financial information guide. Paper No. 3. Marine Resources Branch, Ministry of Environment, Province of British Columbia. 177 p.

QUAYLE, D. B. 1979. Report on Pacific oyster seed plantings on the west coast of Vancouver Island in 1936. Pacific Biological Station, Nanaimo. (manuscript)

Reading List

QUAYLE, D. B. 1969. Pacific oyster culture in B.C. Fish. Res. Board Can. Bull. 169: 192 p.

1969. Paralytic shellfish poisoning in B.C. Fish. Res. Board Can. Bull. 168: 68 p.

1971. Pacific oyster raft culture in B.C. Fish. Res. Board Can. Bull. 178: 134 p.

1949–1957. Oyster bulletins of the Provincial Shellfish Laboratory, Ladysmith, B.C. 8 volumes, 65 numbers, 391 p. British Columbia Department of Fisheries. (mimeo)

1960. The intertidal bivalves of British Columbia. Provincial Museum, Victoria, B.C. 104 p.

QUAYLE, D. B., AND F. BERNARD. 1976. Purification of basket-held Pacific oysters in the natural environment. Proc. Nat. Shellfisheries Assoc., Vol. 66: 69–75.

SHELLFISHERIES NEWSLETTER. Fisheries Division, Ministry of Agriculture and Fisheries. P.O. Box 297, Wellington, New Zealand.

SINDERMAN, C. J. 1977. Disease diagnosis and control in North American Aquaculture. Developments in aquaculture and fisheries science, Vol. 6: 329 p. Elsevier Scientific Publishing Company.

VENTILLA, R. F. 1984. Recent developments in the Japanese oyster culture industry. Advances in marine biology, Vol. 21: 57 p. Academic Press.

WALNE, P. R. 1974. Culture of bivalve molluscs. Fishing News (Books) Ltd. London. 173 p.

WESTWATER RESEARCH CENTRE. 1979. Policy analysis for coastal resources management.

Valiela, D. Vol. I. Oyster ecology and culture in British Columbia.. Tech. Rep. 19: 34 p.

Fralick, J. E. Vol. II. Economic analysis of the British Columbia oyster industry. Tech. Rep. 20: 117 p.

Valiela, D. Vol. III. Policies for development of the British Columbia oyster industry. Tech. Rep. 21: 68 p.

YONGE, C. M. 1960. Oysters. Collins London. 209 p.

Glossary

adductor	muscle holding two valves together
algae	marine plants or seaweeds which reproduce by spores
amoeba	a primitive unicellular animal
anemone	a tentacled soft bodied member of the coelenterate group of marine invertebrates
anterior	front or head
aragonite	a form of crystalline calcium carbonate
auricle	chamber of the heart into which blood is received from the body
bioassay	a test in which the quantity of strength of material is determined by the reaction to it of a living animal
biofouling	the living component of the fouling complex
bloom	a sudden and massive increase in the numbers of micro-organisms great enough to discolour water
B.O.D.	abbreviation for biochemical oxygen demand; the amount of oxygen absorbed by a putrefying waste
box	pair of empty oyster valves
branchial	pertaining to the respiratory organ or gill of an aquatic animal
bryozoa	also known as polyzoa or moss animals. Usually microscopic in size and most often encrusting

bushel	8 dry U.S. gallons or 1.245 cubic feet
byssus	filaments used by a mollusc to attach itself to a substrate
caecum	a blind pouch usually associated with the alimentary canal
capillary	a thin-walled blood vessel of small diameter
cerebral	pertaining to the brain
chitin	a relatively inert skeletal material found mainly in insects
chomata	small denticles on both sides of the hinge in the genus *Ostrea*
cilia	a hair like process with a rhythmic beat which in molluscs produces a water current
cloaca	a posterior chamber into which open the anal, urinary, and genital ducts
coelenterate	one of a group of marine invertebrates with a simple single entrance to the stomach and often characterized by the possession of stinging hair cells. Jellyfish, hydroids and corals
coliform	a bacterium found in the colon of the digestive tract
conchologist	a person who studies molluscan shells
conchyolin	a horny substance found in molluscan shells
condition factor	a measure of plumpness or fatness of an oyster
Crown land	land under governmental ownership or control
crustacea	a group of aquatic animals characterized by jointed legs, i.e. crabs, shrimps
crystalline style	a gelatinous rod-like organ of certain molluscs concerned with digestive processes
cultch	material used to collect oyster spat
demibranch	single plate or leaf of a molluscan gill
depuration	a term used for the purification of molluscan shellfish
detritus	term given to fragmented organic material from plant and animal remains
diatom	a one celled primitive plant enclosed in a siliceous container
dinoflagellate	a motile one celled organism with some animal and some plant characteristics
diploid	the normal number of chromosomes
diverticulum	lateral outgrowth of the stomach cavity
dorsal	pertaining to the back or part of an animal away from the ground
drill	snail preying upon other molluscs whose shell it penetrates with a drilling apparatus
ecology	the relationship between an animal and its environment
echinoderm	the group of spine covered invertebrates such as sea urchins and sea stars
eel grass	a green bladelike marine plant which reproduces with seeds; a true plant and not a seaweed
enzyme	chemicals produced by living cells which aid, but do not take part in chemical reactions
exhalant	emitting or discharge area
exotic	introduced from a foreign country
faeces	indigestible residues remaining in the alimentary canal after digestion
ferrocement	a re-enforced cement often used in boat building
fertilization	the union of egg and sperm
flatworm	a group of flat leaf-like worms, many of which are parasitic
fluting	curved plate-like outgrowths on the surface of a molluscan shell
follicle	a small sac-like structure
folliculina	a microscopic vase shaped protozoan, usually green in colour, visible in numbers to the naked eye. Oyster shells are a favourite settlement site

foreshore	land below the high tide mark
gangeon	a short length of cord
ganglion	an aggregation of nerve cells
gaper	a bivalve mollusc, dead or in the process of dying, with the valves gaping but some meat left within
gastric	pertaining to the stomach
gastropod	a member of a group of molluscs usually characterized by a coiled spiral shell
gazetted	in this instance when the information is published in the official "British Columbia Gazette"
gene pool	a population with similar genetic characteristics
gill	a leaf-like appendage of an aquatic animal and concerned with respiration
glycogen	an animal starch
gonad	the sex gland which produces either eggs or sperms
gribble	a small wood-boring crustacean
halocline	the area of sharp vertical salinity change
hardening	with oysters the process of acclimation to longer and longer periods out of water
hatchery	for shellfish a complex of arrangements for holding and conditioning breeding stock, growing larval food, spawning, larval growing and setting tanks
hybrid	the offspring of the union between two different species or races
hydroid	a small, often bush-like member of the coelenterate group of marine animals
incubate	to hold eggs during development
indigenous	native, not imported
inhalant	the inward direction of a current flow
intracellular	within the cell
invertebrate	an animal without a backbone
labial	pertaining to the lips
lamella	a leaf or plate-like structure
larva	an immature stage between the egg and adult form
lease	rented foreshore or deep water area
lessee	one who obtains a lease
lessor	owner who rents the lease
ligament	fibrous spring-like material joining two valves
lobe	a rounded flap-like projection
long line	an apparatus similar in function to a raft. Consists of a length of rope or cable anchored at each end and supported by a series of buoys, rope or cable anchored at each end and supported by a series of buoys
mantle	a soft fold enclosing the body of a bivalve and which secretes part of the shell
mean	average
microgram	one thousandth of a gram
mouse unit	paralytic shellfish poison measurement unit
mucoprotein	conjugate proteins containing a carbohydrate group
nacre	iridescent calcareous substance forming the innermost layer of a molluscan shell
neap	a series of tides with a relatively small tidal range
nemertean	a thin unsegmented marine worm
nudibranch	a soft bodied shell-less mollusc
oesophagus	junction canal between the mouth and stomach
Order-in-Council	government order or regulation promulgated by Cabinet order

organoleptic	by actual taste
ostia	mouth-like apertures
ovum	egg
palp	a sensory appendage
parasite	an organism which lives in or on another organism and derives sustenance from it without rendering it any service in return
pedal	pertaining to the foot
pericardium	the space or membrane surrounding the heart
peritoneal	pertaining to the body cavity
pest	a predator or parasite
pinworm	marine wood-boring crustacean (*Limnoria*) also called the gribble
plankton	floating or weakly swimming aquatic animals and plants
pleural	pertaining to the pulmonary cavity
plica	a fold-like structure
poach	a mild term for theft
pore	a small aperture
polychaete	a segmented marine worm
PVC	polyvinyl chloride. A synthetic plastic capable of being moulded into a variety of shapes
posterior	the rear, away from the head
predator	an animal which kills and consumes another animal for food
prodissoconch	larval shell of a mollusc
promyal	in front of the adductor muscle
pseudofaeces	false faeces; waste material not taken into the digestive tract
race	a group of animals generally segregated geographically and differing slightly from the typical members of the species
relay	another term for ''transplanting'' shellfish from one bed to another
ren	Japanese term for shellstring
sac	bag or pouch-like structure
salinity	in oceanography the salt content of sea water usually measured in parts per thousand (‰)
seed	a young oyster with no specific definition as to size
set	the accumulated settlement or spatfall of oyster larvae
shackle	a ''u'' piece of iron with a screwable pin for joining two pieces of rope or cable
shell stock	general term for unopened oysters in the shell
shipworm	a mollusc capable of burrowing into wood. Also known as teredos
shuck	to open and remove the oyster meat from the shell
sink float	a raft, usually of logs with the floor sunk below the surface of the water
sinus	in molluscs a loosely defined blood vessel
siphon	in molluscs the special tubes designed to carry water in and out of the shell cavity. Also known as the neck
spat	a newly settled or attached young oyster; a post larval oyster
spatfall	the en masse settlement of oyster larvae
spawn	common term for eggs and sperm
species	term used to denote a group of closely allied, mutually fertile individuals
spring tide	series of tides with a relatively large tidal range
statocyst	an organ for the perception of the position of the body in space
sulphate	a pulp mill process (alkaline) where sodium hydroxide and sodium sulphate are used for cooking the pulp
sulphite	a pulp mill process (acidic) where sulphurous acid with a calcium or magnesium base is used for cooking the pulp
suprabranchial	also epibranchial, meaning above the branchium or gill

thermocline	the area of sharp vertical temperature change
thermograph	an instrument for recording temperature
thimble	a metal insert in a small loop at the end of a rope to prevent abrasion
transplant	another term for relay; to move oysters from one bed or location to another
tubule	a small pipe-like structure
tunicate	a marine invertebrate with a soft body enclosed in a test or tunic
triploidy	triple the haploid number of chromosomes
turbidity	the amount of suspended small particles in a liquid
typhlosole	a longitudinal inwardly projection fold of the wall of the digestive tract
umbo	(pleural umbones) the beak-like projecting which represents the oldest part of a bivalve shell
valve	one of the two parts of a bivalve shell. Two valves make up one shell
veliger	the secondary larval stage of most molluscs, characterized by the presence of a velum
velum	the ciliated locomotor organ of the molluscan veliger larva
ventral	pertaining to that aspect or side of an animal facing the ground
ventricle	the main contractile chamber of the heart
vesicle	a small bladder-like sac or pouch
vexar	a plastic mesh available in a variety sizes
visceral	pertaining to organs within the body
waferworm	a small flat worm belonging to the polyclad group of flatworms
xenomorphism	the characteristic of the left valve of an oyster to follow the contour of the substrate to which it is attached
year-class	a group of animals all spawned in a single year

Metric Conversion

1 barrel = 200 lbs = 91 kg
1 bushel = 8 gallons = 36.3 litres = 1.25 cubic feet
1 Imperial gallon = 4.55 litres
1 U.S. gallon = 3.8 litres
1 cubic foot = 28 litres
1 pound = 454 grams
1 kilogram = 2.2 pounds
1 metric ton (tonne) = 2204.6 pounds

1 mile = 1.6 kilometres
1 metre = 39.4 inches
1 foot = 30 centimetres

1 Acre = 4840 square yards = 4150 square metres
1 Acre = 0.405 hectares
1 Acre = 4050 square metres
1 hectare = 2.47 Acres

1 square yard = 0.84 square metres
1 square foot = 900 square centimetres

1 string shell cultch = 1 bag shell cultch (100 shells)
200 pounds of oysters in the shell = 2.5 gallons = 11.4 litres of meat (approximately)

degrees Centigrade (Celsius) (°C) = 5/9 (°F – 32) = °F
degrees Fahrenheit (°F) = (9/5°C) + 32 = °C

Appendix A — Pendrell Sound Reserve

THAT under the provisions of Section 94 of the "Land Act", being Chapter 175, Revised Statutes of British Columbia, 1948, and amendments, all that area known as Pendrell Sound, Redonda Island, New Westminster District, lying north of a straight line between Walter Point and Durham Point, be reserved and set apart for the use, recreation and enjoyment of the public;

AND TO FURTHER RECOMMEND THAT a copy of this Minute, if approved, be forwarded to the Deputy Minister, Department of Fisheries, Victoria, B.C.

DATES, this 21st day of November A.D. 1950.

Appendix B — Protection of Oyster Lease from Theft

EXCERPTS FROM CRIMINAL CODE OF CANADA — PART VII OFFENCES AGAINST RIGHTS OF PROPERTY

THEFT

269 (1) Every one commits theft who fraudulently and without colour of right takes, or fraudulently and without colour of right converts to his use or to the use of another person, anything whether animate or inanimate, with intent,

(a) to deprive, temporarily or absolutely, the owner of it or a person who has a special property or interest in it, of the thing or of his property or interest in it.

270 (1) Where oysters and oyster brood are in oyster beds, layings or fisheries that are the property of any person and are sufficiently marked out or known as the property of that person, he shall be deemed to have a special property or interest in them.

(2) An indictment is sufficient if it describes an oyster bed, laying or fishery by name or in some other way, without stating that it is situated in a particular territorial division.

PUNISHMENT FOR THEFT

280 Except where otherwise prescribed by law, every one who commits theft is guilty of an indictable offence and is liable

(a) to imprisonment for ten years, where the property stolen is a testamentary instrument or where the value of what is stolen exceeds fifty dollars, or

(b) to imprisonment for two years, where the value of what is stolen does not exceed fifty dollars.

Appendix C — Sanitation Agreement with the United States

PROPOSED MEMORANDUM OF AGREEMENT BETWEEN THE CANADIAN DEPARTMENT OF NATIONAL HEALTH AND WELFARE AND THE UNITED STATES PUBLIC HEALTH SERVICE

In order to bring about improvements in the sanitary practices prevailing in the shellfish industries in Canada and the United States and to facilitate the exchange of information regarding such practices with references to endorsement of shellfish certifications issued by State, Provincial, and other shellfish regulatory authorities, it is agreed as follows:

1. The "Manual of Recommended Practice for Sanitary Control of the Shellfish Industry" recommended by the United States Public Health Service in 1946 and published as

Public Health Bulletin No. 295, is accepted as the statement of the sanitary principles which govern the certification of shellfish shippers.

2. The degree of compliance attained by each state is to be reported to the Canadian Department of Health and Welfare by the United States Public Health Service and the degree of compliance attained by provinces and other regulatory authorities in Canada is to be reported by the Canadian Department of National Health and Welfare to the United States Public Health Service.
3. Whenever inspections of shellfish handling facilities or of growing areas are desired by either party, the other party facilitate such inspections.
4. This agreement may be terminated by either party on thirty days notice.

Appendix D — Regulations

Among the multiplicity of regulations governing the function of an oyster operation, none is probably more important than the Fish Inspection Regulations of British Columbia (B.C. Reg. 12/78). These are supported federally by Fish Inspection Regulations under the Canada Fish Inspection Act. In addition, the Federal Fisheries Act covers Sanitary Control of Shellfish Fisheries Regulations.

The British Columbia Regulations include a list of definitions.

"Depuration" means the removal, in a controlled environment, of micro-organisms of public health significance from live molluscs.

"Molluscs" for the purpose of these regulations means invertebrates of the Phylum Mollusca, including oysters, clams, geoduck, abalone, mussels, and cockles, either shucked or in the shell, and any edible product thereof.

"Plumping" means the holding of shucked molluscs in water of salinity less than that in which they were grown.

Also included is a section on the selling and shipping of oysters and clams.

OYSTERS AND CLAMS

52. No person shall sell oysters and clams in the shell in other than approved containers, except oysters for replanting or relaying.
53. Oysters and clams in the shell shall be alive, individual, undamaged, and free from mussels, limpets, stones, excess mud, and other extraneous material.
54. Each container of oysters and clams in the shell shall be legibly marked in such a manner that the area from which the oysters and clams were harvested can be determined to the satisfaction of an Inspector.
55. (1) Shucked oysters and clams shall be packed, shipped, and sold only in containers approved by the Minister.
    (2) Shucked oysters and clams for *retail trade* shall be packed, shipped, and sold in single-service watertight containers made of clean impervious materials positively sealed so that tampering with the containers can be detected.
    (3) The packer's certificate number, preceeded by "B.C." shall be impressed, embossed, lithographed, or otherwise permanently recorded on every shipping container.
    (4) Every container shall be legibly marked in such a manner that the day, month, and year of packaging can be determined by an Inspector.
56. Where oysters and clams are shipped in bulk, each shipment shall be accompanied by a statement or invoice indicating the species, quantity, the certificate number of the shipper, his name and address, the name and address of the consignee, and, either coded or uncoded, the earliest date the oysters or clams were taken from the water.

In the "Schedules" section are requirements for Mollusc Shucking and Packing establishments.

SCHEDULE A — PART V — MOLLUSC SHUCKING AND PACKING ESTABLISHMENTS

1. (1) Rooms in which shucked molluscs are further processed shall be separated from the shucking-room.
   (2) Delivery windows shall be installed in the wall between the shucking-room and the shucked-mollusc processing-area, which shall be provided with a shelf of smooth metal or concrete, tile, or other approved material, drainage toward the shucking-room, and through which all shucked stock shall be passed to the processing-area.
2. (1) The tops of the shucking-benches or tables and the walls immediately adjacent thereto, to a height of 2 feet above the benches or tables, shall be of smooth concrete, non-corrodible metal, or such other material as the Minister may approve, free from cracks or crevices, and so constructed that drainage is complete and rapid.
   (2) Shucking-blocks shall be constructed of material approved by the Minister.
3. Lockers, or a separate room, shall be provided and maintained in a sanitary condition for storing employee's outer garments.
4. All shucking-pails, opening-knives, slicers, blowers, skimmers, tanks, tubs, trays, measures, colanders, paddles, or other equipment or utensils used in processing of molluscs shall be or noncorrodible material, other than wood, and shall have smooth surfaces free from cracks and crevices.
5. Adequate facilities for refrigeration, capable of cooling shucked molluscs to an internal temperature of 7.2°C or less, within two hours after packing, shall be provided.
6. If shucked molluscs are to be frozen, freezing facilities shall be capable of freezing packaged molluscs at an ambient air temperature of −25°C or less, with packages frozen solid within 12 hours after the start of freezing.

SCHEDULE B — PART V — MOLLUSC SHUCKING AND PACKING ESTABLISHMENTS

1. Employees engaged in mollusc-processing operations shall wear clean outer garments and headgear of a type approved by the Minister.
2. All shell-stock, prior to processing, shall be alive and clean.
3. Shell-stock in dry storage shall be at all times adequately protected from contamination.
4. Floating and wet storage shall not be used unless written approval is obtained each year from the Minister.
5. Molluscs shall be shucked in such a manner that they are not subject to contamination.
6. Shells from which meats have been extracted shall be removed promptly from the shucking-room.
7. (1) All shucked stock shall be thoroughly washed with cold water for a period not exceeding three minutes.
   (2) The practice of plumping is not permitted.
8. An accurate daily record pertaining to every lot of processed molluscs shall be kept on file at the processing establishment to indicate
   (a) the date, quantity, and species of molluscs delivered to the establishment;
   (b) the day and the area from which the molluscs were harvested;
   (c) the date of processing and packaging, either coded or uncoded;
   (d) the date on which the lot is sold and the name of the dealer to whom it is sold.

Appendix E — Fisheries Act

The *Fisheries Act* regulates a number of activities related to oyster culture. These include foreign importations of mature live oysters, seed, or larvae and movement of oysters or equipment between certain specified areas within the province. The harvesting of oysters from vacant Crown foreshore is also regulated and a permit is required for commercial purposes. Shellfish management areas such as Pendrell Sound are also defined.

B.C. Reg. 140/76 Filed February 18, 1976
O.C. 523/76

Fisheries Act

FISHERIES ACT REGULATIONS

[Consolidated July 31, 1983]

PART 3 — CULTURE AND HARVESTING OF SHELLFISH

Interpretation

7. For the purpose of this Part,

''approved area'' means a shellfish area where, in the opinion of the minister and the Minister of National Health and Welfare, shellfish are safe for direct consumption;

''contaminated area'' means the areas designated by Schedule 1 of the Pacific Shellfish Regulations (P.C. 1977–2397, Canada) and areas declared contaminated under the Sanitary Control of Shellfish Fisheries Regulations (P.C. 1972–2406, Canada);

''culture'' means the rearing, improvement, development and production of shellfish;

''Crown land'' means such ungranted Crown or public land or Crown domain as is within the Province and belonging to Her Majesty in right of the Province, and whether or not any waters flow over or cover the same, and includes any right, title or interest therein of the Crown;

''depuration'' means the process of removing micro-organisms that may be dangerous to humans from live shellfish in a controlled environment;

''lease'' means Crown land that has, by a lease granted under the *Land Act* and registered under the *Land Title Act*, been leased for the cultivation of shellfish;

''licence'' means a licence of occupation to occupy and use temporarily Crown land for shellfish culture, subject to such terms, conditions and reservations deemed necessary by the Minister of Environment and Parks, but not exceeding a term of 10 years;

''licensed area'' means Crown land in respect of which a licence of occupation under the *Land Act*, registered under the *Land Title Act*, has been granted for the cultivation of shellfish;

''relaying'', with respect to shellfish, means the moving of shellfish from a contaminated area to an approved area for the purpose of natural biological cleansing;

''safe'' for direct consumption, with respect to shellfish, means that the shellfish are not unwholesome, as that word is defined in Part 4 and may be used for food without being subjected to depuration or relaying;

''shellfish'' means oysters, clams, mussels and other bivalve molluscs, in all their stages of development, and includes parts of them;

''shellfish area'' means an area in which shellfish are found;

''shellfish management area'' means an area set aside under the authority of the *Land Act*, reserved from alienation to protect the culture of oysters, oyster seed production, and to preserve the ecological integrity of the area;

''transplanting'', with respect to shellfish, means the moving of shellfish from one shellfish area to another shellfish area for any purpose other than natural biological cleansing.

[am. B.C. Reg. 13/78, s. 1.]

Oysters

8. (1) No person shall, except by special permission of the minister,
   (a) plant or introduce, or
   (b) use or cause to be used any equipment to plant or introduce into an oyster bed or any waters of the Province oysters, oyster seed, oyster cultch or oyster shells from outside the Province.

(2) No person shall, except by special permission of the minister, transport or cause to be transported from any area described below any oysters, oyster seed, oyster cultch, oyster shell, marine organisms adversely affecting oysters or any tools, boats, scows or other material used in connection with oyster culture or harvesting:
   (a) *Boundary Bay Area*, which comprises all tidelands in Boundary Bay and Mud Bay inside (north) and of the International Boundary line between Point Roberts and the Mainland (Blaine);
   (b) *Crofton Area*, which comprises all tidelands surrounding the Shoal Islands between the Osborne Bay Lighthouse and the Bare Point Lighthouse at Chemainus;
   (c) *Thetis Island Area*, which comprises
      (i) all tidelands in North Cove inside (southeast) of a line drawn between Fraser Point and Pilkey Point,
      (ii) all tidelands, surrounding Hudson Island and Scott Island, and
      (iii) all tidelands in Telegraph and Preedy Harbours inside (east) of a line drawn between Crescent Point and Active Point on Thetis Island;
   (d) *Ladysmith Harbour Area*, which comprises all tidelands inside (west) of a line drawn between Sharpe Point and the white navigation beacon on Holland Bank;
   (e) *Henry Bay Area*, which comprises all tidelands in Henry Bay inside (east) of a line drawn 169° true from Longbeak Point, the northerly tip of Denman Island;
   (f) *Comox Harbour Area*, which comprises all tidelands inside (northeasterly) of a line drawn from Goose Spit to the outer end of the government dock in Comox Harbour;
   (g) *Repealed*. [B.C. Reg. 13/78, s. 2.]

(3) No person shall, except by special permission of the minister, transport, or cause to be transported into any area described below, any oysters, oyster seed, oyster cultch, oyster shell, marine organisms adversely affecting oysters or any tools, boats, scows or other material used in connection with oyster culture or harvesting;
   (a) *Pendrell Sound Area*, which comprises all the waters and tidelands of Pendrell Sound inside (northerly) of a straight line joining Durham Point and Walter Point;
   (b) *Hotham Sound Area*, which comprises all the waters and tidelands of Hotham Sound inside (northerly) of a straight line joining Culloden Point and Foley Head.

(4) No person shall market oysters taken from a contaminated area unless they have been held
   (a) on a noncontaminated lease or licensed area for 2 weeks or for a time specified by the Minister of Fisheries for Canada, or
   (b) in a licensed depuration plant for 48 hours or for such longer time as may be specified by the minister immediately before marketing.

[am. B.C. Reg. 13/78, s. 2.]

Harvesting of Oysters from Vacant Crown Foreshore

9. (1) (a) No person shall take or have in his possession oysters for commercial purposes other than from a registered oyster lease or a registered oyster licensed area,

except as authorized by a permit issued to him by the minister. A permittee under this subsection may only dispose of the oysters so taken to a registered oyster lease holder.

(b) A person who gathers oysters from Crown land for any purpose other than domestic consumption unconnected with any kind of commercial transaction, shall be conclusively deemed, for the purposes of these regulations, to have taken them for commercial purposes.

(c) A person who gathers oysters from Crown land shall be conclusively deemed, for the purposes of these regulations, to have taken them for commercial purposes unless on the day he gathers them

(i) he shucks them on the foreshore and their volume when shucked is 500 mL or less, or

(ii) he does not shuck them on the foreshore and they are 15 or fewer.

(d) A person who has in his possession more oysters than

(i) 30 oysters in the shell, or

(ii) 1 litre of shucked oysters shall, where the oysters are from Crown land, be conclusively deemed, for the purposes of these regulations, to have them in his possession for commercial purposes.

(e) Paragraph (d) does not apply where a person acquires oysters that were not gathered in contravention of these regulations.

(2) The minister may issue permits for the purpose of this section, but each permit so issued

(a) shall require application on a form supplied by the minister, 30 days prior to the applied permit date,

(b) shall cover a specified operation over a specified area of Crown foreshore and shall be for a period not to exceed 30 days,

(c) shall exclude all dispositions made under the provisions of the *Land Act* either prior to or subsequent to the issuance of the permit,

(d) shall be under such terms and conditions as the minister may prescribe,

(e) shall be subject to the payment of a prescribed fee of $10 for each area from which oysters are to be taken,

(f) shall be subject to the riparian rights of upland owners fronting on the Crown foreshore covered by the permit,

(g) shall require the permittee to leave the foreshore in a clean, safe and sanitary condition to the satisfaction of the minister,

(h) shall require the permittee to save the Crown harmless from any claims which may be made against the Province of British Columbia by reason of anything done, or left undone, by the permittee, and

(i) shall be subject to the Sanitary Control of Shellfish Fisheries Regulations (P.C. 1972–2406, Canada).

(3) Every person who is granted a permit to harvest oysters under this section shall render a statement, on a form supplied by the minister, within 10 days of the expiry date of such permit. This statement shall record the production and disposition of shellstock under the permit and shall be submitted to the Aquaculture and Commercial Fisheries Branch.

(4) Every person who harvests oysters under a permit granted under this section shall jointly with the statement in subsection (3) above pay to the Minister of Finance a royalty equal to $5 per ton on all oysters (shellstock) taken.

(5) Any person who violates any provision of this section is liable, on summary conviction, to a fine of not less than $25 and not more than $1000.

(6) Under this section, permits granted to take oysters commercially from foreshore fronting Indian reserves must require written upland owner's consent. Nothing in the foregoing shall prevent, restrict or control the harvesting of oysters in the above mentioned areas by an Indian, as defined in the *Indian Act*, for personal food purposes.

(7) Pursuant to this section, a permit issued for the harvesting of oysters from contaminated vacant Crown foreshore must comply with the Sanitary Control of Shellfish Fisheries Regulations, P.C. 1972–2406.

(8) Pursuant to this section, all harvesters must hold a valid personal commercial fishing licence.

[am. B.C. Reg. 13/78, ss. 3, 4.]

10. *Repealed.* [B.C. Reg. 11/83.]

Shellfish Management Areas

11. The area known as Pendrell Sound, Redonda Island, New Westminster District, lying north of a straight line between Walter Point and Durham Point, is reserved from alienation to protect the culture of oysters, oyster seed production and to preserve the ecological integrity of the area:

(1) Pursuant to this section, all activities defined as oyster culture, conducted in the area known as Pendrell Sound, must have the special permission of the minister;

(2) Pursuant to this section, the part of Pendrell Sound that lies north of a line joining position 50°15′12″N., 124°43′50″W., and position 50°14′47″N., 124°43′21″W., is subject to Part II of Schedule C of the Boating Restriction Regulations, *Canada Shipping Act*, P.C. 1973–770.

[am. B.C. Reg. 246/76.]

Appendix F — Native Oysters

The native oyster is able to change sex during the summer breeding season probably spawning first as a male, then as a female and possibly once more as a male. The breeding season is fairly long and can begin as early as May. Females discharge ova into the mantle cavity and not externally as the Pacific oyster. Males however, discharge sperm externally into the open water and eggs in the mantle cavity of females are fertilized by sperm brought in on the inhalant (respiratory and feeding) current. Larvae are incubated in the mantle cavity and a female in the early stages is said to be "white sick". When the larvae are ready to be released, still in the straight hinged stage, they are quite dark and the oyster is termed "black sick". Thus the developmental stages, lasting about 10 days, can be determined by opening female oysters and observing the colour of the larvae.

Once in the plankton, larvae behave much as those of Pacific oysters, but being already quite well developed they have a shorter planktonic existence of about a week to 10 days.

The settlement process and types of cultch for native oysters are similar to that of the Pacific oyster.

In Pendrell Sound, adult native oysters are rarely seen, being mainly under rocks, but larval production is quite high and good settlement occurs. There is little or no survival because of cultural practices used for Pacific oysters. To collect native oyster spat, the procedure would be to expose cultch in Pendrell Sound early in summer before Pacific oyster breeding time.

The industry in British Columbia was essentially a fishery rather than a culture and it wasn't until the late 1930's that culture attempts were made. Small dikes were constructed in Boundary Bay and Ladysmith Harbour at that time (Fig. 161). Oysters were raked and loaded into scows or on top floats for transfer to shucking houses. Market sized oysters were culled from the smaller oysters and debris, largely made up of dead oyster shell valves. Only small oysters were returned to the beds, which were thus deprived of natural cultch provided by dead oysters. Shucking native oysters is a tedious job for they grow no larger than 5 cm. Most of marketable size are closer to 3 cm and require about 425 for a litre compared to approximately 40 Pacific oysters.

FIG. 161. Typical native oyster dikes.

Centers of the industry were Boundary Bay which abounds in sloughs, a suitable habitat for native oysters, and Ladysmith Harbour. In the early 1940's population size dropped drastically for no single perceptible reason. Probably a combination of deleterious environmental factors was too much for the fishery decimated populations to overcome. Native oysters are now quite rare in both Boundary Bay and Ladysmith Harbour but small populations exist in the Gorge near Victoria, in areas of Barkley Sound, Blunden Harbour and many inlets in the Bella Bella area of the central coast.